Selected Topics in Materials Science and Engineering

Volume I

Selected Topics in Materials Science and Engineering
Volume I

Edited by **Ricky Peyret**

NY RESEARCH
PRESS

New York

Published by NY Research Press,
23 West, 55th Street, Suite 816,
New York, NY 10019, USA
www.nyresearchpress.com

Selected Topics in Materials Science and Engineering: Volume I
Edited by Ricky Peyret

International Standard Book Number: 978-1-63238-410-2 (Hardback)

Printed in China.

Contents

Preface

Materials science and engineering is an arena that involves the classification of the physical and chemical properties of various solid materials, like metals and alloys, magnetic materials, ceramics, optical materials, polymers, superconductors, semiconductors, and composites for the objective of using, changing, as well as enhancing intrinsic properties to create or improve materials. This field examines the microstructure of a material and how it can be changed to affect the strength, optical properties, electrical conductivity or magnetic properties of a material. The discipline of materials science and engineering includes all man-made and natural materials as well as their extraction, characterization, synthesis, properties, processing and development for practical technological applications. This field is essentially multidisciplinary and covers chemical, electrical, mechanical, civil, bio-medical and aerospace engineering as well as physics and chemistry. Students in this field need to develop a foundational understanding of materials at the micro, nano and macro scales and this focus can lead to specialization in branches such as chemical and electrochemical materials science and engineering, electronic, magnetic and optical materials, biomaterials, computational materials science and engineering along with structural materials. Thus, advancements in materials science and engineering enable new technological breakthroughs in almost all branches of the engineering discipline and benefit us on various fronts.

This book is an attempt to compile and collate all available research on materials science and engineering under one umbrella. I am grateful to those who put their hard work, effort and expertise into these researches as well as those who were supportive in this endeavour.

Editor

Morphology and *In Vitro* Behavior of Electrospun Fibrous Poly(D,L-lactic acid) for Biomedical Applications

Toshihiro Inami,[1] Yasuhiro Tanimoto,[2] Masayuki Ueda,[1] Yo Shibata,[3] Satoshi Hirayama,[4] Masaru Yamaguchi,[1] and Kazutaka Kasai[1]

[1] *Department of Orthodontics, Nihon University School of Dentistry at Matsudo, 2-870-1 Sakaecho Nishi, Matsudo, Chiba 271-8587, Japan*
[2] *Department of Dental Biomaterials, Nihon University School of Dentistry at Matsudo, 2-870-1 Sakaecho Nishi, Matsudo, Chiba 271-8587, Japan*
[3] *Department of Oral Biomaterials and Technology, Showa University School of Dentistry, 1-5-8 Hatanodai, Shinagawa-ku, Tokyo 142-8555, Japan*
[4] *Department of Dental Caries and Aesthetic Dentistry, Nihon University School of Dentistry at Matsudo, 2-870-1 Sakaecho Nishi, Matsudo, Chiba 271-8587, Japan*

Correspondence should be addressed to Yasuhiro Tanimoto; tanimoto.yasuhiro@nihon-u.ac.jp

Academic Editor: Peter Majewski

This work describes the fabrication, optimization, and characterization of electrospun fibrous poly(D,L-lactic acid) (PDLLA) for biomedical applications. The influences of the polymer concentration of the electrospinning solution (5, 10, or 15 wt%) and the solution flow rate (0.1, 0.5, 1.0, or 2.0 mL/h) on the morphology of the obtained fibrous PDLLA were evaluated. The *in vitro* biocompatibility of two types of PDLLA, ester terminated PDLLA (PDLLA-R) and carboxyl terminated PDLLA (PDLLA-COOH), was evaluated by monitoring apatite formation on samples immersed in Hanks' balanced salt (HBS) solution. 15 wt% polymer solution was the most beneficial for preparing a fibrous PDLLA structure. Meanwhile, no differences in morphology were observed for PDLLA prepared at various flow rates. Apatite precipitate is formed on both types of PDLLA only 1 day after immersion in HBS solution. After 7 days of immersion, PDLLA-COOH showed greater apatite formation ability compared with that of PDLLA-R, as measured by thin-film X-ray diffraction. The results indicated that the carboxyl group is effective for apatite precipitation in the body environment.

1. Introduction

Owing to its biodegradability and biocompatibility, poly(lactic acid) (PLA) has been demonstrated to be a suitable biodegradable polymer to produce carriers for drug delivery systems, scaffolds for tissue engineering, and implanted medical devices [1–3]. In particular, the fibrous form of PLA is more preferable than the bulk material for bone tissue engineering, because its high porosity encourages the migration and adhesion of osteoblast-like cells [4].

Electrospinning is a simple and effective method for the production of fibrous structures from polymer solutions [5–7]. The advantage of electrospinning method is that it can be used to produce fibrous structures having fiber diameters ranging from a few micrometers to a few hundred nanometers, with large specific surface areas. A schematic diagram of the electrospinning method is shown in Figure 1. There are basically three components required for the process: a high voltage supply, a capillary tube with a pipette or needle of small diameter, and a metallic collecting screen. In electrospinning, a polymer solution is converted to a fibrous structure via electrostatic forces. The polymer solution is delivered through a needle and a high voltage is applied to induce charges in the fluid. When the applied electric field is strong enough to overcome the surface tension, a tiny jet is ejected from the surface of the droplet and drawn toward the collecting plate. Between the needle tip and the collector, the jet is highly stretched, dried, and deposited as

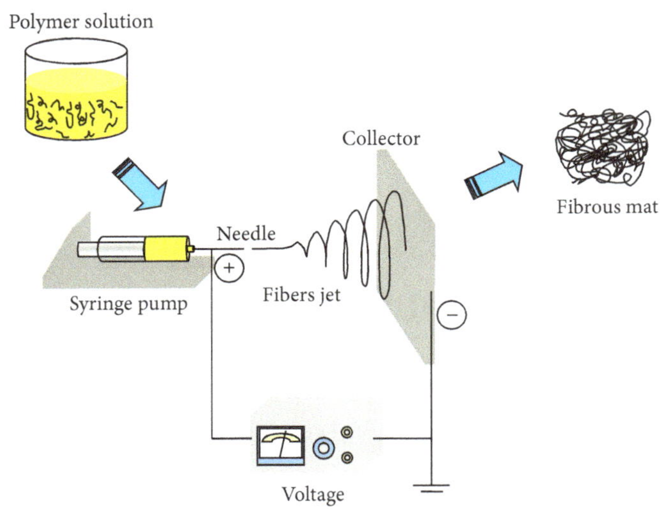

FIGURE 1: Schematic illustration of the electrospinning apparatus.

a nonwoven fiber structure onto the collector. This process is affected by the polymer characteristics, polymer solution parameters, and process parameters. By varying these parameters it is possible to obtain fibrous structures with various morphologies and properties. For example, Hiep and Lee [8] fabricated an electrospun poly(lactic-co-glycolic acid) (PLGA)/polycaprolactone (PCL) blend using various percentages of PLGA in blended PLGA/PCL solutions. The authors concluded that increasing the PLGA concentration affected the mechanical properties of the electrospun membranes and increased the biocompatibility of the resulting fibrous PLGA/PCL mats. Likewise, Ngiam et al. [9] prepared various types of nanofibers, such as pure PLGA and blended PLGA/collagen nanofibers, by electrospinning and investigated their biomineralization ability for nanohydroxyapatite (n-HA) using an alternate calcium phosphate dipping method. The authors reported preferential deposition of n-HA crystallites on the PLGA/collagen nanofibers compared with those on the PLGA nanofibers, suggesting that collagen is a good template material for n-HA nucleation.

The *in vivo* bioactivity, or osteoconductive ability, of a biomaterial is precisely mirrored by its *in vitro* apatite formation ability in simulated body fluid [10]. In particular, there have been several reports that carboxyl (COOH) groups play an effective role in apatite nucleation in the body environment [11–13]. Kitahara et al. [14] prepared three-dimensional (3D) porous PLA scaffolds by freeze-drying and introduced COOH onto their surface, reporting subsequent improvement of the PLA biomaterial for effective and efficient bone regeneration.

With these backgrounds, fibrous PLA with COOH-terminus prepared by electrospinning is considered to be an effective material for apatite formation ability and bone regeneration, because of its high specific surface area and porous structure. However, little is known about biological activity of electrospun fibrous PLA with COOH-terminus.

In the present study, fibrous poly(D,L-lactic acid) (PDLLA) fabricated by electrospinning for biomedical applications was optimized and characterized. Moreover, as an evaluation of *in vitro* biocompatibility, the influence of fibrous PDLLA with or without COOH-terminus on apatite formation in simulated body fluid was investigated.

2. Materials and Methods

2.1. Materials. In this study, two types of PDLLA, ester terminated PDLLA (PDLLA-R) with an inherent viscosity of 0.37 dL/g and carboxyl terminated PDLLA (PDLLA-COOH) with an inherent viscosity of 0.32 dL/g, were supplied by DURECT Corporation, AL, USA. Both PDLLA-R and PDLLA-COOH were dissolved in 1,1,1,3,3,3-hexafluoro-2-propanol (HFIP, Wako, Osaka, Japan) at concentrations of 5, 10, and 15 wt% at room temperature (RT) overnight until the solutions became homogeneous.

2.2. Electrospinning. As mentioned in Figure 1, the electrospinning set-up (IMC-1639, Imoto, Kyoto, Japan) consisted of a high voltage power supply (Matsusada Precision Inc., Shiga, Japan), a syringe pump (KDS-100, ISIS Co. Ltd., Osaka, Japan) with needle, and a stainless steel collector plate. The solution was loaded into a 30 mL plastic syringe (Terumo Corp., Tokyo, Japan) having a 21G stainless steel needle (Terumo Corp.). The PDLLA solution was continuously supplied at flow rates of 0.1, 0.5, 1.0, and 2.0 mL/h through the needle using a syringe pump. The distance between the needle tip and the stainless steel collector was 10 cm. Upon application of a high voltage of 15 kV between the needle tip and the collector, the PDLLA solution was evaporated, and fibrous PDLLAs were deposited on the collector plate at RT. The obtained fibrous PDLLA was vacuum dried overnight at RT.

2.3. Field-Emission Scanning Electron Microscopy. Vacuum drying and platinum sputtering of the specimen surface were carried out, and the appearance of the fibrous PDLLA was then observed under a field-emission scanning electron

FIGURE 2: FE-SEM images showing the morphology of fibrous PDLLA prepared with varying solution concentrations. (a) PDLLA-R; (b) PDLLA-COOH.

TABLE 1: Concentrations of inorganic ions in the HBS solution.

Ion	Na^+	K^+	Mg^{2+}	Ca^{2+}	Cl^-	HPO_4^{2-}	SO_4^{2-}	HCO_3^-
Concentration (mmol/L)	142.	5.81	0.811	1.26	145.	0.788	0.811	4.17

microscope (FE-SEM; JSM-6340F, JEOL, Tokyo, Japan) at an accelerating voltage of 5 kV.

2.4. Characterization of Calcium Phosphate Formed on Fibers in Body Simulated Fluid. PDLLA (10 mm × 10 mm) samples were immersed in 20 mL of simulated body fluid, comprising Hanks' balanced salt (HBS) solution without organic species (pH = 7.4) [15], at 37°C in a Teflon-sealed polystyrene bottle for 1 or 7 days. The composition of the HBS solution used is shown in Table 1. The solution and bottle were changed every day to expose the specimens to fresh solution. After immersion in the HBS solution for 1 or 7 days, the fibrous PDLLAs were rinsed with double-distilled water and then immediately dried in a vacuum desiccator. The morphology of the precipitates on the PDLLA surface was observed by FE-SEM at an accelerating voltage of 5 kV. The precipitates were also characterized by thin-film X-ray diffraction (TF-XRD; RINT2000, Rigaku Corp., Tokyo, Japan) using a Cu Kα X-ray source, power of 50 kV × 100 mA, and scanning range of 20–40°.

3. Results and Discussion

Electrospinning is a versatile polymer processing technique in which a stream of polymer solution or melt is subjected to a high electric field, resulting in formation of a fibrous structure. In the present study, fibrous PDLLAs were prepared by electrospinning for biomedical applications.

The morphology of fibrous materials fabricated by electrospinning is dependent on variables such as fabrication parameters and composition [16, 17]. First, the influence of the solution concentration on the morphology of the electrospun fibrous PDLLA was evaluated. Figure 2 shows FE-SEM images of the structures of fibrous PDLLAs prepared from solution concentrations of 5, 10, and 15 wt% at a flow rate of 2.0 mL/h. In general, fiber morphology varied with PDLLA solution concentration. At a concentration of 5 wt%, many beads or beaded bulks were observed in both PDLLA-R and PDLLA-COOH. At 10 wt%, numerous beads or beaded fibers connected together were still observed throughout the material in both PDLLA-R and PDLLA-COOH. In contrast, fibrous structures were observed for both PDLLA-R and PDLLA-COOH at a concentration of 15 wt%, and the beads completely disappeared. Generally, the amount of the beads decreases with increasing polymer concentration [17]. The shape of the PDLLA changed from bead to fiber when the polymer concentration was increased from 10 to 15 wt%. Thus, it was suggested that the optimum solution concentration for producing fibrous PDLLA was 15 wt%. Next, the influence of the flow rate on morphology of fibrous PDLLA electrospun from 15 wt% solution was evaluated. Figure 3 shows FE-SEM images of the structures of the obtained fibrous PDLLA for flow rates of 0.1, 0.5, and 1.0 mL/h. As previously mentioned, the structure of PDLLA prepared with a solution concentration of 15 wt% at rate of 2.0 mL/h is shown in Figure 2. No large differences in microstructure were found between the samples prepared with flow rates of 0.1, 0.5, 1.0, and 2.0 mL/h.

Apatite precipitation in biological fluids is an important determinant of the bioactivity of fibrous PDLLA. Therefore, we examined precipitate formation on the surface of the obtained PDLLA in HBS solution. Figure 4 shows FE-SEM images of the surface appearances of fibrous PDLLA samples

(a)

(b)

FIGURE 3: FE-SEM images showing the morphology of fibrous PDLLA prepared with varying flow rates. (a) PDLLA-R; (b) PDLLA-COOH.

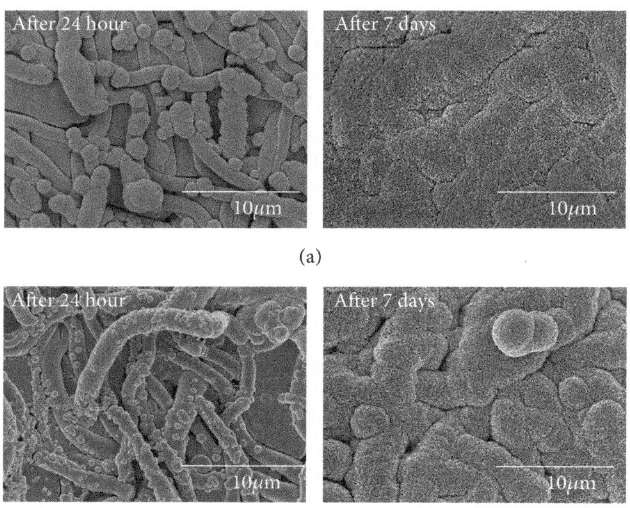

(a)

(b)

FIGURE 4: FE-SEM photographs showing the surface appearances of the fibrous PDLLA after immersion in HBS solution. (Original magnification 5,000x.) (a) PDLLA-R; (b) PDLLA-COOH.

after immersion in HBS solution for 1 or 7 days. Precipitated small globules were observed on both the fibrous PDLLA-R and PDLLA-COOH after 1-day immersion. The surfaces of both the PDLLA-R and the PDLLA-COOH were completely covered with globules after 7-day immersion. The amount of precipitation on the surface of PDLLA-COOH appeared to be greater than that observed on PDLLA-R. However, it is difficult to find evidence to account for the difference in the amount of precipitation from FE-SEM observation alone. Figure 5 shows TF-XRD patterns for the precipitates on the fibrous PDLLAs after immersion in HBS solution for 7 days. For the fibrous PDLLAs immersed for 7 days, two peaks assignable to apatite (JCPDS Card no. 09-0432) were observed at 25.86 and 31.76°. Moreover, the TF-XRD pattern of PDLLA-COOH showed relatively stronger apatite peak intensities compared with those observed for PDLLA-R. Generally, carboxyl (COOH) groups are known to induce apatite nucleation. Tanahashi and Matsuda [12] investigated the dependence of apatite formation in simulated body fluid on the surface functional group. The authors explained that biomineralization such as apatite formation predominantly occurs by calcium ion adsorption on the artificial material. Likewise, Sato et al. [13] investigated the apatite crystal nucleation mechanism on organic monolayers in a simulated body environment. The authors indicated that interfacial interaction between carboxyl groups and calcium ions was an important factor with respect to apatite formation in a simulated body environment, causing hydroxyapatite nucleation. On that basis, the present fibrous PDLLA-COOH is predicted to have a higher *in vivo* apatite formation ability than PDLLA-R. This finding within the limitation speculates that the electrospun fibrous PDLLA, having high specific surface area and porous structure, has more effectiveness of COOH to biological activity in comparison with bulk material. However, more detailed analysis of the fibrous PDLLA and its biological activity such as for cultured cell growth and its tissue response will be investigated further.

4. Conclusions

In this study, fibrous PDLLA was successfully prepared by electrospinning for use as a scaffold. The following is a summary of the authors' interpretation of the results.

(1) A 15 wt% PDLLA solution was the most beneficial for preparing a fibrous structure. In contrast, solutions of 5 or 10 wt% did not yield a completely fibrous structure.

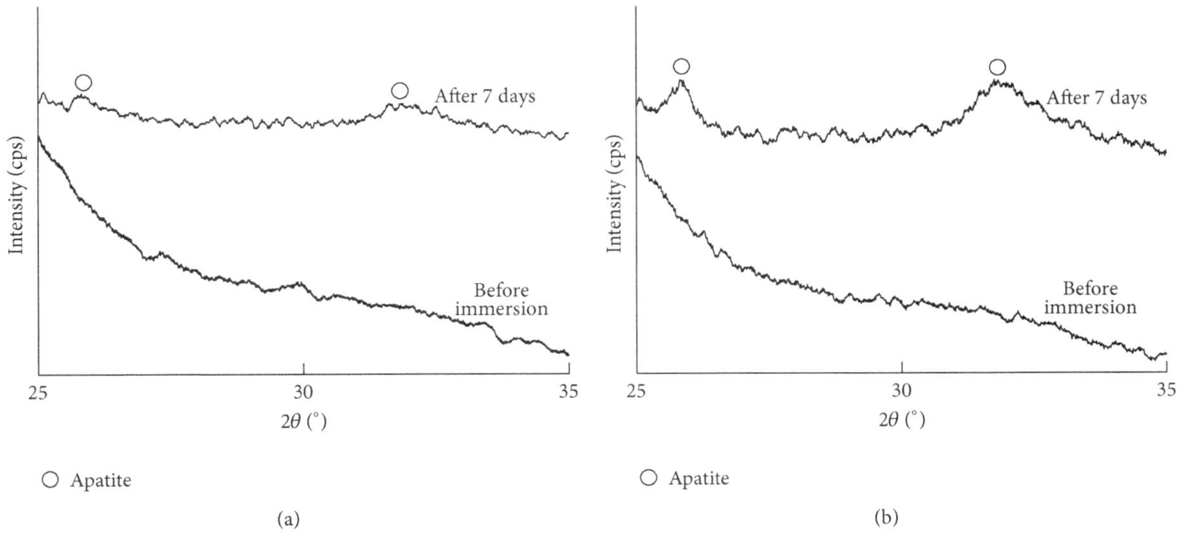

Figure 5: TF-XRD patterns of the PDLLA. Both PDLLA-R and PDLLA-COOH exhibit apatite peaks at 25.86° and 31.76°. (a) PDLLA-R; (b) PDLLA-COOH.

(2) PDLLAs prepared with various flow rates showed no differences in the morphology of the obtained fibrous structure.

(3) Apatite precipitate was formed on the PDLLA only 1 day after immersion in HBS solution. After 7 days of immersion, the apatite formation ability of the PDLLA-COOH was greater than that of the PDLLA-R, indicating that the carboxyl (COOH) group is effective for apatite activity in HBS solution.

Conflict of Interests

The authors declare that no competing interests exist.

Acknowledgment

This work was supported in part by a Grant-in-Aid for Young Scientists (B) (no. 21791954) (24792155) from the Japan Society for the Promotion of Science.

References

[1] A. A. Ignatius, O. Betz, P. Augat, and L. E. Claes, "*In vivo* investigations on composites made of resorbable ceramics and poly(lactide) used as bone graft substitutes," *Journal of Biomedical Materials Research B*, vol. 58, no. 6, pp. 701–709, 2001.

[2] Q. Lu, A. Simionescu, and N. Vyavahare, "Novel capillary channel fiber scaffolds for guided tissue engineering," *Acta Biomaterialia*, vol. 1, no. 6, pp. 607–614, 2005.

[3] Y. Tanimoto and N. Nishiyama, "Preparation and in vitro behavior of a poly(lactic acid)-fiber/ hydroxyapatite composite sheet," *Advances in Materials Science and Engineering*, vol. 2009, Article ID 827241, 2009.

[4] S. I. Jeong, E. K. Ko, J. Yum, C. H. Jung, Y. M. Lee, and H. Shin, "Nanofibrous poly(lactic acid)/hydroxyapatite composite scaffolds for guided tissue regeneration," *Macromolecular Bioscience*, vol. 8, no. 4, pp. 328–338, 2008.

[5] D. Li, Y. Wang, and Y. Xia, "Electrospinning nanofibers as uniaxially aligned arrays and layer-by-layer stacked films," *Advanced Materials*, vol. 16, no. 4, pp. 361–366, 2004.

[6] M. M. Munir, F. Lskander, R. Khairur, and K. Okuyama, "High performance electrospinning system for fabricating highly uniform polymer nanofibers," *Review of Scientific Instruments*, vol. 80, Article ID 026106, 2009.

[7] W. Cui, X. Li, S. Zhou, and J. Weng, "In situ growth of hydroxyapatite within electrospun poly(DL-lactide) fibers," *Journal of Biomedical Materials Research A*, vol. 82, no. 4, pp. 831–841, 2007.

[8] N. T. Hiep and B. T. Lee, "Electro-spinning of PLGA/PCL blends for tissue engineering and their biocompatibility," *Journal of Materials Science: Materials in Medicine*, vol. 21, no. 6, pp. 1969–1978, 2010.

[9] M. Ngiam, S. Liao, A. J. Patil, Z. Cheng, C. K. Chan, and S. Ramakrishna, "The fabrication of nano-hydroxyapatite on PLGA and PLGA/collagen nanofibrous composite scaffolds and their effects in osteoblastic behavior for bone tissue engineering," *Bone*, vol. 45, no. 1, pp. 4–16, 2009.

[10] Y. Tanimoto, T. Hayakawa, T. Sakae, and K. Nemoto, "Characterization and bioactivity of tape-cast and sintered TCP sheets," *Journal of Biomedical Materials Research - Part A*, vol. 76, no. 3, pp. 571–579, 2006.

[11] T. Kokubo, M. Hanakawa, M. Kawashita et al., "Apatite formation on non-woven fabric of carboxymethylated chitin in SBF," *Biomaterials*, vol. 25, no. 18, pp. 4485–4488, 2004.

[12] M. Tanahashi and T. Matsuda, "Surface functional group dependence on apatite formation on self-assembled monolayers in a simulated body fluid," *Journal of Biomedical Materials Research*, vol. 34, no. 3, pp. 305–315, 1997.

[13] K. Sato, Y. Kumagai, and J. Tanaka, "Apatite formation on organic monolayers in simulated body environment," *Journal of Biomedical Materials Research*, vol. 50, no. 1, pp. 16–20, 2000.

[14] K. Kitahara, M. Fuse, and M. Fukumoto, "Effect of the introduction of a carboxylic acid group onto the surface of

a three-dimensional porous poly (lactic acid) scaffold on apatite deposition," *International Journal of Oral-Medical Sciences*, vol. 10, no. 3, pp. 179–186, 2011.

[15] T. Hanawa, M. Kon, H. Ukai, K. Murakami, Y. Miyamoto, and K. Asaoka, "Surface modification of titanium in calcium-ion-containing solutions," *Journal of Biomedical Materials Research*, vol. 34, no. 3, pp. 273–278, 1997.

[16] Y. Wan, X. Cao, S. Zhang, S. Wang, and Q. Wu, "Fibrous poly(chitosan-g-dl-lactic acid) scaffolds prepared via electro-wet-spinning," *Acta Biomaterialia*, vol. 4, no. 4, pp. 876–886, 2008.

[17] Z. M. Huang, Y. Z. Zhang, M. Kotaki, and S. Ramakrishna, "A review on polymer nanofibers by electrospinning and their applications in nanocomposites," *Composites Science and Technology*, vol. 63, no. 15, pp. 2223–2253, 2003.

2

Photocatalytic Degradation of Dichlorvos in Visible Light by Mg^{2+}-TiO_2 Nanocatalyst

T. Siva Rao,[1] Teshome Abdo Segne,[1] T. Susmitha,[1] A. Balaram Kiran,[1] and C. Subrahmanyam[2]

[1] *Department of Inorganic and Analytical Chemistry, School of Chemistry, Andhra University, Visakhapatnam 530003, India*
[2] *Department of Chemistry, Indian Institute of Technology Hyderabad, Hyderabad 500007, India*

Correspondence should be addressed to T. Siva Rao, raost@yahoo.com

Academic Editor: Guohua Jiang

Photocatalytic activity of TiO_2 was studied by doping with magnesium (Mg^{2+}-TiO_2) with varying magnesium weight percentages ranging from 0.75–1.5 wt%. The doped and undoped samples were synthesized by sol-gel method and characterized by X-ray diffraction (XRD), N_2 adsorption-desorption (BET), X-ray photoelectron spectroscopy (XPS), UV-visible diffuse reflectance spectroscopy (DRS), and scanning electron microscopy (SEM). The XRD data has shown that anatase crystalline phase in Mg^{2+}-TiO_2 catalysts, indicating that Mg^{2+} ions did not influence the crystal patterns of TiO_2. The presence of magnesium ions in TiO_2 matrix has been determined by XPS spectra. DRS spectra showed that there is a significant absorption shift towards the visible region for doped TiO_2. The SEM images and BET results showed that doped catalyst has smaller particle size and highest surface area than undoped TiO_2. The photocatalytic efficiency of the synthesized catalysts was investigated by the photocatalytic degradation of aqueous dichlorvos (DDVP) under visible light irradiation, and it was found that the Mg^{2+}-doped catalysts have better catalytic activity than undoped TiO_2. This can be attributed that there is a more efficient electron-hole creation in Mg^{2+}-TiO_2 in visible light, contrary to undoped TiO_2 which can be excited only in UV irradiation. The effect of dopant concentration, pH of solution, dosage of catalysts, and initial pesticide concentration has been studied.

1. Introduction

Advanced oxidation process (AOP) is an alternative way of treating undesirable organic pollutants, including pesticides. According to the literature [1, 2] among AOPs, heterogeneous photocatalysis seems to be an attractive method and has been successfully employed for the degradation or transform into less harmful substances of various families of pollutants [3–5].

The photoinduced redox reactions have received an attention after the discovery of first photoelectrochemical splitting of water on titanium dioxide electrode by Fujishima and Honda [6]. Later, many novel redox reactions of organic and inorganic substrates have been induced by band-gap irradiation of a variety of semiconductor particles [7, 8]. There are a lot of different semiconductor materials which are readily available (e.g., TiO_2, ZnO, Fe_2O_3, CdS, and ZnS), but only a few are suitable for sensitizing the photomineralization of a wide range of organic pollutants. A sensitizer for reaction must be [9, 10] (i) photoactive, (ii) able to utilize visible and/or near the UV light, (iii) biologically and chemically inert, and (iv) photostable.

Among various semiconductor photocatalysts available, TiO_2 has attracted a great deal of study for its high photocatalytic activity, nontoxic property, chemical stability, and highest oxidation rate of many photoactive metal oxides investigated [11]. TiO_2 is capable of decomposing a wide range of organic, inorganic, and toxic pollutants [12].

Although TiO_2 is superior to other semiconductors for many practical uses, two types of defects limit its photocatalytic activity. Firstly, TiO_2 has high bandgap energy (3.2 eV), which is only 4-5% of the overall solar spectrum. Thus this restricts the use of visible light. Secondly, the high rate of electron-hole recombination at TiO_2 particles results in low photoquantum efficiency [13]. In order to overcome these limitations of TiO_2, many attempts have been made to prepare different types of TiO_2 by depositing of noble metals, mixing metal oxides, surface derivatization, and doping selective metal ions into TiO_2 [14, 15], and also Graphene-based semiconductor photocatalysts are prepared [16].

The addition of low percentage of metal by doping was often proposed to improve the photocatalytic activity of TiO_2. The advantage of doping the metal ions into TiO_2 is the temporary trapping of the photogenerated charge carriers by the dopant and the inhibition of their recombination during migration from the inside of the material to the surface, by modifying the bandgap of the photocatalyst. A wide range of metal ions have been studied for doping TiO_2 and their effects on the properties of TiO_2 and on its photocatalysis [17, 18], but reports on alkaline-earth metal ions doping into TiO_2 and their photocatalytic properties have seldom been seen.

Sol-gel process is one of the versatile, simple, and easy means of synthesizing nanosize materials in contrast to the other conventional preparation methods which do not usually produce homogeneous, high-surface-area materials. The main advantage of the sol-gel method in preparing catalytic materials is its excellent control over the properties of the product via a host of parameters that are accessible in all four key processing steps: formation of a gel, aging, drying, and heat treatment. Although titania catalysts have been extensively studied, the procedure of preparing photocatalytic metal-doped titania is still of great interest.

Waste waters generated from the agricultural application of dichlorvos contain the residue of dichlorvos which has been evaluated in a wide range of toxicology assays including bioassays for carcinogenicity and mutagenicity (genotoxicity) [19, 20]. It is also more likely to contaminate surface and ground waters due to its high solubility in water. This could have an adverse effect on the aquatic environment, and procedure has to be found for its degradation. Electron transfer process occurring between pesticide and semiconductor, especially TiO_2, has been examined and found to have practical potential [21].

The present paper has been focused to develop a visible light response catalyst via magnesium-doped titanium dioxide in varying percentages of metal. Characterization has been carried out using XRD, BET, XPS, DRS, and SEM. The photocatalytic activities of the synthesized samples have been evaluated by the degradation of a representative model pesticide pollutant, dichlorvos.

2. Experimental

2.1. Preparation of Photocatalysts. Titanium tetra-n-butoxide [$Ti(O\text{-}Bu)_4$] and magnesium nitrate were obtained from E. Merck (Germany) and used as titanium and magnesium sources for preparing anatase TiO_2 and Mg^{2+}-TiO_2 photo catalysts. Aqueous dichlorvos solution was used as a model compound for degradation. All other chemicals used in this work are of analytical grade, and doubly distilled water was used for the solution preparation.

Mg^{2+}-TiO_2 samples were prepared by sol-gel method in which initially a solution containing 40 mL of absolute alcohol, 6 mL of H_2O, and magnesium nitrate with required percentages (0.75 wt.%, 1.00 wt.%, 1.25 wt.% and 1.50 wt.%) was prepared (solution I). Another solution was prepared by taking 21 mL of $Ti(O\text{-}Bu)_4$ in 40 mL of absolute alcohol with stirring for 10 minutes, and then 3 mL of HNO_3 was added dropwise under continuous stirring for 30 minutes (solution II). Solution I was added to solution II slowly from the burette with vigorous stirring at room temperature until the transparent sol was obtained, and the resulting sol was further stirred for 1 h. The gel was prepared by aging the solution for 48 h at room temperature. The derived gel was dried at 100°C in an oven and ground. The catalyst powder was calcined at 400°C in furnace for 2 h. Undoped TiO_2 is also prepared with same procedure without metal nitrate. The doping concentrations are expressed as weight percentage of titanium atom. The powders are stored in black coated air-tight glass containers.

2.2. Characterization of Photocatalysts. The crystal phase composition of the prepared photocatalysts (TiO_2, Mg^{2+}-TiO_2) was determined by X-ray diffraction measurement carried out at room temperature using a PANalytical, D/Max-III A diffractometer with CuK_α radiation (λ = 0.15148 nm) with a liquid nitrogen gas-cooled germanium solid-state detector. The accelerating voltage of 35 kV and emission current of 30 mA were used and studied in the range of 2°–90° 2θ with a step time of 2° min^{-1}. The Brunauer-Emmett-Teller (BET) surface area was determined from the N_2 adsorption-desorption isotherm at 77.3 K by using a Quantachrome Nova 2200 E system. The sample was outgassed for 3 h at 300°C prior to the adsorption. The specific surface area was determined by using the standard BET method. To study the valence state of the photocatalysts, X-ray photoelectron spectroscopy (XPS) was recorded with the PHI quantum ESCA microprobe system, using the AlK_α line of a 250 W X-ray tube as a radiation source with the energy of 1486.6 eV, 16 mA × 12.5 kV, and a working pressure lower than 1×10^{-8} Nm^{-2}. As an internal reference for the absolute binding energies, the C 1s peak of hydrocarbon contamination was used as reference to 284.8 eV. The fitting of XPS curves was analyzed with MultiPak 6.0 A software. The DRS were recorded with a Shimadzu 3600 UV-visible NIR spectrophotometer equipped with an integrating sphere diffuse reflectance accessory, using $BaSO_4$ as reference scatter. Powder samples were loaded into a quartz cell, and spectra were recorded in the range of 200–900 nm. The morphology and size of particles was characterized using SEM (JSM-6610 LV) spectrophotometer and operated at 20 kV. A Horiba Jobin FluoroMax-4 instrument has been used for the PL, spectral data analysis, with a PMT voltage was 150 V and with slit set both at 2.5 nm.

2.3. Photocatalytic Activity of the Catalysts. Photocatalytic degradation studies of DDVP were carried out in a modified photoreactor system, in which 100 mL of reaction mixture illuminated with 400 W high-pressure mercury vapor lamp (Osram, India). The distance between the light source and the reaction beaker was 20 cm. Prior to irradiation, the solution with catalyst was stirred in the dark for 45 min to ensure establishment of adsorption-desorption equilibrium of dichlorvos on catalyst surface. Aliquots of the samples were withdrawn from the solution by using Millipore syringe (0.45 μm) at certain time intervals and analyzed for dichlorvos concentration. The quantitative determination of DDVP was performed by measuring the absorption of phosphate ion formed at 465 nm with a Milton-Roy spectronic-1201 UV-visible spectrometer. The IR (>700 nm) and UV (<360 nm) radiations were filtered by water and UV filters, respectively.

Previous studies have shown that the photocatalytic degradation of pesticides by irradiating with UV light source and using TiO_2 or ZnO as catalyst leads to the formation of H^+, Cl^-, PO_4^{3-}, and CO_2 as final products, and their formation during the progress of the degradation of DDVP was confirmed by simple qualitative analysis tests. PO_4^{3-} was determined calorimetrically by the molybdenum blue method:

$$(CH_3O)_2POOCHCCl_2 + \left(\frac{9}{2}\right)O_2 \longrightarrow PO_4^{3-} + 4CO_2 + 2\,Cl^- + 5\,H^+ + H_2O. \quad (1)$$

Hence, in order to study the extent of mineralization of DDVP over illuminated Mg^{2+}-doped TiO_2, the absorbance measurements of PO_4^{3-} were carried out. The percentage of degradation was calculated from the following equation:

$$\%\ \text{Phosphate ion formation} = \left[\frac{A_t}{A_0}\right] \times 100, \quad (2)$$

where A_0 is calculated concentration of phosphate ion of the corresponding concentration and A_t is absorbance at time t.

3. Results and Discussions

3.1. XRD Diffraction Study and BET Results. The crystalline phases of the synthesized TiO_2 and Mg^{2+}-TiO_2 were examined by XRD, and the diffractograms are given in Figure 1. The XRD patterns of calcined (400°C) TiO_2 and the samples of Mg^{2+}-TiO_2 (Mg/Ti = 0.75, 1.0 and 1.25 wt.%) show only anatase form, indicating that Mg^{2+} ions in TiO_2 did not influence the crystal patterns of TiO_2 particle. Further, examination of XRD patterns of both pure and Mg^{2+}-doped TiO_2 illustrates the existence of peaks at 25.3, 37.7, 47.7, and 54.2 (2θ). The intensity of anatase TiO_2 peaks increased due to doping, and peaks corresponding to $MgNO_3$ and $MgCO_3$ were not detected. This may be due to the well dispersion of Mg^{2+} content in TiO_2 particles.

Since Mg^{2+} is more electropositive, the electronic cloud in each TiO_2 might be loosely held, favoring the formation of less dense anatase phase. In other words, the tight packing arrangement required for rutile phase formation is fully suppressed by the addition of magnesium nitrate in water which enhances the polarity of water, thus facilitating the formation of anatase phase exclusively. In addition, the presence of residual alkyl groups can also reduce the rate of crystallization of TiO_2 which favored the formation of less dense anatase phase.

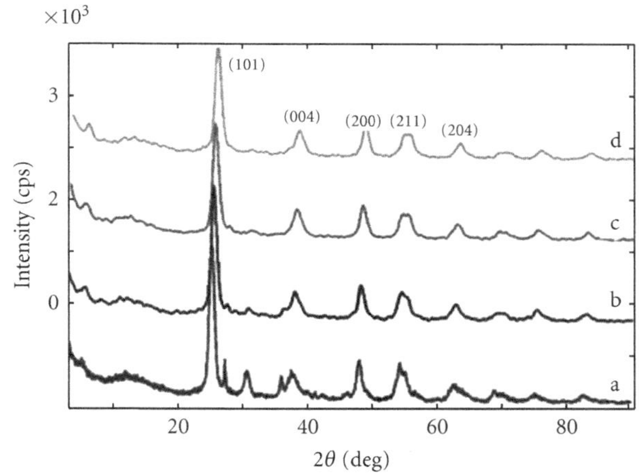

FIGURE 1: XRD patterns of (a) undoped TiO_2, (b) 0.75 wt.% of Mg^{2+}-doped TiO_2, (c) 1.0 wt.% of Mg^{2+}-doped TiO_2, and (d) 1.25 wt.% of Mg^{2+}-doped TiO_2.

The BET surface area results have shown that there is a much increase in surface area of doped catalysts than undoped one. With the incorporation of Mg^{2+} dopant during the sol-gel preparation technique, there was crystal growth suppression, favoring the formation of smaller TiO_2 crystallite. This effect may be attributed to the enhanced lattice strain in the doped TiO_2 network and then decrease grain growth rate.

3.2. Diffuse Reflectance Spectroscopy. The DRS of Mg^{2+}-doped TiO_2 and undoped TiO_2 samples are given in Figure 2. The spectra of undoped TiO_2 showed an absorption peak at 388 nm in the UV region. The absorption peak of the doped TiO_2 with different percentages of Mg^{2+} at 0.75 wt.%, 1.0 wt.%, 1.25 wt.%, and 1.50 wt.% shows considerable shift towards the visible region at around 400–800 nm for all the samples. The tailing absorption peaks can be considered as the extra tail states in the bandgap because Mg^{2+} was added to the TiO_2 matrix. The extension of absorption edge to longer wavelengths for Mg^{2+}/TiO_2 indicates the existence of good contact between TiO_2 and the metal (Mg^{2+}) ion grain.

The UV-Vis absorption edge and bandgap energies of the samples were determined from the reflectance [F(R)] spectra using the Kubelka-Munk (KM) formalism and the Tauc plot method [22]. The extrapolated lines for Mg^{2+}-doped TiO_2 have been given in Figure 3 and used to determine the bandgap energies for Mg^{2+}-doped TiO_2, undoped TiO_2, and Degussa P25 samples. The calculated bandgap energies are given in Table 1. The largest reduction gap is observed for catalyst containing 1.25 wt.% of Mg^{2+}-doped TiO_2. This reduction bandgap may be attributed to the doping of Mg^{2+} as impurity into the host structure (TiO_2) and create extra energy levels situated within the bandgap [23]. The extra energy level created within the bandgap due to shifting of the Fermi energy level of Mg^{2+} towards valence band since metal create P-type semiconductor.

Table 1: Crystallite size, BET surface area, and bandgap energy of each catalyst.

Catalyst no.	wt.% doping of Mg^{2+} in TiO_2	Calcination temp. (°C)	Crystallite size (nm)	BET surface area (m^2/g)	Bandgap energy (eV)
(1)	0.75	400	15.00	85.209	2.90
(2)	1.0	400	12.11	105.987	2.85
(3)	1.25	400	11.25	107.267	2.75
(4)	1.50	400	13.26	101.820	2.95
(5)	Undoped TiO_2	400	11.30	79.382	3.2
(6)	Degussa P25	—	—	69.962	3.1

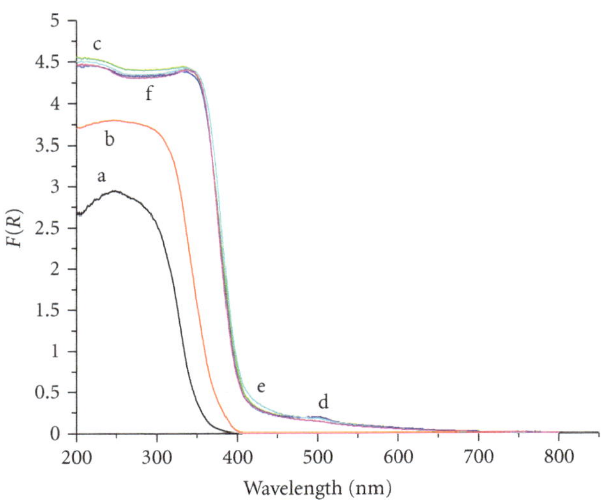

(a) Undoped TiO_2
(b) Degussa P25
(c) 0.75 wt.% of Mg^{2+}
(d) 1 wt.% of Mg^{2+}
(e) 1.25 wt.% of Mg^{2+}
(f) 1.5 wt.% of Mg^{2+}

Figure 2: The DRS-UV-Vis spectra of Degussa P25, undoped, and doped TiO_2 with different % of Mg^{2+}.

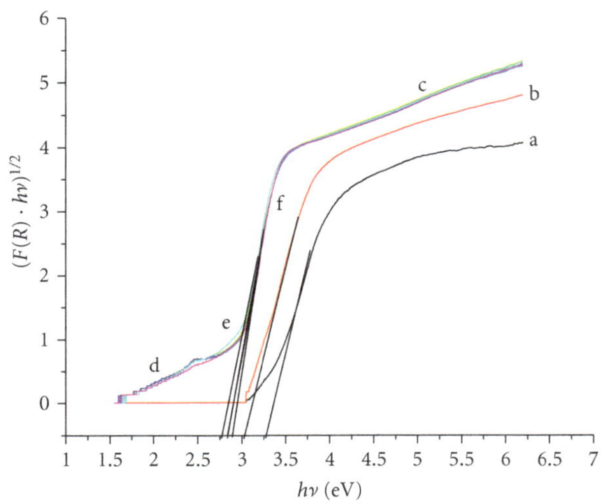

(a) Undoped TiO_2
(b) Degussa P25
(c) 0.75 wt.% of Mg^{2+}
(d) 1 wt.% of Mg^{2+}
(e) 1.25 wt.% of Mg^{2+}
(f) 1.5 wt.% of Mg^{2+}

Figure 3: Plots of transformed Kubelka-Munk functions $[F(R)\cdots h\nu]^{1/2}$ versus $h\nu$ for Degussa P25, undoped TiO_2, and Mg^{2+}-doped TiO_2 samples.

3.3. BET Surface Areas and Pore Distribution. The BET surface areas of the as-prepared Mg^{2+}-TiO_2 sample have been given in Table 1, and nitrogen adsorption and desorption isotherms and corresponding pore size distribution curves (inset) are given in Figure 4. It is observed that at relatively high pressures between 0.075 and 0.15 (P/P_o), the curve exhibit a hysteresis loop indicating the presence of mesopores. The shape of the hysteresis loop is of type H3, associated with particle giving rise to narrow slit shaped pores [24–27]. From Figure 4 the inset figure indicated the pore size distribution range from 20 to 35 nm. The pore volume of the as-prepared catalyst is 0.135 m^3/g. However, the existing poresarise maybe due to removal of nitrate ion and organic moiety from the metal precursors (magnesium and titanium). Due tothis, organized porous structure within the bulk sample of the catalyst may helpful in photocatalysis.

3.4. Analysis of Hydroxyl Radicals ($OH^{\bullet}$). The analysis of hydroxyl radicals has been carried out by photoluminescence technique [28]. 0.1 g of Mg^{2+}-TiO_2 in 10 ppm of coumarin solution was illuminated with visible light. The fluorescence emission spectrum excited at 428 nm from the coumarin solution was measured at every 15 min of illumination. Figure 5 shows the induction of fluorescence from 10 ppm coumarin solution. As shown in the figure, gradual increase in the fluorescence at about 500 nm for 7-hydroxycoumarin which is obtained due to reaction of OH radicals and coumarin was observed by continuous illumination of visible light on the Mg^{2+}-TiO_2 solution. The trend observed in the figure indicated that formation of the OH radicals is proportional to the irradiation time.

3.5. XPS Measurements. X-ray photoelectron spectroscopy (XPS) analysis of magnesium- (II) doped sample was performed, and high-resolution scans are shown in Figures 6(a), 6(b), and 6(c). The XPS observations show that only Mg, Ti and O elements were detected from the samples in spectrum analysis. The binding energy of Mg 2p was found

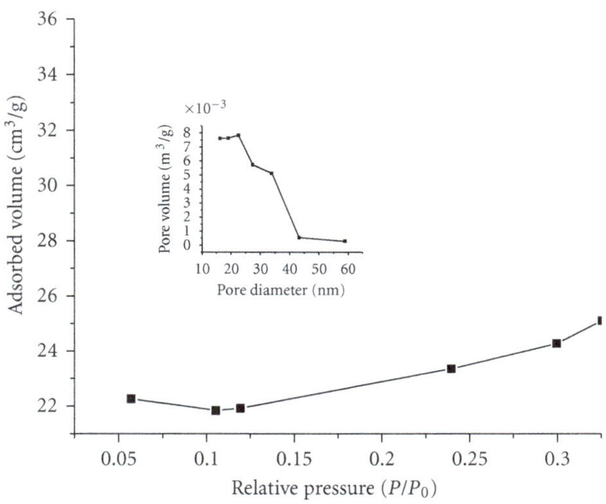

FIGURE 4: Nitrogen adsorption-desorption isotherms and corresponding pore size distribution. Curves (inset) of the Mg^{2+}-TiO_2 prepared at 1.25 wt.% of Mg^{2+}.

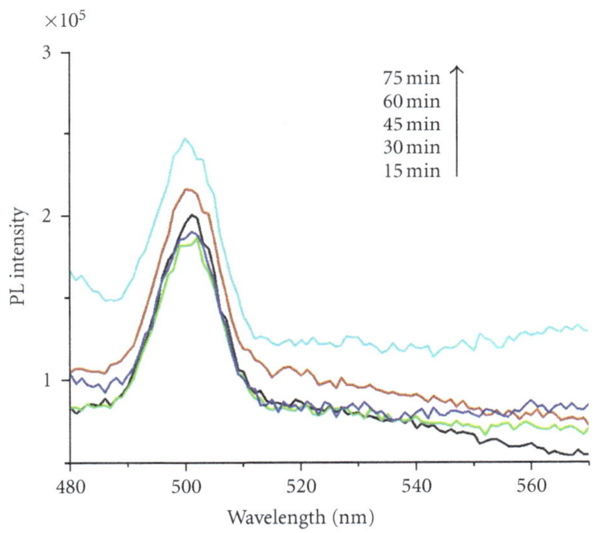

FIGURE 5: PL spectral changes with irradiation time on the undoped TiO_2 and 10 ppm coumarin.

to be 50.4 eV (Figure 4(a)) which is a typical of Mg^{2+} that bonded to oxygen and was assigned to the compound MgO. The binding energies of Ti $2p_{3/2}$ and Ti $2p_{1/2}$ were found to be 458.4 and 464.0 eV, and these bands could belong to Ti^{4+} (Figure 6(b)) and there was no fitting peak for Ti^{3+}. Hence the XPS-spectra confirmed the chemical composition of magnesium-doped sample to be MgO and TiO_2.

3.6. Scanning Electron Microscopy. The SEM images of TiO_2 and metal-doped TiO_2 catalysts are shown in Figures 7(a) and 7(b). The image of samples containing 1.25 wt.% of Mg^{2+} ion shows the morphological changes induced by the addition of alkaline earth cation. The catalysts are found to contain irregular-shaped particles which are again aggregates of tiny crystals. Figure 7(a) shows anatase (undoped TiO_2) produced with no addition of magnesium ion. The sample appears as large blocks of coarse material of dimensions from around 1 to 20 μm. Figure 7(b) shows sample containing 1.25 wt.% of Mg^{2+}, with reduction in the particle size to 1–10 μm range. This clearly illustrates the altered morphology of the catalyst powders, which consists of a large portion of submicron-sized particles.

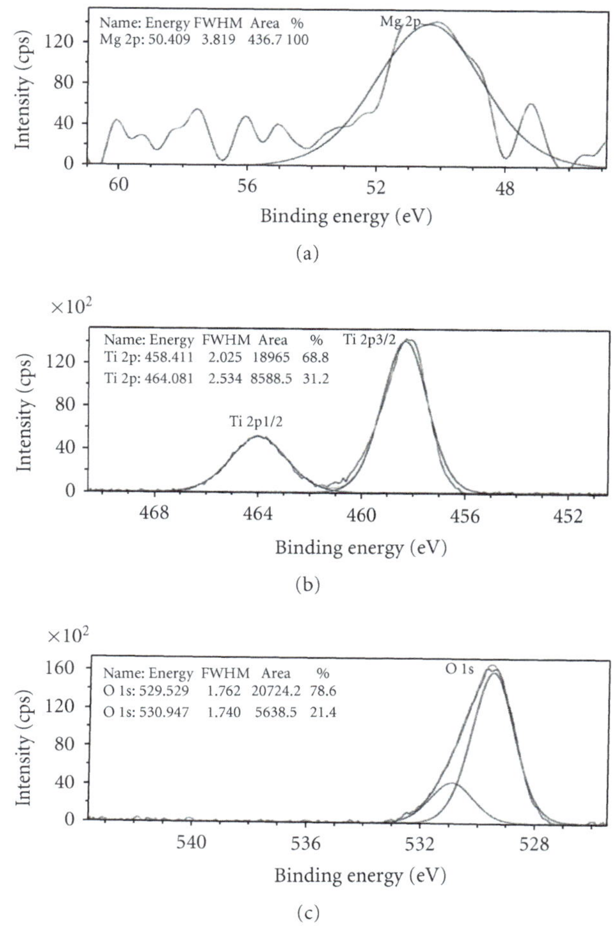

FIGURE 6: (a) High resolution of Mg 2p spectrum, (b) high resolution of Ti 2p spectrum, and (c) high resolution of O 1s spectrum.

3.7. FT-IR Spectral Study. FT-IR spectra of undoped TiO_2 and 1.25 wt.% of Mg^{2+}-TiO_2 given in Figures 8(a) and 8(b) have shown peaks corresponding to stretching vibrations of O–H and bending vibrations of adsorbed water molecules around 2322–3317 cm^{-1} and 1613 cm^{-1}, respectively. The intensity of the peaks became less in Mg^{2+}-TiO_2, indicating the removal of large portion of adsorbed water from TiO_2.

The broad band below 456.06 cm^{-1} of undoped TiO_2 is due to Ti–O–Ti vibration. This peak has been shifted to 470.63 cm^{-1} in Mg-doped TiO_2 which may be because of Ti–O–Mg vibration. Hence the FT-IR spectral study along with XPS established the substitution of Mg^{2+} into TiO_2 lattice.

FIGURE 7: SEM images of (a) undoped TiO_2 and (b) 1.25 wt.% of Mg-doped TiO_2.

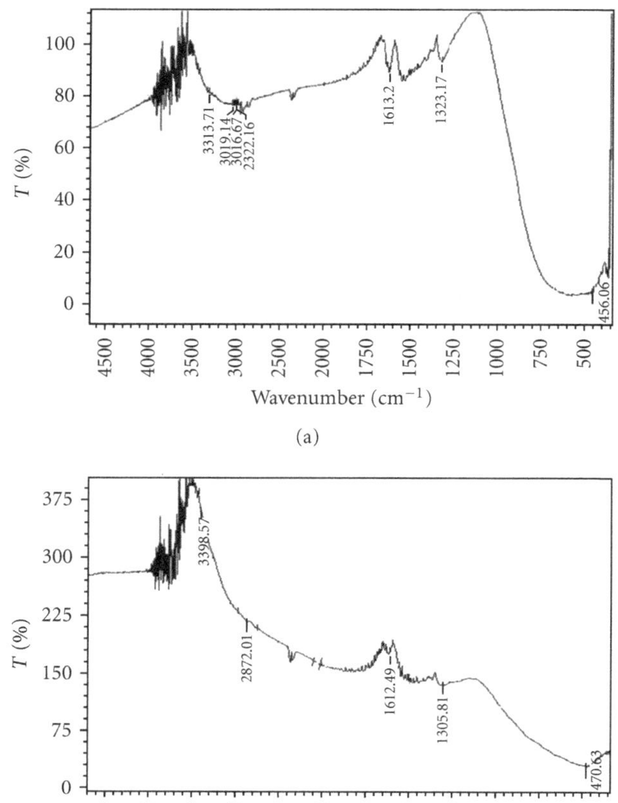

FIGURE 8: FT-IR spectra of (a) undoped TiO_2 and (b) 1.25 wt.% Mg^{+2}-TiO_2.

3.8. Photocatalytic Degradation of Dichlorvos. To understand the photocatalytic activity of the prepared catalysts, degradation of dichlorvos pesticide under the irradiation of visible light has been carried out, in the presence and absence of catalysts. The percentage of DDVP degradation is significantly less in the absence of catalyst (0.05% for 9 hours irradiation) when compared to that of undoped or doped titanium dioxide. A blank experiment in the absence of irradiation along with catalysts demonstrated that no significant change in the DDVP concentration was observed. The efficiency of photocatalytic degradation process depends on various experimental parameters such as effect of dopant concentration, pH, catalyst dosage, and initial concentration of pollutant. Hence, it is essential to optimize these parameters to achieve higher degradation efficiency of photocatalyst.

3.8.1. Effect of Mg Dopant Concentration on Photocatalytic Degradation of DDVP. Figure 9 shows the results of the photocatalytic degradation of DDVP by irradiation of visible light over Degussa P25, undoped TiO_2 catalyst, and Mg-doped TiO_2 photocatalyst with different doping concentrations of Mg in TiO_2. Among various concentrations of Mg, 1.25 wt.% of Mg-TiO_2 catalyst exhibits the highest photocatalytic activity. This may be attributed to that increase in dopant concentration leads to increase in number of trapped charge carriers per particle [15, 29] and is the highest in 1.25 wt.% of Mg^{2+}-TiO_2 which is the optimum concentration showing maximum photocatalytic activity.

The results of the experiment show that the rate of phosphate ion formation increased with increase in the concentration of magnesium ions, and it was optimum at 1.25 wt.%. However, magnesium ions' doping became detrimental when the dopant concentration was above 1.25 wt.%. It also demonstrates that, even though the concentration of the doped magnesium ions is small, it still gives much influence on the photocatalytic activity of TiO_2 particles.

3.8.2. Effect of pH. The adsorption properties of catalysts can be greatly changed at different pH values; hence, the catalyst-assisted photodegradation of DDVP has also been monitored by in situ measurements of pH of the aqueous solution with irradiation time, at a fixed weight of catalyst and DDVP pesticide concentration at different pH values. Solution pH influences adsorption and desorption of the substrate, catalyst surface charge, oxidation potential of the valence band, and other physicochemical properties.

The effect of pH on the photocatalytic degradation of DDVP is shown in Figure 10. The degradation plot for DDVP has been plotted between rate constant and pH of the solution. Based on the results, it is observed that the rate of degradation decreased with increase in pH of the solution. The optimum pH of the solution for the degradation of DDVP is 2. Generally in acidic pH the surface of the TiO_2 is in the form of $TiOH_2^+$. The photogenerated electrons can be captured by the adsorbed H^+ to form $H^{\bullet}_{ads}$. At higher pH,

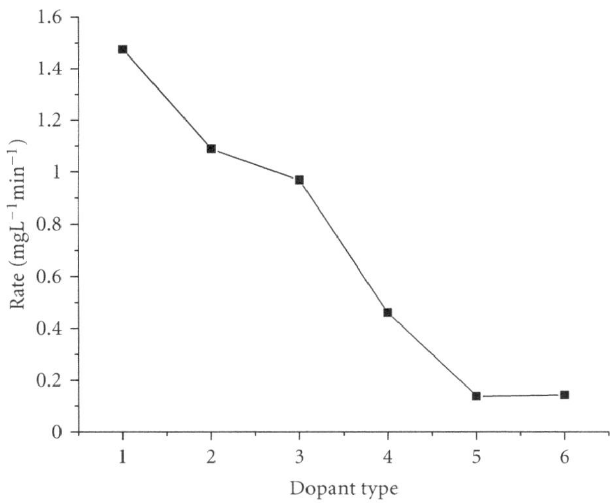

(1) 1.25 wt.% of Mg^{2+}
(2) 1 wt.% of Mg^{2+}
(3) 1.5 wt.% of Mg^{2+}
(4) 0.75 wt.% of Mg^{2+}
(5) Undoped TiO_2
(6) Degussa P25

FIGURE 9: Effect of dopant concentration on the degradation of dichlorvos. Catalyst dosage = 0.10 g, pH = 3, and [DDVP] = 150 ppm.

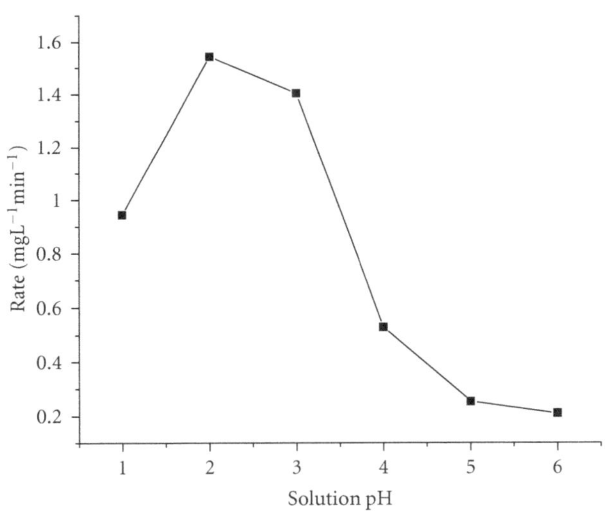

FIGURE 10: The effect of pH on the degradation of dichlorvos by Mg^{2+}-doped TiO_2. Catalyst dosage = 0.10 g, [DDVP] = 150 ppm.

the surface of catalyst has a net negative charge as TiO_2^-, and hence the degradation rate was found to be lower:

$$\begin{aligned} TiOH + H^+ &\longrightarrow TiOH_2^+, \quad pH < 6.25,\\ TiOH &\longrightarrow TiO^- + H^+, \quad pH > 6.25. \end{aligned} \tag{3}$$

3.8.3. Effect of Catalyst Dosage. The effect of the amount of catalyst on the photodegradation rate was investigated at a fixed pH and initial concentration of the dichlorvos, experiments were performed with varying amounts of Mg^{2+}-TiO_2

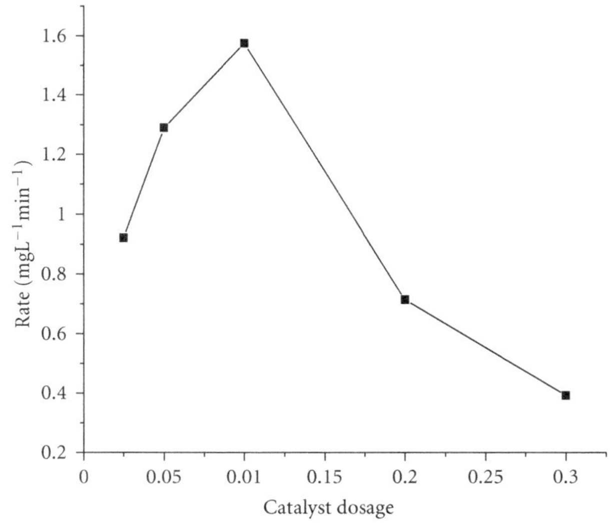

FIGURE 11: The effect of catalyst dosage on the degradation of dichlorvos by Mg^{2+}-doped TiO_2, pH = 2, [DDVP] = 150 ppm.

from 0.025 g to 0.3 g in 100 mL DDVP pesticide aqueous solution. The degradation pattern from such experiments has been depicted in Figure 11, which reveals that the rate of degradation increases linearly with increase in the amount of catalyst up to 0.1 g and then decreases. This may be explained by two main reasons. The first one is due to the presence of foreign metal ions (Mg^{2+}) as dopant that helps in inhibiting electron-hole pair recombination and enhancing interfacial charge transfer reactions, which leads to increased catalytic activity.

Secondly, as the amount of catalyst increases, the number of photons absorbed and the number of DDVP molecules adsorbed are increased leading to the increase in DDVP degradation. At higher concentrations of catalyst, although more surface areas are available for constant DDVP molecules, the solution turbidity increases and it interferes with the penetration of light. The deactivation of activated molecules by collision with ground-state molecules may also hinder the photocatalytic efficiency [30]. Hence, above a certain level, additional catalyst amount is not involved in catalysis, and thus the rate levels off.

3.8.4. Effect of Initial Pesticide Concentration. At a fixed weight of Mg^{2+}-TiO_2 and pH, variation of initial pesticide concentration on photo degradation was studied and shown in Figure 12. It was noted that the degradation rate increases with the increase of pesticide concentration up to a certain concentration (150 ppm), and a further increase leads to a decrease in the pesticide degradation rate. This is due to the reduction of generation of radicals ($OH^\bullet$) on the catalyst surface since the active surfaces are covered by pesticide molecules.

3.9. Photocatalytic Mechanism. Based on the experimental results and reports of the previous workers, the following

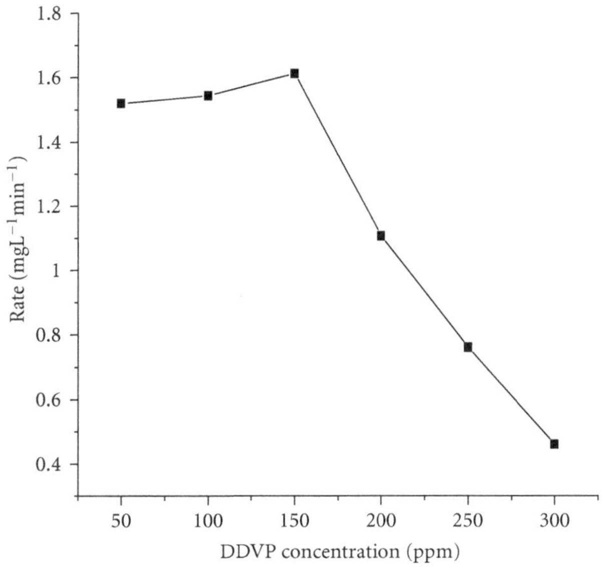

FIGURE 12: The effect of concentration of DDVP on the degradation of dichlorvos by Mg^{2+}-doped TiO_2, pH = 2, catalyst dosage = 0.10 g.

mechanism would be proposed for the photocatalytic reactions of magnesium-doped TiO_2 [31–33].

(a) Upon visible light illumination of photocatalyst, electrons are ejected from the valence band to the conduction band leaving positive holes in the valence band.

(b) When the metal ion doped into TiO_2 lattice, it trapped the ejected electrons, holding up the recombination process.

(c) The trapped electrons can be further scavenged by molecular oxygen, which is adsorbed on the TiO_2 surface to generate superoxide radical, and this in turn produces hydrogen peroxide (H_2O_2), hydroperoxy ($HO_2^{\bullet}$), and hydroxyl ($^{\bullet}OH$) radicals.

(d) The positive holes in the valence band act as good oxidizing agents available for degradation of pesticide in the solution. Thus, the DDVP is attacked by the hydroxyl radicals formed both by trapped electrons and hole in the VB as given in the above equations, to generate phosphate ions, organic radicals, or other intermediates, which further undergo degradation.

4. Conclusions

The DRS analysis shows that doping of Mg^{2+} into TiO_2 shifts the absorbance band of TiO_2 from UV to visible region, and the bandgap energy is reduced for all doped catalysts (from 3.2 eV to 2.75 eV); the largest reduction gap was observed for doped catalyst of 1.25 wt.% of Mg. The incorporation of Mg^{2+} ions into the TiO_2 was evidenced by XPS analysis. The BET results also show that there is an increase in surface area of the catalyst, which enhances the photocatalytic degradation of DDVP in visible light. The present work imparts the significance of magnesium ion doping into titania, in dichlorvos degradation. It also establishes the doping of magnesium by sol-gel as a best alternative method. The experimental results show that the photocatalytic activity of undoped TiO_2 is lower than magnesium-doped TiO_2.

Magnesium doping causes better separation of electron and holes on the modified TiO_2 surface allowing more efficient channeling of the produced electrons into useful redox reactions, thus enhancing the DDVP degradation in the presence of visible light, when compared to undoped TiO_2 which can be excited only in less available UV light. The degradation of DDVP was maximum at 150 ppm DDVP concentration, 0.1 g of Mg^{2+}-TiO_2 catalyst, and at a pH of 2 using 1.25 wt.% of magnesium-doped titania catalyst.

Acknowledgments

T. Rao and Mrs. T. Susmitha are thankful to UGC, New Delhi, for financial assistance as major research project to carry out research work and also thankful to Dr. Sridhar, Scientist, IICT, Hyderabad, for providing XPS spectral data. T. A. Segne is also thankful to the Ethiopian Government for their financial support for doing Ph.D. at the Andhra University, Visakhapatnam, India.

References

[1] O. Legrini, E. Oliveros, and A. M. Braun, "Photochemical processes for water treatment," *Chemical Reviews*, vol. 93, no. 2, pp. 671–698, 1993.

[2] D. Chen and A. K. Ray, "Photocatalytic kinetics of phenol and its derivatives over UV irradiated TiO_2 ," *Applied Catalysis B*, vol. 23, no. 2-3, pp. 143–157, 1999.

[3] A. L. Linsebigler, G. Lu, and J. T. Yates, "Photocatalysis on TiO_2 surfaces: principles, mechanisms, and selected results," *Chemical Reviews*, vol. 95, no. 3, pp. 735–758, 1995.

[4] C. Minero, E. Pelizetti, P. Pichat, M. Sega, and M. Vincenti, "Formation of condensation products in advanced oxidation technologies: the photocatalytic degradation of dichlorophenols on TiO_2 ," *Environmental Science and Technology*, vol. 29, no. 9, pp. 2226–2234, 1995.

[5] P. K. J. Robertson, "Semiconductor photocatalysis: an environmentally acceptable alternative production technique and effluent treatment process," *Journal of Cleaner Production*, vol. 4, no. 3, pp. 203–212, 1996.

[6] A. Fujishima and K. Honda, "Electrochemical photolysis of water at a semiconductor electrode," *Nature*, vol. 238, no. 5358, pp. 37–38, 1972.

[7] R. W. Matthews, "Photooxidation of organic impurities in water using thin films of titanium dioxide," *Journal of Physical Chemistry*, vol. 91, no. 12, pp. 3328–3333, 1987.

[8] M. A. Fox, C. C. Chen, K. Park, and J. N. Younathan, "Controlled organic redox reactivity on irradiated semiconductor surfaces," *ACS Symposium Series*, vol. 278, pp. 69–78, 1985.

[9] A. Mills, R. H. Davies, and D. Worsley, "Water purification by semiconductor photocatalysis," *Chemical Society Reviews*, vol. 22, no. 6, pp. 417–425, 1993.

[10] S. Peng, Y. Li, F. Jiang, G. Lu, and S. Li, "Effect of Be^{2+} doping TiO_2 on its photocatalytic activity," *Chemical Physics Letters*, vol. 398, no. 1–3, pp. 235–239, 2004.

[11] M. R. Hoffmann, S. T. Martin, W. Choi, and D. W. Bahnemann, "Environmental applications of semiconductor photocatalysis," *Chemical Reviews*, vol. 95, no. 1, pp. 69–96, 1995.
[12] Y. Ku, W. Wang, and Y. S. Shen, "Reaction behaviors of decomposition of monocrotophos in aqueous solution by UV and UV/O_3 processes," *Journal of Hazardous Materials*, vol. 72, no. 1, pp. 25–37, 2000.
[13] C. Kormann, D. W. Bahnemann, and M. R. Hoffmann, "Preparation and characterization of quantum-size titanium dioxide," *Journal of Physical Chemistry*, vol. 92, no. 18, pp. 5196–5201, 1988.
[14] J. Moser and M. Grätzel, "Photosensitized electron injection in colloidal semiconductors," *Journal of the American Chemical Society*, vol. 106, no. 22, pp. 6557–6564, 1984.
[15] W. Choi, A. Termin, and M. R. Hoffmann, "The role of metal ion dopants in quantum-sized TiO_2: correlation between photoreactivity and charge carrier recombination dynamics," *Journal of Physical Chemistry*, vol. 98, no. 51, pp. 13669–13679, 1994.
[16] Q. Xiang, J. Yu, and M. Jaroniec, "Graphene based semiconductor photocatalysis," *Chemical Society Reviews*, vol. 41, no. 2, pp. 782–796, 2012.
[17] Y. Bessekhouad, D. Robert, J. V. Weber, and N. Chaoui, "Effect of alkaline-doped TiO_2 on photocatalytic efficiency," *Journal of Photochemistry and Photobiology A*, vol. 167, no. 1, pp. 49–57, 2004.
[18] S. Liu, J. Yu, and M. Jaroniec, "Anatase TiO_2 with dominant high-energy {001} facets: synthesis, properties and applications," *Chemistry of Materials*, vol. 23, no. 18, pp. 4085–4093, 2011.
[19] E. D. Booth, E. Jones, and B. M. Elliott, "Review of the in vitro and in vivo genotoxicity of dichlorvos," *Regulatory Toxicology and Pharmacology*, vol. 49, no. 3, pp. 316–326, 2007.
[20] E. Evgenidou, K. Fytianos, and I. Poulios, "Semiconductor-sensitized photodegradation of dichlorvos in water using TiO_2 and ZnO as catalysts," *Applied Catalysis B*, vol. 59, no. 1-2, pp. 81–89, 2005.
[21] T. Wu, G. Liu, J. Zhao, H. Hidaka, and N. Serpone, "TiO_2 assisted degradation of dyes IX: photooxidation of asquarylium cyanine dye in aqueous dispersions under visible light irradiation," *Environmental Science and Technology*, vol. 33, no. 9, pp. 1379–1387, 1999.
[22] D. Blake, "Bibliography of work on the heterogeneous photocataloytic removal of hazardous compounds from water and air," Tech. Rep. NREL/TP-510-31319, National Technical Information Service, United States, Department of Commerce, 2001.
[23] W. Wang, J. Zhang, F. Chen, D. He, and M. Anpo, "Preparation and photocatalytic properties of Fe^{3+} Ag@TiO_2 core-shell nanoparticles," *Journal of Colloid and Interface Science*, vol. 323, no. 1, pp. 182–186, 2008.
[24] K. S. W. Sing, D. H. Everett, R. A. W. Haul et al., "Reporting physisorption data for gas/solid interface with special reference to the determination of surface area and porosity," *Pure and Applied Chemistry*, vol. 57, pp. 603–319, 1985.
[25] S. J. Gregg and K. S. W. Sing, *Adsorption, Surface Area and Poracity*, Academic Press, New York, NY, USA, 1982.
[26] J. Yu, J. C. Yu, K.-P. Mitch et al., "Effects of acidic and basic hydrolysis catalysts on the photocatlytic activity and structures of bimodal mesoporous titania," *Journal of Catalysis*, vol. 217, no. 1, pp. 69–78, 2003.
[27] J. Yu, S. Liu, and H. Yu, "Microstructures and photoactivity of mesoporous anatase hollow microspheres fabricated by fluoride-mediated self-transformation," *Journal of Catalysis*, vol. 249, no. 1, pp. 59–66, 2007.
[28] Q. Xiang, J. Yu, and P. K. Wong, "Quantitative characterization of hydroxyl radicals produced by various photocatalysts," *Journal of Colloid and Interface Science*, vol. 357, no. 1, pp. 163–167, 2011.
[29] B. Neppolian, H. C. Choi, S. Sakthivel, B. Arabindoo, and V. Murugesan, "Solar/UV-induced photocatalytic degradation of three commercial textile dyes," *Journal of Hazardous Materials*, vol. 89, no. 2-3, pp. 303–317, 2002.
[30] J. M. Herrmann and C. Guillard, "Photocatalytic degradation of pesticides in agricultural used waters," *Comptes Rendus de l'Académie des Sciences IIC*, vol. 3, no. 6, pp. 417–422, 2000.
[31] H. Li, X. Duan, G. Liu, and L. Li, "Synthesis and characterization of copper ions surface-doped titanium dioxide nanotubes," *Materials Research Bulletin*, vol. 43, no. 8-9, pp. 1971–1981, 2008.
[32] K. Chiang, R. Amal, and T. Tran, "Photocatalytic degradation of cyanide using titanium dioxide modified with copper oxide," *Advances in Environmental Research*, vol. 6, no. 4, pp. 471–485, 2002.
[33] Y. Yang, Q. Wu, Y. Guo, C. Hu, and E. Wang, "Efficient degradation of dye pollutants on nanoporous polyoxotungstate-anatase composite under visible-light irradiation," *Journal of Molecular Catalysis A*, vol. 225, no. 2, pp. 203–212, 2005.

3

Simulation of Electronic Structure of Aluminum Phosphide Nanocrystals Using Ab Initio Large Unit Cell Method

Hamad R. Jappor,[1] Zeyad Adnan Saleh,[2] and Mudar A. Abdulsattar[3]

[1] *Department of Physics, College of Education, University of Babylon, P.O. Box 4, Hilla, Iraq*
[2] *Department of Physics, College of Science, Al-Mustansiriyah University, P.O. Box 46010, Baghdad, Iraq*
[3] *Directorate of Materials Research, Ministry of Science and Technology, P.O. Box 8012, Baghdad, Iraq*

Correspondence should be addressed to Mudar A. Abdulsattar, mudarahmed3@yahoo.com

Academic Editor: Markku Leskela

Ab initio restricted Hartree-Fock method coupled with the large unit cell method is used to determine the electronic structure and physical properties of aluminum phosphide (AlP) nanocrystals between 216 and 1000 atoms with sizes ranging up to about 3 nm in diameter. Core and surface parts with different sizes are investigated. Investigated properties include total energy, cohesive energy, energy gap, valence band width, ionicity, and degeneracy of energy levels. The oxygenated (001)-(1 × 1) facet that expands with larger sizes of nanocrystals is investigated to determine the rule of the surface in nanocrystals electronic structure. Results revealed that electronic properties converge to some limit as the size of the large unit cell increases and that the 216 core atoms approaches bulk of Aluminum phosphide material in several properties. Increasing nanocrystals size also resulted in a decrease in lattice constant, increase of core cohesive energy (absolute value), increase of core energy gap, increase of core valence band width and decrease of ionicity. Valence and conduction bands are wider on the surface due to splitting and oxygen atoms. The method also shows fluctuations in the converged energy gap, valence band width and cohesive energy of core part of nanocrystals duo to shape variation.

1. Introduction

Semiconductor nanostructures have been of particular interest because confinement effects lead to large increase in the band gap (blue shift) and efficient light emission, even in Si for which coupling to light is extremely weak in the bulk crystal [1]. In the case of a pure semiconductor, the broken bonds lead to reconstruction of the surface, and in the smallest clusters, there is little resemblance to the bulk structures. On the other hand, terminating the surface by atoms such as hydrogen, which remove the dangling bonds makes the resulted cluster much more like a small, terminated piece of the bulk. Aluminum phosphide (AlP), with one of the largest direct gaps of the III–V compound semiconductors, is undoubtedly the most "exotic" and least studied [2]. At normal conditions, AlP crystallizes in the zinc blende (zb) structure. However, in recent years, it attracted special attention to its incorporation in the AlAs/AlP- and GaP/AlP-based heterostructures. AlP is a subject of extensive theoretical studies ranging from the semiempirical to the first principles methods [3], within the density functional theory (DFT) framework using both pseudopotential [4], and all-electron approaches. For the bulk properties of AlP, theoretical calculations based on the Hartree-Fock [5], and potential model [6] have obtained a very good description of its structural and electronic properties. Recently, Annane et al. [7] investigated the structural and electronic properties of AlAs and AlP compounds using the full potential linearized augmented plane wave plus local orbitals method based on density functional theory.

The present work addresses large nanocrystals that have the size range of 216–1000 total atoms. These atoms are divided between core and surface. The present lower limit (of atoms) is chosen since it provides the first occurrence of idle core part (nearly ideal zincblende structure) that is far from surface reconstruction by more than three surface layers. To the best of our knowledge, no theoretical investigations of cubic zb AlP nanocrystals have appeared in the literature up to the present date.

In the present work, we study the electronic structure and physical properties of AlP nanocrystals core and surface part with different sizes by using an ab initio restricted Hartree-Fock (RHF) method coupled with large unit cell method (LUC-RHF). LUC method was formulated and used before for several kinds of bulk materials including diamond and zinc blend (zb) structured materials [8–11].

2. Computational Details

Ab initio self-consistent Hartree-Fock (RHF) is used to obtain Aluminum phosphide nanocrystal molecular orbitals. Correlation corrections are neglected in the present calculations relying on Koopmans theorem [12]. This theorem states that comparisons of Hartree-Fock closed-shell results (which is the case in the present work) with experimental values suggest that in many cases the energetic corrections due to relaxation effects nearly cancel the corrections due to electron correlation. Simple STO-3G basis [13, 14] is used in the present work to reach higher number of core and surface atoms.

This method uses ($\mathbf{k} = 0$) approximation, that is, one point in the wave vector space. The LUC method is one kind of supercell methods with the above $\mathbf{k} = 0$ restriction. In this method, instead of adding more **k** points, the single central cell is expanded to contain more atoms, which are now called core atoms in the present work. The LUC method was first suggested and applied to covalent semiconductors in the 1970s [10, 15]. The method was found suitable for nanocrystal calculations because the $\mathbf{k} = 0$ approximation retains only one central cluster of atoms surrounded by other atoms to passivate the outer dangling bonds [16]. The method is used to simulate parts of specific symmetry in the nanocrystal (core and surface) in the same way it is used for bulk materials.

We used 3D PBC (Periodic Boundary Condition) method in GUSSIAN 03 code [17] to calculate ab initio restricted Hartree-Fock electronic structure of AlP nanocrystals internal core. The 2D PBC calculations are used to simulate oxygenated (001)-(1 × 1) surface. In the present work, we divided the work into two parts, core and surface parts, which is the traditional method used in microscopic-size solid-state calculations. In LUC core part only the lattice constant is optimized, at the surface part all bond lengths and lattice constant still need to be optimized because of surface reconstruction. The (001) chosen surfaces are terminated by oxygen atoms to passivated dangling bonds. The width of surface part is set to one conventional lattice that is deduced from AlP surface bonds calculations. The oxygenated (001)-(1 × 1) AlP surface in the oxygen double bonding configuration is investigated to obtain the total electronic structure of the AlP nanocrystal with (001)-(1×1) facets. The oxygenated (001)-(1 × 1) AlP surface is chosen since it is one of the least reconstructed surfaces. Unlike the core part, the bond lengths and lattice constant are not unique and a variation of geometrical parameters is needed. Normally surface

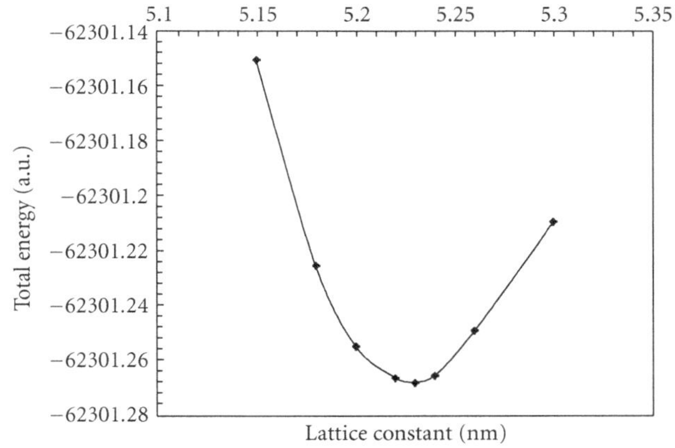

Figure 1: Total energy of 216 atoms aluminum phosphide core LUC as a function of lattice constant.

effects do not penetrate more than four layers (one lattice constant) from the crystal surface [18]. On the other hand, short-range sp^3 bonds do not require more than fourth neighbor's interaction range to conduct electronic structure calculations successfully using molecular orbital methods [16]. The upper two identical conditions are applied in the present calculations.

Two kinds of core LUCs are investigated, namely, cubic and parallelepiped cells. The cubic cells are multiples of zb structure Bravais unit cells, while the parallelepiped cells are multiples of primitive zb structure unit cells [18]. Cubic core cells include 8, 64, and 216 atoms. Parallelepiped cells include 16, 54, 128, and 250 atoms.

3. Calculations and Results

In this section, we present the calculated band structures of aluminum phosphide nanocrystals. Figure 1 shows the total energy for AlP nanocrystal as a function of lattice constant for 216 core atoms, from which we obtained the equilibrium lattice constant. Figure 2 shows the variation of the lattice constant of aluminum phosphide nanocrystals core as a function of number of core atoms. This figure is obtained after minimizing the energy lattice curve of every investigated number of core atoms. From this figure, we see that the lattice constant for the core part of the crystal decrease with increasing number of atoms. The converged lattice length for a high number of core atoms (0.523 nm) is in good agreement with the experimentally reported value (0.545 nm) for bulk [19, 20].

The bulk modulus B can be obtained using the equation of state of Murnghan [21]. Bulk modulus of AlP nanocrystal as a function of number of atoms in the core is shown in Figure 3. This figure shows that the bulk modulus increases with the number of atoms in the core. We note that the bulk modulus for AlP nanocrystal of 97 GPa is in reasonable agreement with the experimental value of 86 [19] of the bulk taking into consideration that Hartree-Fock theory always overestimates bulk modules values [8].

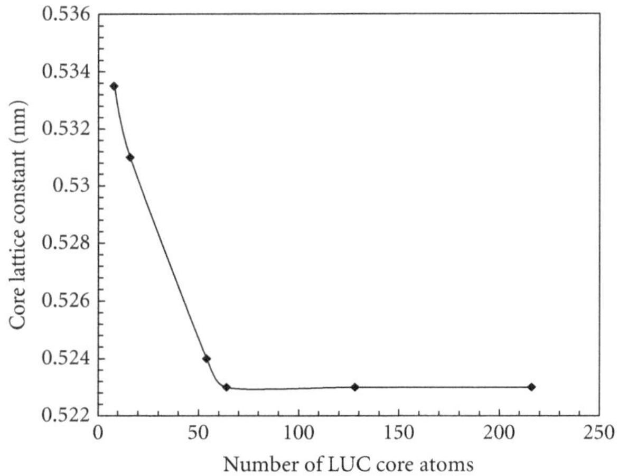

Figure 2: Energetically optimized zb-AlP core lattice constant as a function of the number of atoms in the core.

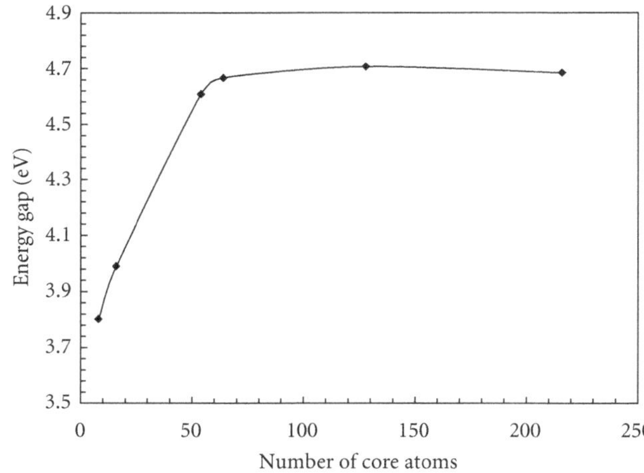

Figure 4: Energy gap of the core part of zb-AlP nanocrystals as a function of the number of core atoms.

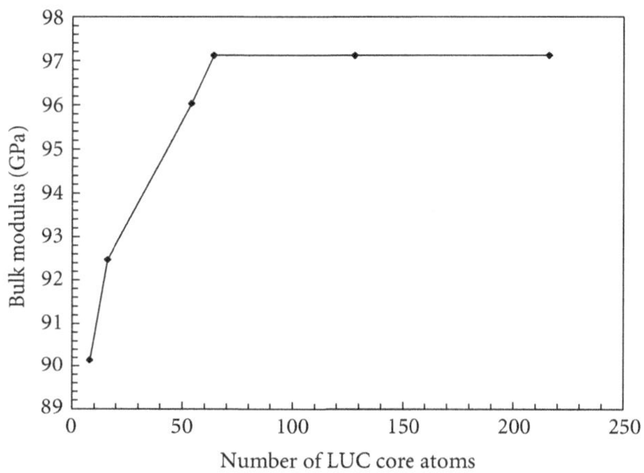

Figure 3: Bulk modulus as a function of the number of core atoms.

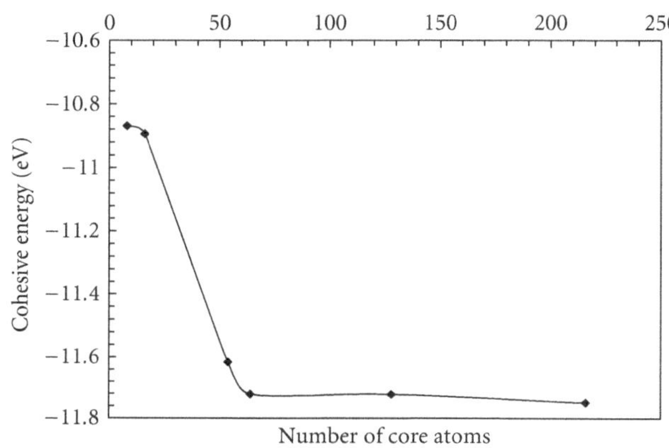

Figure 5: Cohesive energy of the core part of zb-AlP nanocrystals as a function of the number of core atoms.

AlP has an indirect minimum gap between highest occupied molecular orbital (HOMO) and lowest unoccupied molecular orbital (LUMO) with the CBM at X point of the Brillouin zone [22]. Figure 4 shows the energy gap of the core part of the investigated nanocrystals against the number of core atoms. The increase of the energy gap in the first part of the considered range is not a violation of the quantum confinement theory since the energy gap is controlled by the surface part gap [23, 24], which is much smaller gap than the core part. All energies converge quickly for the nanocrystals with more than 64 core atoms with a very wide energy gap variation range approximately (3.80 eV to 4.69 eV), which is higher than the experimentally reported energy gap (2.45–2.5 eV) for bulk zb-AlP [25, 26] but within the trends of nanocrystals blue shift [23, 24, 27, 28]. The relation between the cohesive energy and the number of atoms is shown in Figure 5. A reasonable agreement for the cohesive energy with experimental data (8.34 eV) [29] and other calculations [20, 30] of the bulk value (8.27–9.73 eV). Figure 5 shows that the cohesive energy decreases with increasing number of atoms with small fluctuation after (−11.721 eV) for the cell which contains 64 core atoms.

In the present work, we calculated valence bandwidth which is the difference between HOMO and LOMO levels. The valence bandwidth as a function of number of atoms for AlP nanocrystal core part is shown in Figure 6. One can see from the figure that this band generally increases with increasing number of atoms and becomes nearly stable at the 64 core atoms.

The ionicity charge is calculated as a function of number of core atoms. Figure 7 shows that smaller nanocrystals are more ionic than larger ones.

Figure 8 shows the variation of highest occupied molecular orbital energy (HOMO) and lowest unoccupied molecular orbital energy (LUMO) as the core of nanocrystals grows up in size and changes its shape. This curve fluctuates strongly because of the change in size and shape that produces different surfaces with different properties. HOMO and LUMO variation shows that not all quantities have definite convergence behaviour similar to the upper investigated quantities.

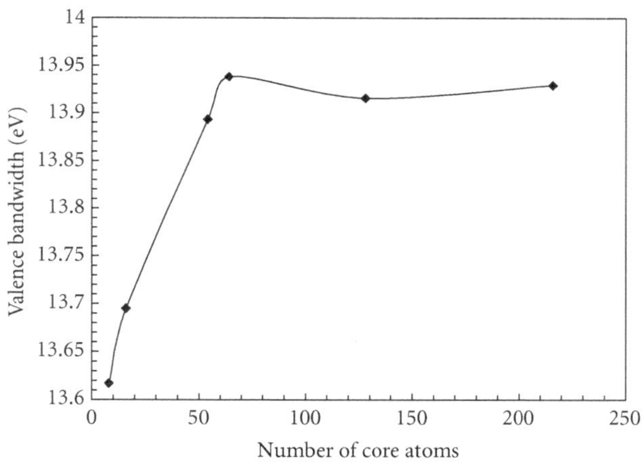

FIGURE 6: Valence bandwidth of the core part of zb-AlP nanocrystals as a function of core atoms.

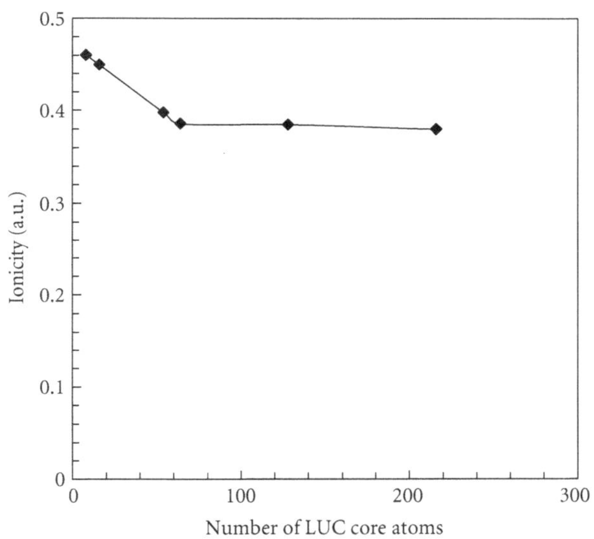

FIGURE 7: Ionicity of the core part of zb-AlP nanocrystals as a function of the number of core atoms.

HOMO and LUMO energy levels plotted in Figure 8 show fluctuating behaviour that continues indefinitely. A negative sign to the approximate ionization potential and affinity, respectively, [31] relates HOMO and LUMO levels. As nanocrystals grow up in size they develop different shapes and surfaces. Each of the developed surfaces has different properties such as the upper mentioned ionization potential and affinity or HOMO and LUMO energy levels. Therefore, these fluctuations start at the nanoscale, continue to the micro-, and bulk scale. A similar behaviour is observed for BP and SiGe nanocrystals [27, 32].

The degeneracy of energy level states for AlP nanocrystal as a function of energy levels for the lowest energy lattice constant is calculated for core and surface parts. Figure 9 shows the degeneracy of states for 8 atoms of core part. Due to high symmetry, the number of degenerate states is high in the core part. The highest number of degenerate states in the valence band is between 6 for the 8 atom cores and 30 for the 216 atom cores. A similar behaviour is observed for diamond nanocrystals ab initio large unit cell calculations [33] and SiC nanocrystals [34].

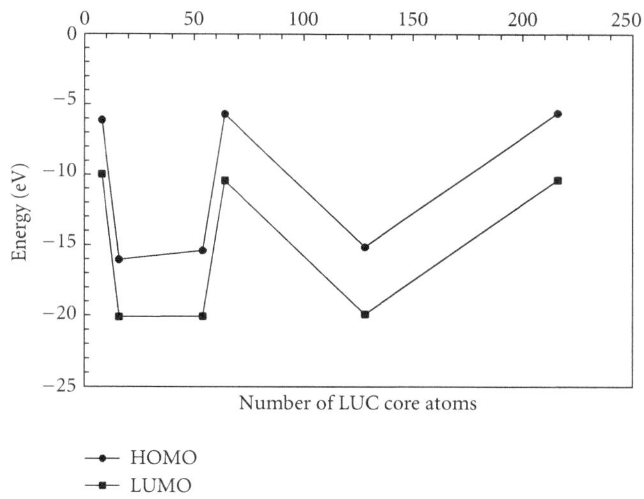

FIGURE 8: HOMO and LUMO energies that correspond to different core sizes of AlP nanocrystals.

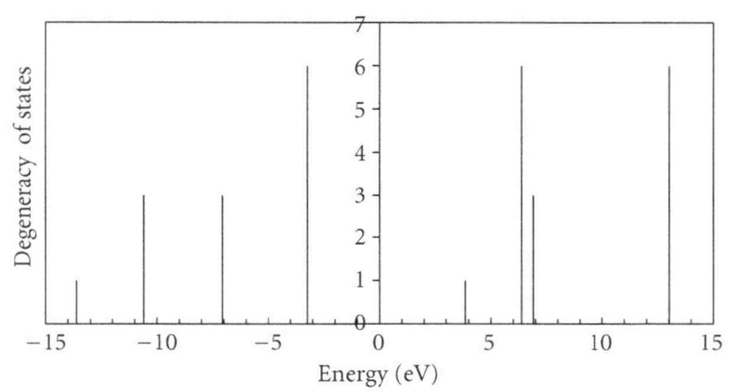

FIGURE 9: Degeneracy of states of 8-atom core part of zb-AlP nanocrystal as a function of level energy. The HOMO level is taken as the reference level.

To investigate nanocrystals surface electronic structure, it is natural to assume that larger nanocrystals have larger surfaces. Three oxygenated oxygen terminated (001)-(1 × 1) zb-AlP surface stoichiometry slabs are investigated, namely, $Al_8O_4P_8$, $Al_{32}O_{16}P_{32}$, and $Al_{72}O_{36}P_{54}$ with three surface areas of a^2, $4a^2$, and $9a^2$, respectively, where (a) is the lattice constant.

The results of the oxygenated (001)-(1 × 1) surface for the lattice constant and the energy gap for the three different areas are shown in Figures 10 and 11, respectively.

The average lattice constants of the three investigated stoichiometries are less than the corresponding core parts, but they follow the same decreasing order. On the other hand, the energy gap of (001)-(1 × 1) oxidized surface shows much smaller gap than that of the core part.

The oxygenated (001)-(1 × 1) surface slab calculations of the nanocrystal converge to approximately the same lattice constant value of the core (0.5195 nm). This good matching

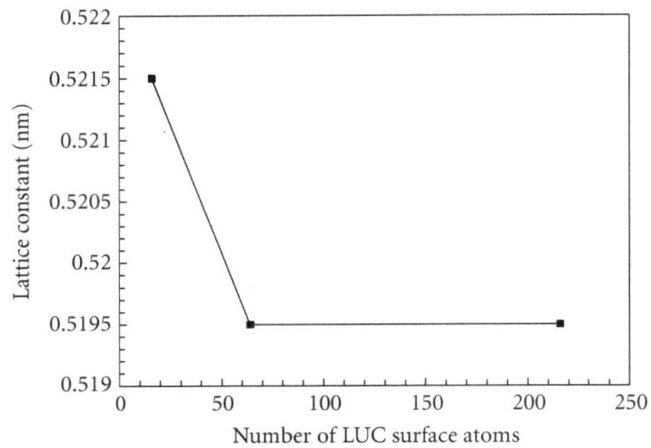

FIGURE 10: Lattice constant variation as a function of nanocrystal oxygenated (001)-(1 × 1) surface face.

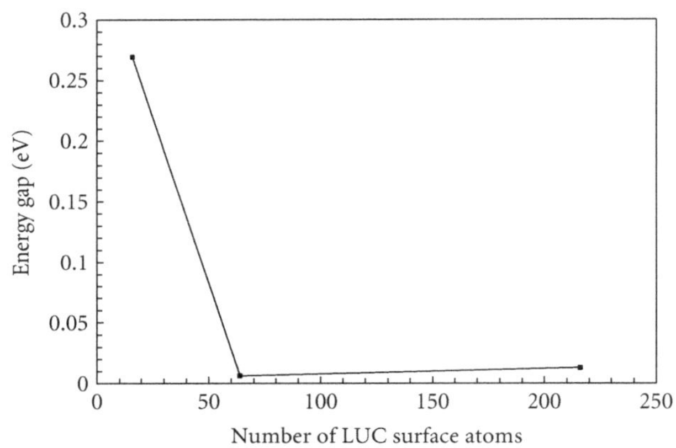

FIGURE 11: Energy gap of AlP nanocrystals oxygenated face as a function of surface area of nanocrystals facet.

between surface and core lattice constants shows that good adhesion is expected between the two parts.

In the surface part, discontinuity of surface imposes level splitting on the degenerate energy levels. The splitting of energy levels does not only affect the number of degenerate states, but also the valence and conduction bandwidths which are wider in surface part. On the other hand, the splitting of states and additional oxygen surface states strongly affect the energy gap and considerably reduce its value.

Figure 12 represents the degeneracy of states for oxygenated (001)-(1 × 1) surface part of $Al_8O_4P_8$ stoichiometry slab nanocrystal surface. This surface has lower number of degenerate states compared with core part of Figure 9. This low degeneracy is due to broken symmetry at the surface discontinuity. The existence of oxygen atoms leads to varying the bond lengths and lattice constant. Because of splitting, energy gap is reduced (see Figure 11).

4. Conclusions

In this paper, we studied some properties of zb-AlP nanocrystals. The total energy, lattice constant, bulk modulus, cohesive energy, energy gap, valence bandwidth, ionicity,

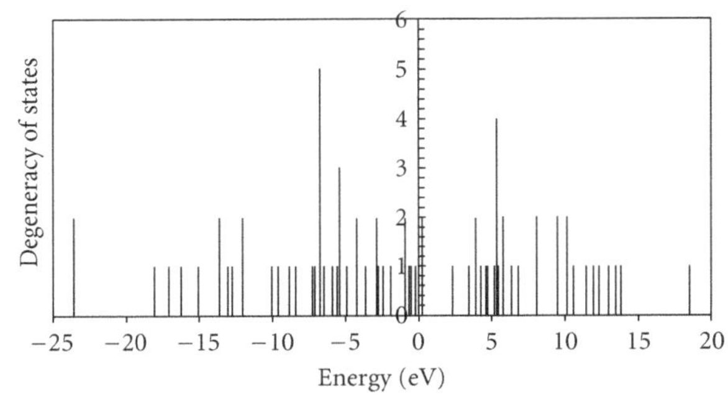

FIGURE 12: Degeneracy of states of surface part of $Al_8O_4P_8$ stoichiometry slab as a function of level energy. The HOMO level is taken as the reference level.

and degeneracy of energy levels have been calculated by the self-consistent restricted Hartree-Fock method coupled with the large unit cell method (RHF-LUC). The lattice constant and ionicity decreases with increasing size of large unit cell for core and surface parts. Cohesive energy of zb-AlP nanocrystals increases (absolute value) with the increase of number of atoms. Valence bandwidth increases with increasing size of LUC for core and surface part. Core energy gap increases with increasing size of LUC; however, surface gap decreases. The Valence bandwidth in surface part is larger than that for core part caused by the existence of oxygen atoms and surface states splitting. The degeneracy of states of core part for all sizes of LUC is larger than the degeneracy of surface part. On the other hand, (001)-(1×1) oxidized surface show much smaller gap and slightly larger lattice constant.

References

[1] R. M. Martin, *Electronic Structure Basic Theory and Practical Methods*, Cambridge University Press, Cambridge, UK, 2005.

[2] I. Vurgaftman, J. R. Meyer, and L. R. Ram-Mohan, "Band parameters for III-V compound semiconductors and their alloys," *Journal of Applied Physics*, vol. 89, no. 11, pp. 5815–5875, 2001.

[3] A. Mujica, A. Rubio, A. Munoz, and R. Needs, "High-pressure phases of group-IV, III–V, and II–VI compounds," *Reviews of Modern Physics*, vol. 75, no. 3, pp. 863–912, 2003.

[4] A. Mujica, P. Rodríguez-Hernández, S. Radescu, R. J. Needs, and A. Muñoz, "AlX (X = As, P, Sb) compounds under pressure," *Physica Status Solidi B*, vol. 211, no. 1, pp. 39–43, 1999.

[5] S. Froyen and M. Cohen, "Structural properties of III–V zinc-blende semiconductors under pressure," *Physical Review B*, vol. 28, no. 6, pp. 3258–3265, 1983.

[6] A. Jivani, H. Trivedi, P. N. Gajjar, and A. Jani, "Total energy, equation of state and bulk modulus of AlP, AlAs and AlSb semiconductors," *Pramana*, vol. 64, no. 1, pp. 153–158, 2005.

[7] F. Annane, H. Meradji, S. Ghemid, and F. El Haj Hassan, "First principle investigation of AlAs and AlP compounds and ordered $AlAs_{1-x}P_x$ alloys," *Computational Materials Science*, vol. 50, no. 2, pp. 274–278, 2010.

[8] M. A. Abdulsattar and K. H. Al-Bayati, "Corrections and parametrization of semiempirical large unit cell method for covalent semiconductors," *Physical Review B*, vol. 75, no. 24, Article ID 245201, 2007.

[9] R. A. Evarestov, M. I. Petrashen, and E. M. Ledovskapa, "The translational symmetry in the molecular models of solids," *Physica Status Solidi B*, vol. 68, no. 1, pp. 453–461, 1975.
[10] A. Harker and F. Larkins, "A large unit cell semiempirical molecular orbital approach to the properties of solids. I. General theory," *Journal of Physics C*, vol. 12, no. 13, pp. 2487–2495, 1979.
[11] A. Harker and F. Larkins, "A large unit cell semiempirical molecular orbital approach to the properties of solids. II. Covalent materials: diamond and silicon," *Journal of Physics C*, vol. 12, no. 13, pp. 2497–2508, 1979.
[12] W. Hehre, L. Radom, P. Schileyer, and J. Pople, *Ab Initio Molecular Orbital Theory*, Wiley, New York, NY, USA, 1986.
[13] W. J. Hehre, R. F. Stewart, and J. A. Pople, "Self-consistent molecular-orbital methods. I. Use of gaussian expansions of slater-type atomic orbitals," *The Journal of Chemical Physics*, vol. 51, no. 6, pp. 2657–2664, 1969.
[14] J. B. Collins, P. V. R. Von Schleyer, J. S. Binkley, and J. A. Pople, "Self-consistent molecular orbital methods. XVII. Geometries and binding energies of second-row molecules. A comparison of three basis sets," *The Journal of Chemical Physics*, vol. 64, no. 12, pp. 5142–5151, 1976.
[15] A. M. Dobrotvorskii and E. A. Evaeestov, "The quasi-molecular large unit cell model in the theory of deep levels in imperfect crystals: point defects in graphitic boron nitride," *Physica Status Solidi B*, vol. 66, no. 1, pp. 83–91, 1974.
[16] M. A. Abdulsattar, "Size effects of semiempirical large unit cell method in comparison with nanoclusters properties of diamond-structured covalent semiconductors," *Physica E*, vol. 41, no. 9, pp. 1679–1688, 2009.
[17] M. J. Frisch, G. W. Trucks, H. B. Schlegel et al., *Gaussian 03, Revision B.01*, Gaussian, Pittsburgh, Pa, USA, 2003.
[18] C. Kittel, *Introduction to Solid State Physics*, John Wiley & Sons, New York, NY, USA, 5th edition, 1976.
[19] S. Froyen and M. L. Cohen, "Structural properties of III–V zinc-blende semiconductors under pressure," *Physical Review B*, vol. 28, no. 6, pp. 3258–3265, 1983.
[20] R. Ahmed, F. Aleem, S. J. Hashemifar, and H. Akbarzadeh, "First-principles study of the structural and electronic properties of III-phosphides," *Physica B*, vol. 403, no. 10-11, pp. 1876–1881, 2008.
[21] F. D. Murnghan, "The compressibility of media under extreme pressures," *Proceedings of the National Academy of Sciences of The United States of America* , vol. 30, no. 9, pp. 244–247, 1944.
[22] W. Martienssen and H. Warlimont, Eds., *Springer Handbook of Condensed Matter and Materials Data*, Springer, Berlin, Germany, 2005.
[23] N. A. Nama, M. A. Abdulsattar, and A. M. Abdul-Lettif, "Surface and core electronic structure of oxidized silicon nanocrystals," *Journal of Nanomaterials*, vol. 2010, Article ID 952172, 2010.
[24] N. H. Aysa, M. A. Abdulsattar, and A. M. Abdul-Lettif, "Electronic structure of germanium nanocrystals core and (001)-(1×1) oxidised surface," *Micro & Nano Letters*, vol. 6, no. 3, pp. 137–140, 2011.
[25] H. C. Casey Jr. and M. B. Panish, *Heterostructure Lasers, Part A, Fundamental Principles and Part B, Materials and Operating Characteristics*, Academic Press, New York, NY, USA, 1978.
[26] M. Shishkin and G. Kresse, "Self-consistent GW calculations for semiconductors and insulators," *Physical Review B*, vol. 75, no. 23, Article ID 235102, 2007.
[27] M. A. Abdulsattar, "Mesoscopic fluctuations of electronic structure properties of boron phosphide nanocrystals," *Electronic Materials Letters*, vol. 6, no. 3, pp. 97–101, 2010.
[28] T. Shimazaki and Y. Asai, "Energy band structure calculations based on screened Hartree-Fock exchange method: Si, AlP, AlAs, GaP, and GaAs," *Journal of Chemical Physics*, vol. 132, no. 22, Article ID 224105, 2010.
[29] M. Causa and A. Zupan, "Density functional LCAO calculation of periodic systems. A posteriori correction of the Hartree-Fock energy of covalent and ionic crystals," *Chemical Physics Letters*, vol. 220, no. 3–5, pp. 145–153, 1994.
[30] S. Lee, J. Kang, and M. Kang, "Structural properties of semiconductors in the generalized gradient approximation," *Journal of the Korean Physical Society*, vol. 31, no. 5, pp. 811–814, 1997.
[31] S. J. Sque, R. Jones, and P. R. Briddon, "Structure, electronics, and interaction of hydrogen and oxygen on diamond surfaces," *Physical Review B*, vol. 73, no. 8, Article ID 085313, pp. 1–15, 2006.
[32] H.M. Abduljalil, M.A. Abdulsattar, and S.R. Al-Mansoury, "SiGe nanocrystals core and surface electronic structure from Ab initio large unit cell calculations," *Micro & Nano Letters*, vol. 6, no. 6, pp. 386–389, 2011.
[33] M. A. Abdulsattar, "Ab initio large unit cell calculations of the electronic structure of diamond nanocrystals," *Solid State Sciences*, vol. 13, no. 5, pp. 843–849, 2011.
[34] I. V. Kityk, A. Kassiba, K. Plucinski, and J. Berdowski, "Band structure of large-sized SiC nanocomposites," *Physics Letters A*, vol. 265, no. 5-6, pp. 403–410, 2000.

4

Reduction of Maximum and Residual Drifts on Posttensioned Steel Frames with Semirigid Connections

Arturo López-Barraza,[1] Edén Bojórquez,[2] Sonia E. Ruiz,[1] and Alfredo Reyes-Salazar[2]

[1] *Coordinación de Mecánica Aplicada, Instituto de Ingeniería, Universidad Nacional Autónoma de México, Ciudad Universitaria, Coyoacán, 04510 México City, Mexico*
[2] *Facultad de Ingeniería, Universidad Autónoma de Sinaloa, Calzada de las Américas y Boulevard Universitarios S/N, Ciudad Universitaria, 80040 Culiacán, Sinaloa, Mexico*

Correspondence should be addressed to Edén Bojórquez; eden@uas.edu.mx

Academic Editor: John W. Gillespie

The aim of this paper is to study the seismic performance of self-centering moment-resisting steel frames with posttensioned connections taking into account nonlinear material behavior, for better understanding of the advantages of this type of structural system. Further, the seismic performance of traditional structures with rigid connections is compared with the corresponding equivalent posttensioned structures with semirigid connections. Nonlinear time history analyses are developed for both types of structural systems to obtain the maximum and the residual interstory drifts. Thirty long-duration narrow-banded earthquake ground motions recorded on soft soil sites of Mexico City are used for the analyses. It is concluded that the structural response of steel buildings with posttensioned connections subjected to intense earthquake ground motions is reduced compared with the seismic response of traditional buildings with welded connections. Moreover, residual interstory drift demands are considerably reduced for the system with posttensioned connections, which is important to avoid the demolition of the buildings after an earthquake.

1. Introduction

Moment-resisting steel frames (MRSFs) with posttensioned connections (PTCs) represent a viable alternative to traditional steel frames with welded beam-column connections (FWCs). The posttensioned frames with semirigid connections (FPTCs) have been recently proposed as an alternative for controlling the structural damage and to improve the behavior of steel structures in seismic areas [1–12]. The system is designed to prevent brittle fractures in the nodes' areas of steel frames, which may cause reduction in the ductility capacity of the frame (as occurred in many cases during the 1994 Northridge earthquake). The beams are posttensioned to the columns by high strength posttensioning strands oriented in parallel to the axis of the beam, which provides capacity of self-centering of the frame, reducing the maximum and particularly the residual drifts and, at the same time, providing energy dissipation capacity through special devices placed in the connections.

While the control of maximum interstory drift is very important because it is one of the main parameters recommended by seismic design codes to assess the performance of structures, the energy dissipation capacity is crucial for structures subjected to long-duration narrow-band ground motions [13–16]. Finally, residual drift demands are an important parameter to decide if a structure is repaired or demolished after the occurrence of an intense earthquake. This problem has been addressed by several authors. For example, MacRae and Kawashima [17] studied the residual displacements of inelastic single-degree-of-freedom (SDOF) systems and illustrated their significant dependence on the postyielding stiffness ratio. Christopoulos et al. [18] studied residual displacements of five SDOF systems using different hysteretic rules and suggested that this type of displacements

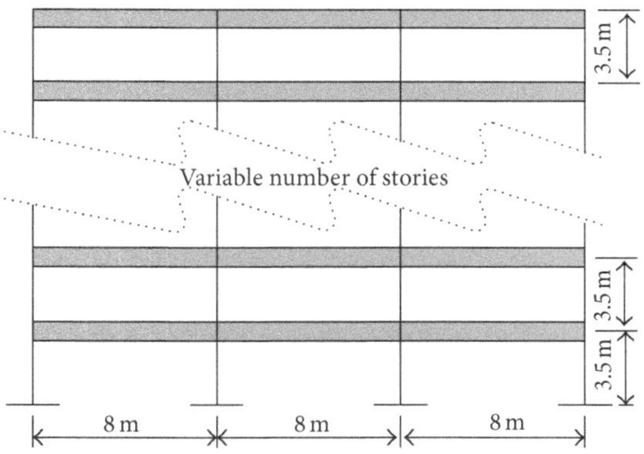

Figure 1: Geometrical characteristics of the frames.

Table 1: Relevant characteristics of the steel frames.

Frame	F6	F8	F10
Number of stories	6	8	10
Internal columns			
Story 1	W30 × 173	W36 × 210	W36 × 280
Story 2	W30 × 173	W36 × 210	W36 × 280
Story 3	W30 × 148	W36 × 194	W36 × 245
Story 4	W30 × 148	W36 × 194	W36 × 245
Story 5	W30 × 124	W36 × 170	W36 × 210
Story 6	W30 × 124	W36 × 170	W36 × 210
Story 7		W36 × 160	W36 × 182
Story 8		W36 × 160	W36 × 182
Story 9			W36 × 150
Story 10			W36 × 150
External columns			
Story 1	W27 × 146	W36 × 194	W36 × 280
Story 2	W27 × 146	W36 × 194	W36 × 280
Story 3	W27 × 129	W36 × 182	W36 × 245
Story 4	W27 × 129	W36 × 182	W36 × 245
Story 5	W27 × 114	W36 × 160	W36 × 210
Story 6	W27 × 114	W36 × 160	W36 × 210
Story 7		W36 × 135	W36 × 182
Story 8		W36 × 135	W36 × 182
Story 9			W36 × 150
Story 10			W36 × 150
Beams			
Story 1	W18 × 71	W21 × 83	W21 × 68
Story 2	W18 × 76	W21 × 93	W21 × 93
Story 3	W18 × 76	W21 × 93	W21 × 101
Story 4	W16 × 67	W21 × 83	W21 × 101
Story 5	W16 × 50	W18 × 71	W21 × 101
Story 6	W16 × 45	W18 × 65	W21 × 93
Story 7		W18 × 55	W21 × 73
Story 8		W18 × 46	W21 × 68
Story 9			W21 × 57
Story 10			W21 × 50

decrease when the postyielding stiffness ratio increases. Ruiz-Garcia and Miranda [19] showed that residual displacements are more sensitive to changes in local site conditions, earthquake magnitude, distance to the source range, and hysteretic behavior than peak displacements. Pampanin et al. [20] studied the seismic response of multidegree-of-freedom (MDOF) systems and highlighted a significant sensitivity of residual drifts to the hysteretic rule, postyield stiffness ratio, and global plastic mechanism. McCormick et al. [21] addressed the issue of discomfort of the occupants of the buildings due to residual deformations; they reported that significant discomfort is felt by occupants of building with residual interstory drift around 0.8%. Based on past experiences of buildings that have undergone differential settlement, they propose a residual drift limit of 0.5% to be considered in performance-based seismic design. They also reported that repairing damaged structures with residual story drifts greater than 0.5% after the 1995 Hyogoken-Nanbu earthquake was not financially viable. Moreover, in a recent study, Bojórquez and Ruiz-García [22] concluded that the residual interstory drift demands in traditional steel buildings are larger than 0.5% when subjected to long-duration earthquake ground motions. Their study was developed considering buildings designed for controlling the maximum interstory drift with new seismic design guidelines.

Motivated by the limitations of the traditional moment-resisting steel frames to achieve satisfactory seismic performance, and because most of the studies to estimate the seismic performance of structures with posttensioned connections are developed using broad-band records [10, 12], long-duration narrow-band earthquake ground motions recorded on very soft soil as those taken from Mexico City are not commonly used. The objective of the present study is to compare the seismic performance of traditional steel frames with posttensioned structures with semirigid connections. For this aim, six moment-resisting steel frame models with different numbers of stories have been subjected to a set of thirty narrow-band earthquake ground motions recorded on soft soil of Mexico City. Three frames are representative of traditional MRSFs with welded connections and the others, of equivalent frames with posttensioned semirigid connections. It is shown that the posttensioned structural systems can reduce considerably the maximum and residual interstory drift demands when subjected to long-duration earthquake ground motions.

2. Structural Models

2.1. Steel Frame Models. Six structural steel frames are analyzed in the present study. Three corresponds to FWC, and the rest are representatives of FPTC. The first group of structures was designed according to the Mexico City Building Code [23]. The buildings are assumed to be for office occupancy. They have 6, 8, and 10 stories and 3 bays; hereafter identified as F6WC, F8WC, and F10WC, respectively. The dimensions of the frames are shown in Figure 1. The beams and columns are A36 steel W sections. A bilinear hysteretic model for accounting the nonlinear material behavior with 3% of postyielding stiffness was considered for the analyses,

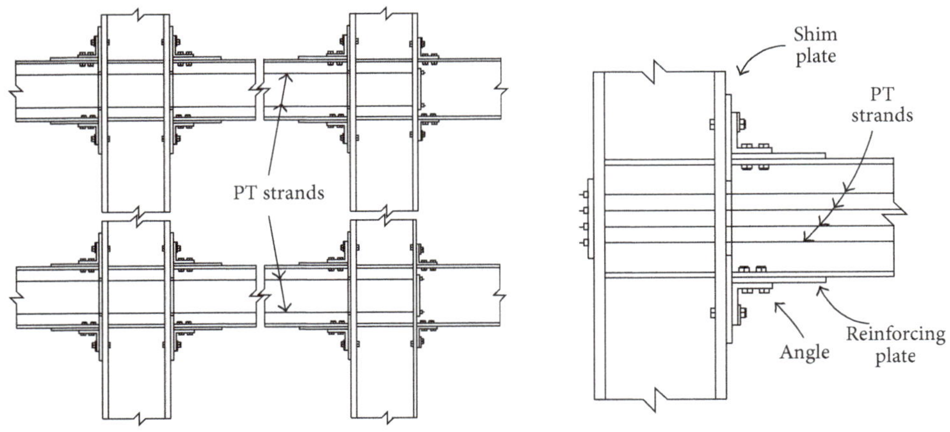

FIGURE 2: Angles and posttensioned strands in FPTC.

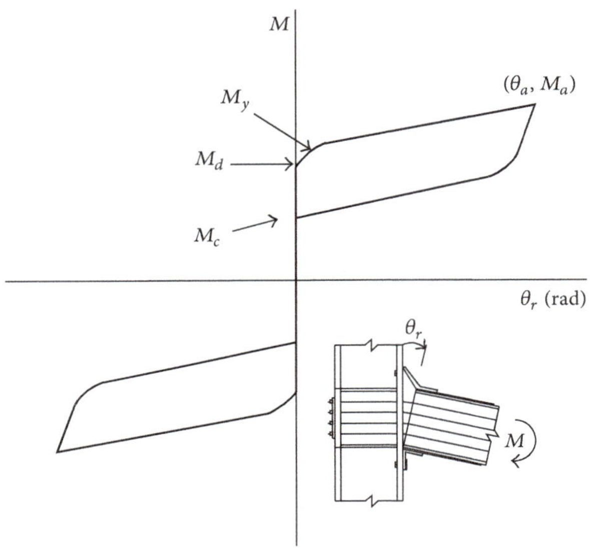

FIGURE 3: Moment-relative rotation hysteretic curve of the posttensioned connection.

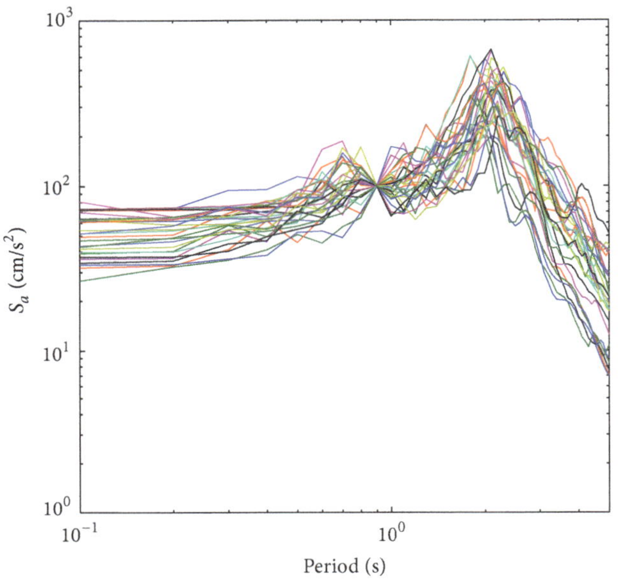

FIGURE 4: Elastic response spectra for the records scaled at the same spectral acceleration $S_a(T_1) = 100\,\text{cm/s}^2$ for period equal to 0.9 s and 3% of critical damping.

and the damping used was 3% of critical. The fundamental periods of vibration (T_1) are 1.07, 1.20, and 1.37 s, respectively. On the other hand, the FPTCs were designed in accordance with the recommendations proposed by Garlock et al. [6], which basically start with the design of the steel frames as usually is done (considering rigid connections), then the semirigid posttensioned connections are designed to satisfy the requirements of the serviceability and resistance conditions. The beam-column connections consist of two angles bolted to the flanges of the beam and to the column flange (top and seat). For the design of posttensioned connections, steel grade 50 was used for the angles. The length of the angles was taken equal to the width of the flange of the beams (b_f). Different angle sizes were tested. In the end, 152×152×13 mm angles were used in all the cases. Posttensioned cables consist of seven wires with an area of 150 mm^2 and withstanding a load of 279 kN; they are parallel to the axis of the beam passing through the interior columns and fixed to the outer face of the columns at the ends of the frame. An initial tension of the cables less than 0.33 of their maximum capacity was used according to the suggestions given by Garlock et al. [6]. The three FPTC models are identified here as F6PTC, F8PTC, and F10PTC, for the frames with 6, 8, and 10 stories, respectively. The FPTC models have fundamental periods of vibration of 1.03, 1.25, and 1.37 s, respectively. The columns of the ground floor are fixed at the base without posttensioned. The beam and column members used for all the frames are illustrated in Table 1. Note that the mechanical characteristics and dimensions of beams and columns are the same for both FWC and FPTC. Figure 2 shows a typical assembly of a posttensioned steel frame, where the posttensioned strands

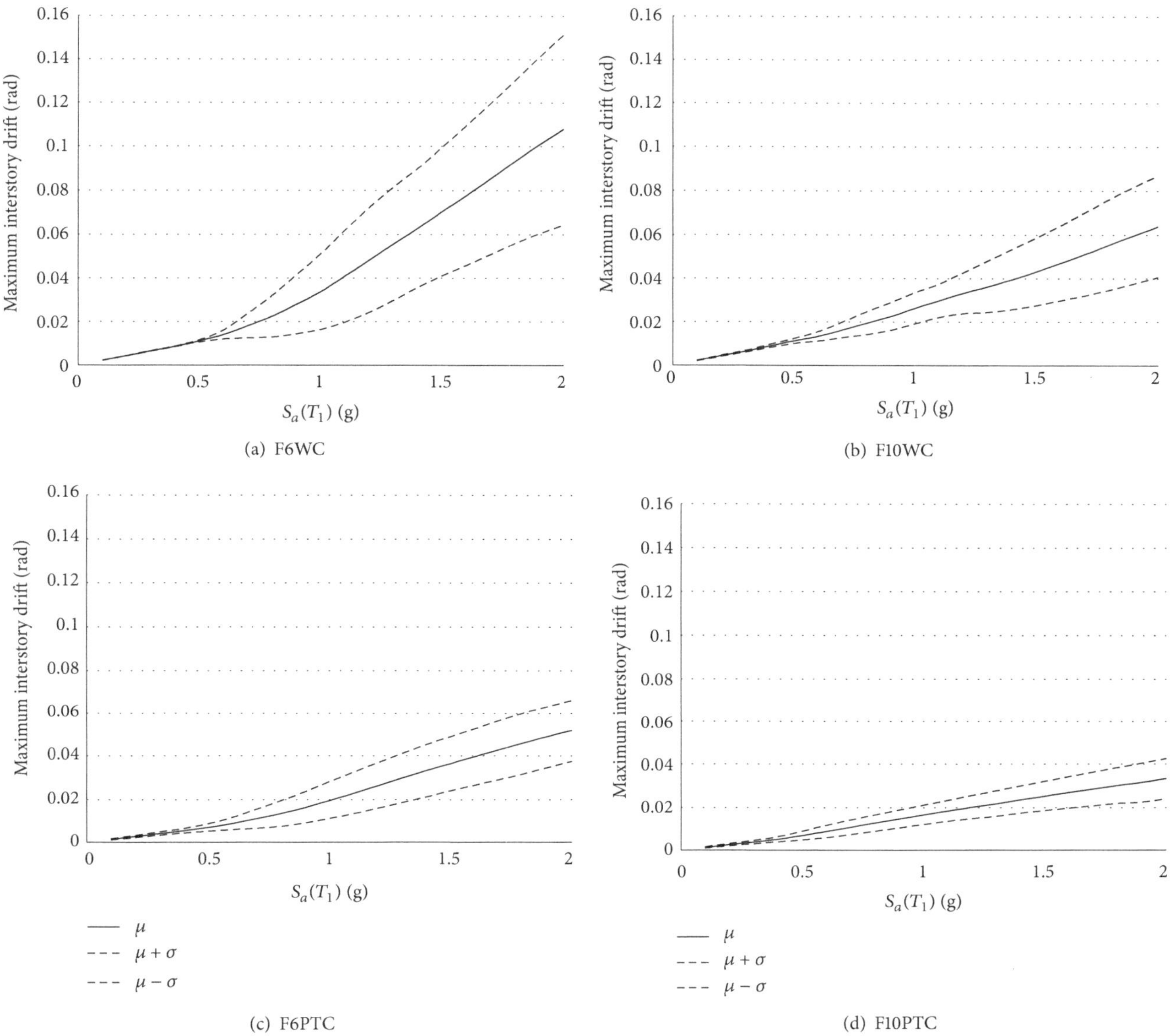

FIGURE 5: Median values of maximum interstory drifts ± one standard deviation for models: (a) F6WC, (b) F10WC, (c) F6PTC, and (d) F10PTC.

can be identified and the energy-dissipating elements (bolted angles) can be observed.

2.2. Connection Model: Nonlinear Behavior. The nonlinear behavior or hysteretic rules that represent the cyclic behavior of the semirigid connections of the posttensioned frames are characterized by moment-rotation curves (M-θr), which usually present shapes similar to a flag. This representation characterizes the nonlinearity, self-centering capability, and energy dissipation capacity of the connection. Experimental tests with isolated angles, subjected to cyclic and monotonic loads conducted by Shen and Astaneh-Asl [24], showed a stable cyclic response and good capability of hysteretic energy dissipation. In general, ultimate strength exceeds 3 times the yield strength and ductility reached values between 8 and 10. The strength and stiffness in bending of the posttensioned connections is provided by the contribution of the angles of the PTC and by posttensioned strands. Wires and angles work as springs in parallel. Posttensioned strands exhibit linear behavior, while connecting angles behave nonlinearly. Figure 3 shows a typical example of a hysteretic curve corresponding to a posttensioned connection. This behavior was modeled by using (1) and (2) which represent the loading and unloading curves, respectively, which in turn were obtained from the superposition of the exponential equation proposed by Richard and Abbott [25] for semirigid connections, considering the linear contribution of the strands, decompression moments (M_d), and the closing moment (M_c) of the connection. The first function, given by (1),

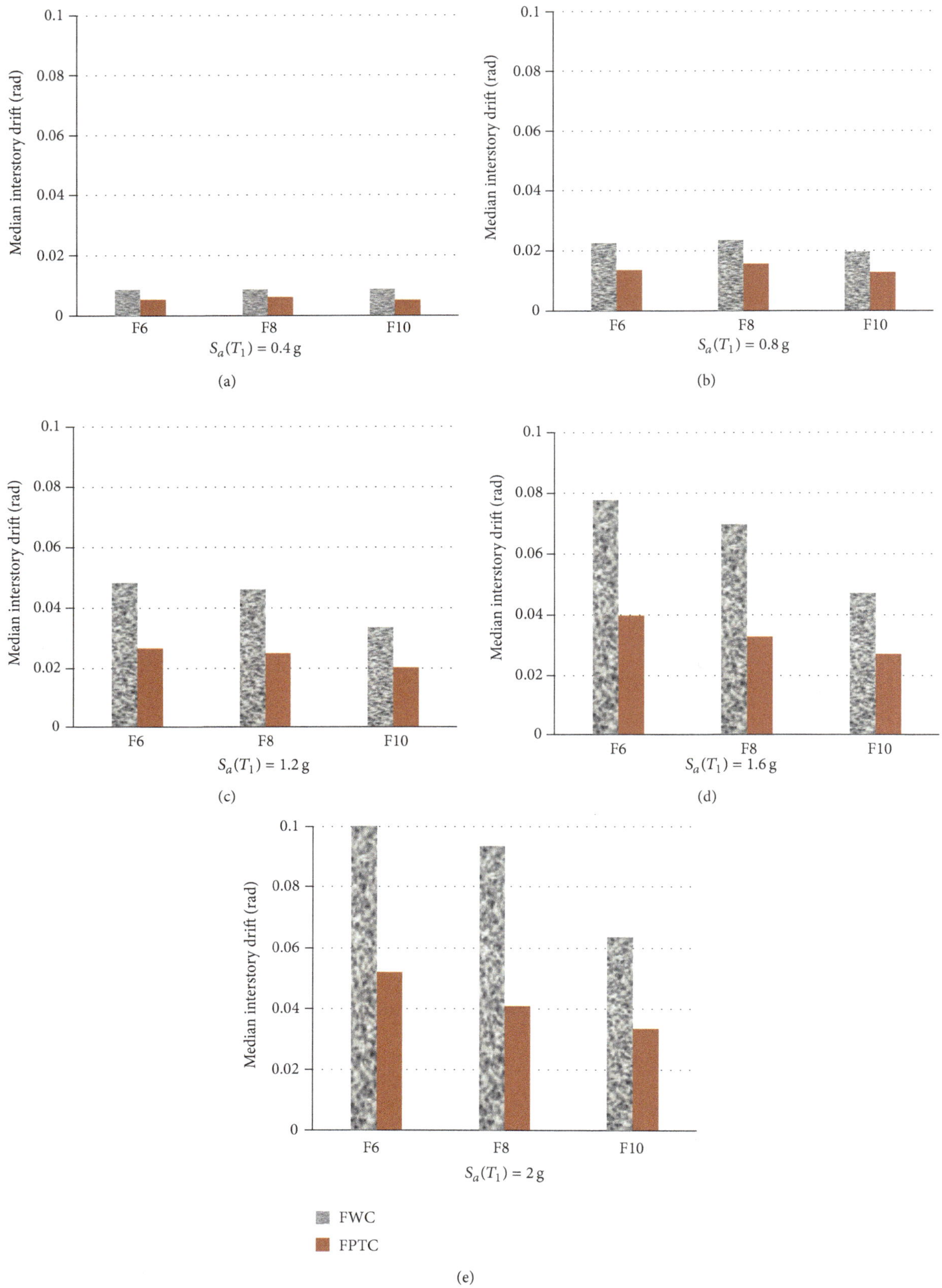

FIGURE 6: Median of maximum interstory drift of the frames FWC and FPTC.

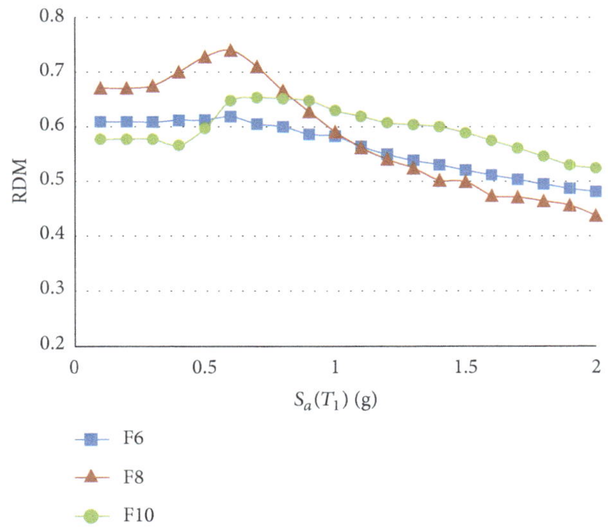

FIGURE 7: Ratio of median values of IDR_{max} for FPTC divided by IDR_{max} for FWC.

corresponds to the initial loading cycle; two types of variables are observed: (1) variables depending only on geometric and physical properties of angles, such as initial (k) and postyield (k_p) stiffness, the reference moment (M_o), and N that defines the curvature in the transition between the linear and plastic behavior, and (2) variables depending on the number and type of tendons such as the rigidity of the posttensioned tendons ($k_{\theta S}$) and the bending moment associated to the opening of the connection (named decompression moment, M_d) which is a function of the resulting initial tension in the tendons. The second function, given by (2), defines the unloading and reloading process in the connection; M_a and θ_a are the maximum values reached in each cycle, and the parameter φ defines the magnitude of the closing moment of the connection (M_c), which must be greater than zero in order to insure complete closure of the connection after getting complete unloading; moreover, this parameter largely defines the E_H dissipation capacity of the connection (enclosed area). The curves obtained with the modified model exhibit good accuracy in comparison with experiment results [5], and they were modeled in RUAUMOKO [26] as the flag-shaped bi-linear hysteresis,

$$M = M_d + \frac{(k - k_p)\,\theta_r}{\left[1 + \left|(k - k_p)\,\theta_r / M_o\right|^N\right]^{1/N}} + (k_p + k_{\theta S})\,\theta_r, \tag{1}$$

$$M = M_a - \frac{(k - k_p)(\theta_a - \theta_r)}{\left[1 + \left|(k - k_p)(\theta_a - \theta_r)/\varphi M_o\right|^N\right]^{1/N}} - (k_p + k_{\theta S})(\theta_a - \theta_r). \tag{2}$$

3. Earthquake Ground Motion Records

A set of 30 long-duration narrow-band ground motions recorded at Lake Zone sites of Mexico City was considered. Particularly, all motions were recorded at sites having soil periods of 2 s, during seismic events with magnitudes near of seven or larger and having epicenters located at distances of 300 km or more from Mexico City. Some important characteristics of the records are summarized in Table 2. While PGA and PGV denote the peak ground acceleration and velocity, respectively, the duration (t_D) was estimated according to Trifunac and Brady [27]. It should be mentioned that sites having soil periods of 2 s are fairly common within the Lake Zone and that the higher levels of shaking (in terms of PGA) have been consistently observed at these sites. Further, the long-duration narrow-band ground motion records demand large amounts of energy to structures compared with those subjected to broad-band ground motion [14]. The ground motions were recorded in sites where most of the damage during the Mexico earthquake of September 19, 1985, occurred.

Figure 4 shows the elastic response spectra for the records under consideration (where S_a denotes pseudoacceleration) corresponding to 3% of critical damping. The narrow-banded records exhibit similar values of parameter N_p, which is an indicator of the characteristics of their spectral shape [28–30]; further information of this parameter can be found in [29]. As a result, there is a strong similarity between their spectral shapes when they are scaled to the same spectral acceleration at first mode of vibration. Figure 4 shows seismic response spectra for all records scaled for an elastic spectral acceleration of 100 cm/s^2 for a period of 0.9 s. The similitude exhibited by all spectra indicates that the spectral acceleration is a good indicator of the damage potential of the ground motions and emphasizes the good correspondence that exists between parameter N_p and the spectral shape of the motions. Note that the spectral acceleration of 100 cm/s^2 was chosen only for illustrative purposes and that the similitude exhibited by the spectra is valid for any level of scaling.

4. Methodology

The comparison of maximum interstory drift (IDR_{max}) and peak residual interstory drift ($RIDR_{max}$) of the traditional and posttensioned steel frames was estimated using incremental nonlinear dynamic analysis [31]. For this aim, the building frame models were subjected to the set of thirty long-duration narrow-band earthquake ground motions, which were scaled at different values of the seismic intensity in terms of spectral acceleration at first mode of vibration of the structure $S_a(T_1)$. The selected seismic intensity values were from 0.1 g to 2.0 g with increments of 0.1 g. The RUAUMOKO [26] computer program was used for the time history nonlinear dynamic analysis.

The seismic performance results considering nonlinear behavior are expressed in terms of maximum and residual interstory drifts at the end of the earthquake motions for each scaling level.

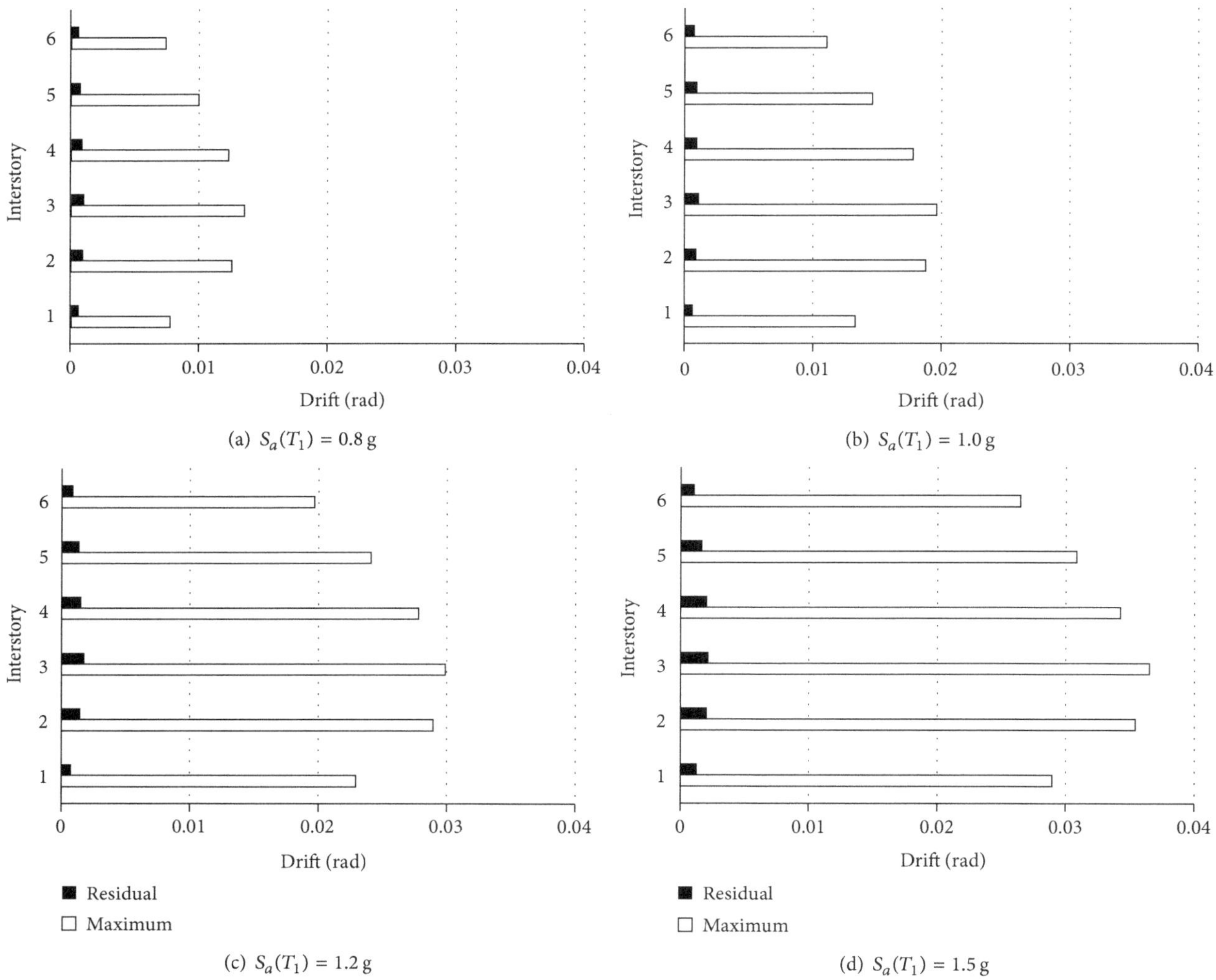

FIGURE 8: Median values of maximum and residual interstory drifts for F6PTC at different spectral acceleration values: (a) 0.8 g, (b) 1.0 g, (c) 1.2 g, and (d) 1.5 g.

4.1. Maximum Interstory Drift Results. The median (μ) and standard deviation (σ) of IDR_{max} for each frame and seismic intensity level are estimated in first place. Figures 5(a), 5(b), 5(c), and 5(d) show the values of IDR_{max} at different pseudospectral acceleration for frames F6WC, F10WC, F6PTC, and F10PTC, respectively. It is observed that for frames of the same height subjected to the same ground motion intensities, the median values of IDR_{max} are smaller for the FPTC than for the FWC. The difference tends to increases as the spectral acceleration also increases. Figures 5(a)–5(d) also indicate that the maximum interstory drift demands are smaller for the structures with fundamental periods equal to 1.37 s than for those with periods around 1.07 s due to the softening phenomenon that these structures suffer; it occurs because as the seismic intensity increases and the structure presents inelastic nonlinear behavior, the effective period of vibration becomes larger, which implies that it becomes closer to the peak ordinate of the spectrum. As a consequence, the structural response grows very rapidly. In other words, the structure is "*softening*" into the peak of the spectrum, as it is shown in Montiel and Ruiz [32]. Note that the uncertainty in the estimation of maximum interstory drift is larger for the steel frames with rigid connections compared with the FPTC (see Figure 5). Moreover, for the FPTC models, IDR_{max} tend to increase almost linearly with $S_a(T_1)$. This can be explained observing that the strands remain elastic which significantly contribute to the stiffness capacity of the connection.

Figure 6 shows the maximum interstory drift for all frames and the ground motion intensities of 0.4 g, 0.8 g, 1.2 g, 1.6 g, and 2.0 g. The figure indicates that the maximum interstory drift of the FPTC models are smaller than those of the corresponding FWC models; this difference increases as the intensity seismic demand increases. It suggests the advantages of using moment-resisting steel frames with posttensioned connections.

The ratio of the maximum interstory drift of FPTC divided by the same parameter for FWC model (here defined as Ratio of Maximum Drifts RMD) is illustrated in Figure 7 at different intensity levels. It is observed that RMD in all the cases is smaller than unity, indicating that the IDR_{max}

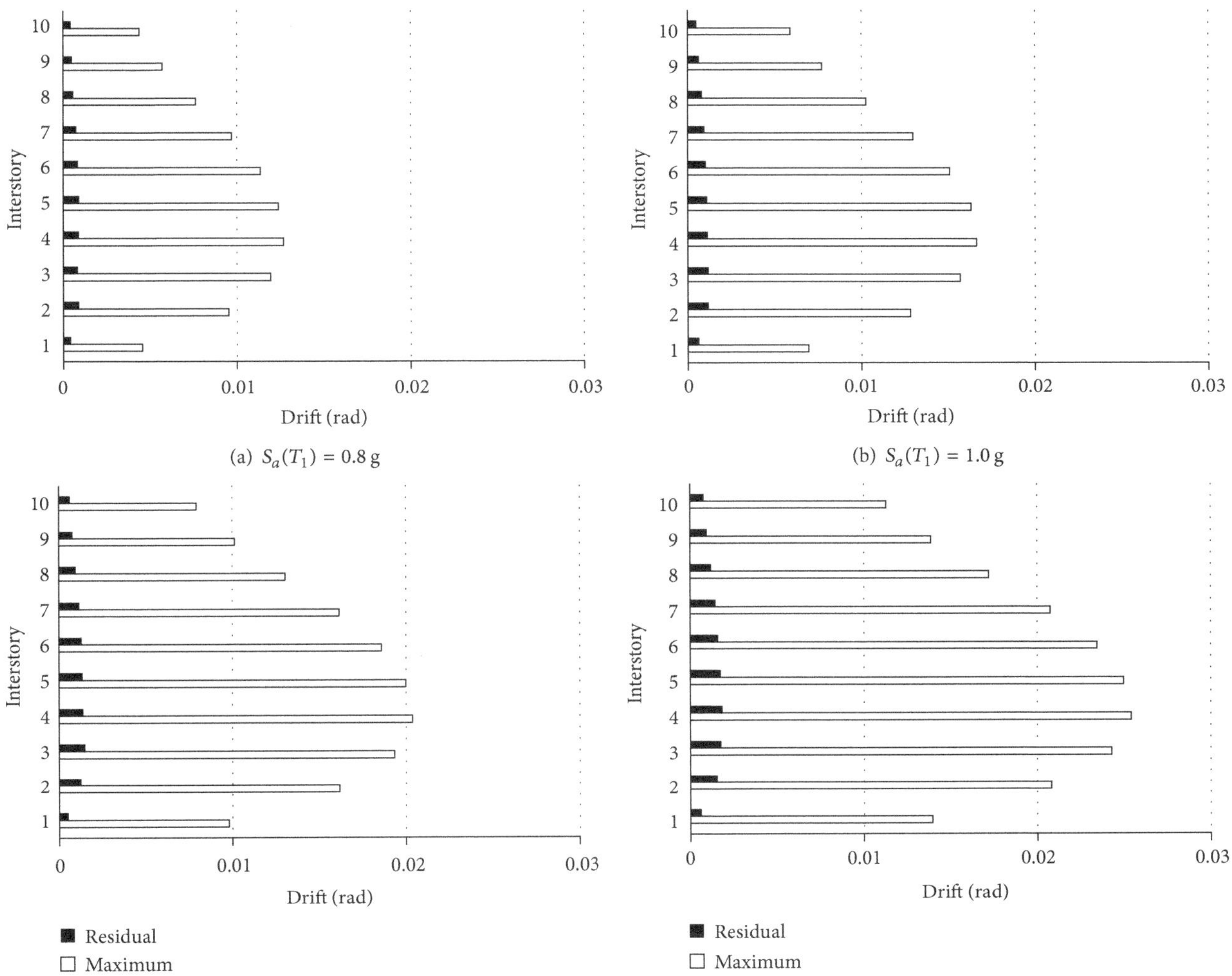

(a) $S_a(T_1) = 0.8$ g (b) $S_a(T_1) = 1.0$ g (c) $S_a(T_1) = 1.2$ g (d) $S_a(T_1) = 1.5$ g

FIGURE 9: Median values of maximum and residual interstory drifts for F10PTC at different spectral acceleration values: (a) 0.8 g, (b) 1.0 g, (c) 1.2 g and (d) 1.5 g.

values of the FPTC systems are smaller than those of the FWC system. Furthermore, for larger values of nonlinear behavior, the ratio is almost the same for all the frames analyzed with magnitude about 50%, which represent an important reduction of the maximum seismic demands compared with the traditional structural steel frame models.

4.2. Comparison of Maximum and Residual Interstory Drift Demands. Maximum and residual interstory drift demands are compared at different intensity levels in terms of spectral acceleration in this section. The results of median values of the selected engineering demand parameters are compared in Figures 8 and 9 considering ground motion intensities of 0.8 g, 1.0 g, 1.2 g, and 1.5 g, for the structural models F6PTC and F10PTC. As it was expected, the magnitudes of IDR_{max} tend to increase for larger values of spectral acceleration; on the other hand, the $RIDR_{max}$ slightly increases when $S_a(T_1)$ becomes larger. Moreover, the value of $RIDR_{max}$ is smaller than 0.5% for $S_a(T_1)$ equal to 1.5 g, indicating that the buildings with posttensioned steel connections do not necessarily must be demolished after an earthquake. Note that the value of 0.5% of residual interstory drift was proposed by McCormick et al. [21] as the limit for feasible financially structures and to guarantee the comfort of the occupants of the buildings. Furthermore, the larger seismic demands in terms of maximum and residual deformations occur in the intermediate stories (between 0.3 and 0.6 of the height).

Figure 10 shows the median values of $RIDR_{max}$ for the frames with 6, 8, and 10 stories and spectral acceleration values from 0.1 to 2.0 g. It can be observed that the magnitude of $RIDR_{max}$ increases when $S_a(T_1)$ becomes larger. For $S_a(T_1) \geq 0.5$ g, the $RIDR_{max}$ of the FPTC models are smaller than those of the FWC systems. The difference increases when $S_a(T_1)$ also increases. In the case of FWC, the residual interstory drift strongly depends on the fundamental period of the frames being smaller for frames with larger fundamental period of vibration. Note that the results of the incremental dynamic analyses of the FPTC suggests that

TABLE 2: Selected long-duration narrow-band ground motions.

Records	Date	Magnitude	Station	PGA (cm/s^2)	PGV (cm/s)	t_D (s)
1	19/09/1985	8.1	SCT	178.0	59.5	34.8
2	21/09/1985	7.6	Tlahuac deportivo	48.7	14.6	39.9
3	25/04/1989	6.9	Alameda	45.0	15.6	37.8
4	25/04/1989	6.9	Garibaldi	68.0	21.5	65.5
5	25/04/1989	6.9	SCT	44.9	12.8	65.8
6	25/04/1989	6.9	Sector Popular	45.1	15.3	79.4
7	25/04/1989	6.9	Tlatelolco TL08	52.9	17.3	56.6
8	25/04/1989	6.9	Tlatelolco TL55	49.5	17.3	50.0
9	14/09/1995	7.3	Alameda	39.3	12.2	53.7
10	14/09/1995	7.3	Garibaldi	39.1	10.6	86.8
11	14/09/1995	7.3	Liconsa	30.1	9.62	60.0
12	14/09/1995	7.3	Plutarco Elías Calles	33.5	9.37	77.8
13	14/09/1995	7.3	Sector Popular	34.3	12.5	101.2
14	14/09/1995	7.3	Tlatelolco TL08	27.5	7.8	85.9
15	14/09/1995	7.3	Tlatelolco TL55	27.2	7.4	68.3
16	09/10/1995	7.5	Cibeles	14.4	4.6	85.5
17	09/10/1995	7.5	CU Juárez	15.8	5.1	97.6
18	09/10/1995	7.5	Centro urbano Presidente Juárez	15.7	4.8	82.6
19	09/10/1995	7.5	Córdoba	24.9	8.6	105.1
20	09/10/1995	7.5	Liverpool	17.6	6.3	104.5
21	09/10/1995	7.5	Plutarco Elías Calles	19.2	7.9	137.5
22	09/10/1995	7.5	Sector Popular	13.7	5.3	98.4
23	09/10/1995	7.5	Valle Gómez	17.9	7.18	62.3
24	11/01/1997	6.9	CU Juárez	16.2	5.9	61.1
25	11/01/1997	6.9	Centro urbano Presidente Juárez	16.3	5.5	85.7
26	11/01/1997	6.9	García Campillo	18.7	6.9	57.0
27	11/01/1997	6.9	Plutarco Elías Calles	22.2	8.6	76.7
28	11/01/1997	6.9	Est. no. 10 Roma A	21.0	7.76	74.1
29	11/01/1997	6.9	Est. no. 11 Roma B	20.4	7.1	81.6
30	11/01/1997	6.9	Tlatelolco TL08	16.0	7.2	57.5

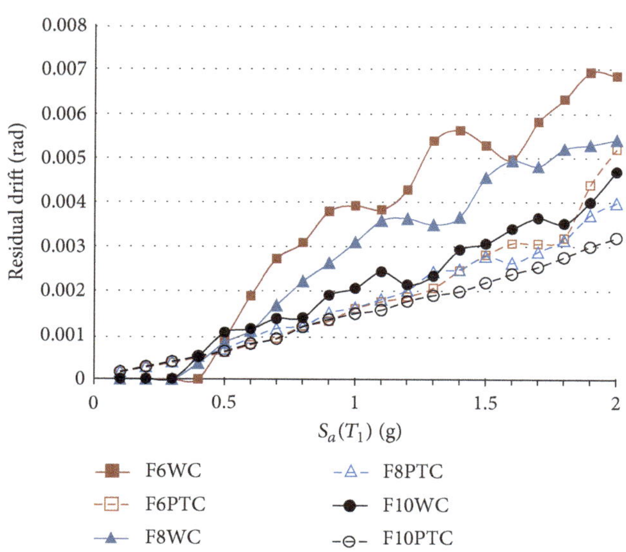

FIGURE 10: Median values of maximum residual interstory drift of the frames FWC and FPTC.

there is a lack of dependence on the fundamental period of structural vibration in the values of $RIDR_{max}$ at different intensity scaling levels, in such a way that for FPTC with different stories the residual demands are similar for the same spectral acceleration intensity.

The ratio of residual interstory drift demands for the FWC and FPTC structural systems defined as RRD is introduced to compare the structural performance of both types of systems. Figure 11 shows the mean values of RRD for all the frames under consideration at different intensity scaling levels starting from 0.5 g; for smaller values, the structures exhibit elastic behaviour. Note that for a specific frame the ratio tends to be constant for all the intensity values under consideration; also, the ratio increases when the structural fundamental period is larger. This is summarized in Table 3, where the mean and standard deviation of RRD are presented for the frames obtained for all the scaling levels. Results indicate that the mean value increases and the standard deviation (σ) decreases as the fundamental structural period increases, which implies a large difference in mean and in dispersion of the results corresponding to the frames under

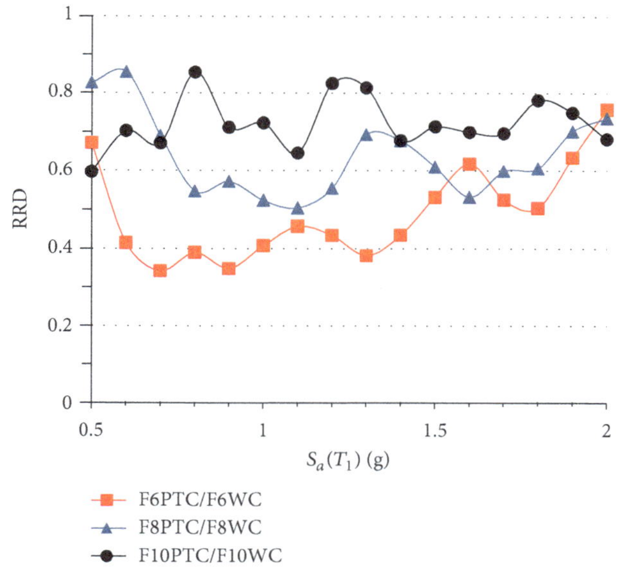

FIGURE 11: Ratios of mean values of $RIDR_{max}$ for FPTC divided by $RIDR_{max}$ for FWC.

TABLE 3: Mean and standard deviation of RRD values.

Ratio of $RIDR_{max}$	Mean	Standard deviation
F6PTC/F6WC	0.49	0.123
F8PTC/F8WC	0.64	0.106
F10PTC/F10WC	0.72	0.069

study. For the 6 stories frame the mean RRD is about 0.49; this implies that the average of the residual interstory drift for F6PTC is about 51% smaller than the F6WC residual interstory drift. In the same way, the reductions are of 36% and 28% for the 8-story and 10-story frames, respectively. The results show the advantage of using posttensioned structural systems with energy dissipating connections.

5. Conclusions

The study compares the seismic performance of buildings in terms of maximum and residual interstory drifts' demands of posttensioned self-centering moment-resisting steel frames with traditional welded connection steel frames. Incremental nonlinear dynamic analysis was used to estimate the seismic assessment of six structural steel frames which were subjected to thirty long-duration ground-motion records scaled at different values of $S_a(T_1)$. For the analyses of the frames, nonlinear material behavior was considered. The numerical study indicates that in all the cases, the maximum interstory drifts of the FPTC are smaller than those of the FWC; in fact, reductions in the order of 47% were obtained when the FPTC was employed as structural system compared with the traditional system. Moreover, the residual interstory drifts for the FPTC systems are smaller than 0.5% which is the limit for feasible financially structures and to guarantee the comfort of the occupants of the buildings, and this conclusion is valid for different intensity levels. Furthermore, the maximum residual interstory drift was reduced at values as large as 51% when the frames have posttensioned connections. For this reason, the results suggest that PT connections in structural frame buildings can control significantly the seismic performance in terms of peak and residual drift demands for structures subjected to long-duration narrow-band earthquake ground motions, and this is valid for frames with different story levels and at different intensity of the ground motion. Note that most of the previous studies were developed using low-rise buildings which were subjected only to broad-band records.

Finally, it is important to say that the reduction of the maximum interstory drift implies less structural damage in the buildings and, on the other hand, the reduction of the maximum residual interstory drift is very important to reduce the economic consequences of repairing and interruption of the functionality of the buildings.

Akcnowledgments

The support given by the Universidad Nacional Autónoma de México under Project PAPIIT-IN109011-3 and PROMEP is appreciated. Financial support also was received from El Consejo Nacional de Ciencia y Tecnología to E. Bojórquez under Grant CB-2011-01-167419 and the Universidad Autónoma de Sinaloa under Grant PROFAPI 2012. Finally, the authors wish to express their gratitude to the anonymous reviewers for their suggestions to improve the quality of the paper.

References

[1] J. M. Ricles, R. Sause, M. Garlock, and C. Zhao, "Post-tensioned seismic-resistant connections for steel frames," *Journal of Structural Engineering*, vol. 127, no. 2, pp. 113–121, 2001.

[2] J. M. Ricles, R. Sause, S. W. Peng, and L. W. Lu, "Experimental evaluation of earthquake resistant posttensioned steel connections," *Journal of Structural Engineering*, vol. 128, no. 7, pp. 850–859, 2002.

[3] C. Christopoulos, A. Filiatrault, and C. M. Uang, "Self-centering post-tensioned energy dissipating (PTED) steel frames for seismic regions," Tech. Rep. SSRP-2002/06, University of California, 2002.

[4] C. Christopoulos and A. Filiatrault, "Seismic response of post-tensioned energy dissipating moment resisting steel frames," in *Proceedings of the 12th European Conference on Earthquake Engineering*, no. 61, London, UK, 2002.

[5] M. Garlock, J. Ricles, and R. Sause, "Experimental studies on full-scale post tensioned steel connections," *Journal of Structural Engineering*, vol. 131, no. 3, pp. 438–448, 2005.

[6] M. Garlock, R. Sause, and J. Ricles, "Behavior and design of post-tensioned steel frames system," *Journal of Structural Engineering*, vol. 133, no. 3, pp. 389–399, 2007.

[7] M. M. Garlock, J. M. Ricles, and R. Sause, "Influence of design parameters on seismic response of post-tensioned steel MRF systems," *Engineering Structures*, vol. 30, no. 4, pp. 1037–1047, 2008.

[8] P. Rojas, J. M. Ricles, and R. Sause, "Seismic performance of post-tensioned steel moment resisting frames with friction

devices," *Journal of Structural Engineering*, vol. 131, no. 4, pp. 529–540, 2005.

[9] C.-C. Chou and J. H. Chen, "Column restraint in post-tensioned self-centering moment frames," *Earthquake Engineering and Structural Dynamics*, vol. 39, no. 7, pp. 751–774, 2010.

[10] C.-C. Chou and J. H. Chen, "Tests and analyses of a full-scale post-tensioned RCS frame subassembly," *Journal of Constructional Steel Research*, vol. 66, no. 11, pp. 1354–1365, 2010.

[11] C.-C. Chou and J.-H. Chen, "Seismic design and shake table test of a steel post-tensioned self-centering moment frame with a slab accommodating frame expansion," *Earthquake Engineering and Structural Dynamics*, vol. 40, pp. 1241–1261, 2011.

[12] C. C. Chou and J. H. Chen, "Analytical model validation and influence of column bases for seismic responses of steel post-tensioned self-centering MRF systems," *Engineering Structures*, vol. 33, no. 9, pp. 2628–2643, 2011.

[13] E. Bojórquez and S. E. Ruiz, "Strength reduction factors for the valley of Mexico taking into account low cycle fatigue effects," in *Proceedings of the 13th World Conference on Earthquake Engineering*, no. 516, Vancouver, Canada, 2004.

[14] A. Teran-Gilmore and J. O. Jirsa, "Energy demands for seismic design against low-cycle fatigue," *Earthquake Engineering and Structural Dynamics*, vol. 36, no. 3, pp. 383–404, 2007.

[15] E. Bojórquez, A. Reyes-Salazar, A. Terán-Gilmore, and S. E. Ruiz, "Energy-based damage index for steel structures," *Steel and Composite Structures*, vol. 10, no. 4, pp. 343–360, 2010.

[16] E. Bojórquez, A. Terán-Gilmore, S. E. Ruiz, and A. Reyes-Salazar, "Evaluation of structural reliability of steel frames: inter-story drifts versus plastic hysteretic energy," *Earthquake Spectra*, vol. 27, no. 3, pp. 661–682, 2011.

[17] G. A. Macrae and K. Kawashima, "Post-earthquake residual displacements of bilinear oscillators," *Earthquake Engineering and Structural Dynamics*, vol. 26, no. 7, pp. 701–716, 1997.

[18] C. Christopoulos, S. Pampanin, and M. J. N. Priestley, "Performance-based seismic response of frame structures including residual deformations. Part I: single-degree of freedom systems," *Journal of Earthquake Engineering*, vol. 7, no. 1, pp. 97–118, 2003.

[19] J. Ruiz-García and E. Miranda, "Residual displacement ratios for assessment of existing structures," *Earthquake Engineering and Structural Dynamics*, vol. 35, no. 3, pp. 315–336, 2006.

[20] S. Pampanin, C. Christopoulos, and M. J. N. Priestley, "Performance-based seismic response of frame structures including residual deformations. Part II: multi-degree of freedom systems," *Journal of Earthquake Engineering*, vol. 7, no. 1, pp. 119–147, 2003.

[21] J. McCormick, H. Aburano, M. Ikenaga, and M. Nakashima, "Permissible residual deformation levels for building structures considering both safety and human elements," in *Proceedings of the 14th Conference on Earthquake Engineering*, Beijing, China, October 2008.

[22] E. Bojórquez and Ruiz-García, "Residual drift demands in moment-resisting steel frames subjected to narrow-band earthquake ground motions," *Earthquake Engineering and Structural Dynamics*. In press.

[23] Mexico City Building Code, "Reglamento de Construcciones para el Distrito Federal (In Spanish)," 2004.

[24] J. Shen and A. Astaneh-Asl, "Hysteretic behavior of bolted-angle connections," *Journal of Constructional Steel Research*, vol. 51, no. 3, pp. 201–218, 1999.

[25] R. M. Richard and B. J. Abbott, "Versatile elastic plastic stress-strain formula," *Journal of Engineering Mechanics*, vol. 101, no. 4, pp. 511–515, 1975.

[26] A. Carr, *RUAUMOKO Inelastic Dynamic Analysis Program*, University of Canterbury, Department of Civil Engineering, 2011.

[27] M. D. Trifunac and A. G. Brady, "A study of the duration of strong earthquake ground motion," *Bulletin of the Seismological Society of America*, vol. 65, no. 3, pp. 581–626, 1975.

[28] E. Bojórquez, I. Iervolino, and G. Manfredi, "Evaluating a new proxy for spectral shape to be used as an intensity measure," in *Proceedings of the Seismic Engineering International Conference Commemorating the 1908 Messina and Reggio Calabria Earthquake (MERCEA '08)*, 2008.

[29] E. Bojórquez and I. Iervolino, "Spectral shape proxies and nonlinear structural response," *Soil Dynamics and Earthquake Engineering*, vol. 31, no. 7, pp. 996–1008, 2011.

[30] E. Bojórquez, I. Iervolino, A. Reyes-Salazar, and S. E. Ruiz, "Comparing vector-valued intensity measures for fragility analysis of steel frames in the case of narrow-band ground motions," *Engineering Structures*, vol. 45, pp. 472–480, 2012.

[31] D. Vamvatsikos and C. A. Cornell, "Incremental dynamic analysis," *Earthquake Engineering and Structural Dynamics*, vol. 31, no. 3, pp. 491–514, 2002.

[32] M. A. Montiel and S. E. Ruiz, "Influence of structural capacity uncertainty on seismic reliability of buildings under narrow-band motions," *Earthquake Engineering and Structural Dynamics*, vol. 36, no. 13, pp. 1915–1934, 2007.

5

Optimum Design of Fibre Orientation in Composite Laminate Plates for Out-Plane Stresses

Rasoul Khandan, Siamak Noroozi, Philip Sewell, John Vinney, and Mehran Koohgilani

School of Design, Engineering and Computing, Bournemouth University, Dorset Bournemouth BH12 5BB, UK

Correspondence should be addressed to Rasoul Khandan, ras_khandan@yahoo.com

Academic Editor: Philip Harrison

Previous studies have shown that composite fibre orientations can be optimised for specific load cases such as longitudinal or in-plane loading. However, the methodologies utilised in these studies cannot be used for general analysis of such problems. In this research, an extra term is added to the optimisation penalty function in order to consider the transverse shear effect. This modified penalty function leads to a new methodology whereby the thickness of laminated composite plate is minimized by optimising the fibre orientations for different load cases. Therefore, the effect of transverse shear forces is considered in this study. Simulated annealing (SA) is used to search for the optimal design. This optimisation algorithm has been shown to be reliable as it is not based on the starting point, and it can escape from the local optimum points. In this research, the Tsai-Wu failure and maximum stress criteria for composite laminate are chosen. By applying two failure criteria at the same time the results are more reliable. Experimentally generated results show a very good agreement with the numerical results, validating the simulated model used. Finally, to validate the methodology the numerical results are compared to the results of previous research with specific loading.

1. Introduction

The demand for high strength, high modulus and low density industrial materials has generated an increased number of applications for fibre laminated composite structures in many different fields such as in submarines, sport equipment, medical instruments, civil engineering, enabling technologies, primary and secondary marine and aerospace structures, astronavigation and many more industries [1]. Composite constructions are usually multilayer produced structures, mostly made of flat and curved panels, built up from several layers or laminae, which are bonded together [2].

In the last half century, the use of composite materials has grown rapidly. These materials are ideal for structural applications that require high strength and low weight. They have good fatigue characteristics and are resistant to corrosion. They provide some flexibility in design through the variation of the fibre orientation or stacking sequence of fibre and matrix materials [3, 4]. Another advantage of fibre laminate composites is the capability to design the physical structure and mechanical properties prior to manufacture. The mechanical behaviour of laminates strongly depends on the orientation of fibres and thickness of lamina. Accordingly, the lamina should be designed to satisfy the specific requirements of each particular application in order to obtain the maximum advantages from the directional properties of materials. Accurate and efficient structural analysis, design sensitivity analysis, and optimisation procedures for size and shape and the orientation of fibres within the material are also required. This provides a good opportunity to tailor the material properties to the specific application [5, 6]. However, it increases the complexity of the design problem. This complexity exists, not only because of numerous design variables, but also because of having a multimodal and variable-dimensional optimisation problem with unattainable or costly derivatives [7].

Optimum strength designs of continuous fibre-reinforced composite laminates have been used since the early days of these materials. The first research to investigate the fibre orientation of a unidirectional lamina yielding maximum strength under in-plane stress conditions has

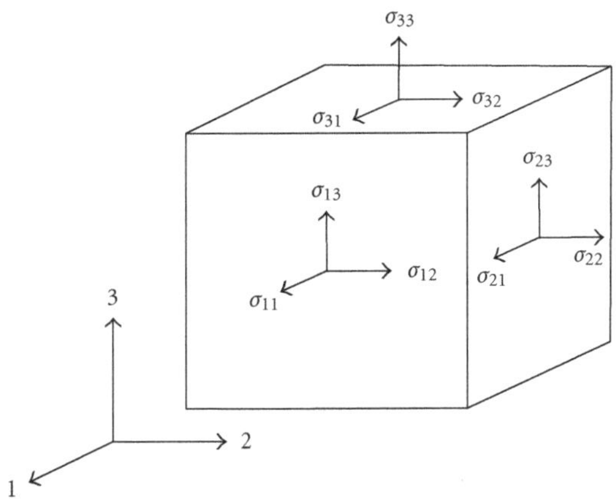

FIGURE 1: State of stress at a point of a continuum [36].

been carried out by Sandhu [8]. Brandmaier found that the strength of a unidirectional lamina under in-plane stresses could be maximised analytically with respect to the fibre orientation [9]. The results based upon Tsai-Hill failure criterion indicated that the optimum fibre orientation depended upon the stress state and the relative value of the transverse and in-plane shear strengths of the lamina. When the strength of a multidirectional composite laminate is to be maximised, more complicated and explicit optimisation techniques are needed [10]. Chao et al. were the pioneers that sought the optimum strength design of multidirectional laminates using a search technique [11]. Latterly, many studies have been devoted to the optimum strength design of multidirectional laminate. Among these are the works by Park [12], Fukunaga and Chou [13], Miravete [14], Fukunaga and Vanderplaats [15], Gurdal et al. [16, 17], Spallino et al. [18], Weaver [19], Chattopadhyay et al. [20], Luersen and Riche. [21], and Ghiasi et al. [22].

Previous studies have shown that composite fibre orientation angles can be optimised by different optmisation methods for specific load cases such as longitudinal or in-plane loading [23–35].

In this paper, the thickness of laminated composite plates is minimised by optimising the fibre orientation angles for different load cases. The novelty of the research presented in this paper is that the effect of transverse shear forces, and therefore, the induced twist angle are considered.

2. Governing Equation

2.1. Analysis of a Laminate Composite Plate. The state of stress at a point in a general continuum can be represented by nine stress components σ_{ij} $(i, j = 1, 2, 3)$ acting on the sides of an elemental cube with sides parallel to the axes of a reference coordinate system (Figure 1).

In the most general case the stress and strain components are related by the generalised Hook's law as follows:

$$\begin{bmatrix}\sigma_{11}\\ \sigma_{22}\\ \sigma_{33}\\ \sigma_{23}\\ \sigma_{31}\\ \sigma_{12}\\ \sigma_{32}\\ \sigma_{13}\\ \sigma_{21}\end{bmatrix} = \begin{bmatrix} C_{1111} & C_{1122} & C_{1133} & C_{1123} & C_{1131} & C_{1112} & C_{1132} & C_{1113} & C_{1121}\\ C_{2211} & C_{2222} & C_{2233} & C_{2223} & C_{2231} & C_{2212} & C_{2232} & C_{2213} & C_{2221}\\ C_{3311} & C_{3322} & C_{3333} & C_{3323} & C_{3331} & C_{3312} & C_{3332} & C_{3313} & C_{3321}\\ C_{2311} & C_{2322} & C_{2333} & C_{2323} & C_{2331} & C_{2312} & C_{2332} & C_{2313} & C_{2321}\\ C_{3111} & C_{3122} & C_{3133} & C_{3123} & C_{3131} & C_{3112} & C_{3132} & C_{3113} & C_{3121}\\ C_{1211} & C_{1222} & C_{1233} & C_{1223} & C_{1231} & C_{1212} & C_{1232} & C_{1213} & C_{1221}\\ C_{3211} & C_{3222} & C_{3233} & C_{3223} & C_{3231} & C_{3212} & C_{3232} & C_{3213} & C_{3221}\\ C_{1311} & C_{1322} & C_{1333} & C_{1323} & C_{1331} & C_{1312} & C_{1332} & C_{1313} & C_{1321}\\ C_{2111} & C_{2122} & C_{2133} & C_{2123} & C_{2131} & C_{2112} & C_{2132} & C_{2113} & C_{2121} \end{bmatrix} \begin{bmatrix}\epsilon_{11}\\ \epsilon_{22}\\ \epsilon_{33}\\ \epsilon_{23}\\ \epsilon_{31}\\ \epsilon_{12}\\ \epsilon_{32}\\ \epsilon_{13}\\ \epsilon_{21}\end{bmatrix}, \quad (1)$$

or

$$\sigma_{ij} = C_{ijkl}\epsilon_{kl} \ (i, j, k, l = 1, 2, 3), \quad (2)$$

where C_{ijkl} is the stiffness components [30].

By considering the symmetry of the stress and strain tensors and the energy relations, it is proven that the stiffness matrices are symmetric. Thus, the state of stress (strain) at a point can be described by six components of stress (strain), and the stress-strain equations are expressed in terms of 21 independent stiffness constants [36].

2.2. In-Plane Stress. The simplest equivalent single-layer (ESL) laminated plate theory, based on the displacement field, is the classical laminated plate theory (CLPT) [37–43]. The two-dimensional classical theory of plates was initiated by Kirchhoff [44] in the 19th century, and then was continued by Love [45] and Timoshenko [46] during the early 20th century. The principal assumption in CLPT is that normal lines to the midplane before deformation remain straight and normal to the plane after deformation. Although this assumption leads to simple constitutive equations, it is the main deficiency of the theory. The effect of the transverse shear strains on the deformation of the elastic two-dimensional structure are ignored and some of the deformation mode constraints by reducing the model to a single degree of freedom (DoF) results are neglected. This is a consequence of the basic assumptions made. It is also worth mentioning that neglecting shear stresses leads to a reduction or removal of the three natural boundary conditions that should be satisfied along the free edges. These boundary conditions being the normal force, bending moment, and twisting couple [47].

For solving in-plane stress normally the classical laminate theory is used. It is assumed that plane stress components are

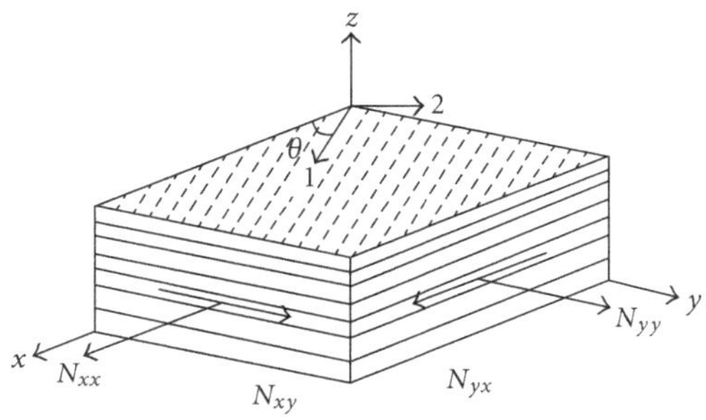

Figure 2: A scheme of composite plate under in-plane stress [35].

taken as zero. With respect to the coordinate system shown in (Figure 2) the in-plane stress components are related to the strain components as

$$\begin{bmatrix} \sigma_{xx} \\ \sigma_{yy} \\ \tau_{xy} \end{bmatrix}_k = \begin{bmatrix} \overline{Q}_{11} & \overline{Q}_{12} & \overline{Q}_{16} \\ \overline{Q}_{12} & \overline{Q}_{22} & \overline{Q}_{26} \\ \overline{Q}_{16} & \overline{Q}_{26} & \overline{Q}_{66} \end{bmatrix} \begin{bmatrix} \epsilon_{xx} \\ \epsilon_{yy} \\ \gamma_{xy} \end{bmatrix}, \quad (3)$$

where k is the lamina number, $\overline{Q}_{ij}$ are the off-axis stiffness components, which can be explained in terms of principal stiffness components, Q_{ij}, using the tensor transformation rules [48] as

$$\begin{aligned}
\overline{Q}_{11} &= Q_{11}\cos^4\theta + 2(Q_{12} + 2Q_{66}) \\
&\quad \times \sin^2\theta\cos^2\theta + Q_{22}\sin^4\theta, \\
\overline{Q}_{22} &= Q_{11}\sin^4\theta + 2(Q_{12} + 2Q_{66}) \\
&\quad \times \sin^2\theta\cos^2\theta + Q_{22}\cos^4\theta, \\
\overline{Q}_{12} &= (Q_{11} + Q_{22} - 4Q_{66})\sin^2\theta\cos^2\theta \\
&\quad + Q_{12}\left(\sin^4\theta + \cos^4\theta\right), \\
\overline{Q}_{16} &= (Q_{11} - Q_{12} - 2Q_{66})\sin\theta\cos^3\theta \\
&\quad + (Q_{12} - Q_{22} + 2Q_{66})\sin^3\theta\cos\theta, \\
\overline{Q}_{26} &= (Q_{11} - Q_{12} - 2Q_{66})\sin^3\theta\cos\theta \\
&\quad + (Q_{12} - Q_{22} + 2Q_{66})\sin\theta\cos^3\theta, \\
\overline{Q}_{66} &= (Q_{11} + Q_{22} - 2Q_{12} - 2Q_{66}) \\
&\quad \times \sin^2\theta\cos^2\theta + Q_{66}\left(\sin^4\theta + \cos^4\theta\right).
\end{aligned} \quad (4)$$

The principal stiffness terms, Q_{ij}, are related to elastic properties of the material along the principal directions, E_1, E_2, G_{12}, ϑ_{12}, and ϑ_{21} [48]. The effect of transverse shear stress is not considered by previous work [35] because in their work the laminate is only subject to in-plane loads. Therefore, strain components defined with respect to x-y-z coordinate system are the same for each ply regardless of the fibre orientation. For the same reason, the mechanical response of the laminate is independent of the stacking sequence. Stress resultants, or forces per unit length of the cross section, are obtained as

$$\begin{bmatrix} N_{xx} \\ N_{yy} \\ N_{xy} \end{bmatrix} = \int_{-h/2}^{h/2} \begin{bmatrix} \sigma_{xx} \\ \sigma_{yy} \\ \tau_{xy} \end{bmatrix} dz = 2\sum_{k=1}^{m} n_k t_0 \begin{bmatrix} \sigma_{xx} \\ \sigma_{yy} \\ \tau_{xy} \end{bmatrix}_k . \quad (5)$$

Here, m is the number of distinct laminae, n_k is the number of plies in the kth lamina. Here, lamina is meant to be a group of plies with the same orientation angle. Substituting the stress-strain relation given by (3) into (5) we have

$$\begin{bmatrix} N_{xx} \\ N_{yy} \\ N_{xy} \end{bmatrix} = \begin{bmatrix} A_{11} & A_{12} & A_{16} \\ A_{12} & A_{22} & A_{26} \\ A_{16} & A_{26} & A_{66} \end{bmatrix} \begin{bmatrix} \epsilon_{xx} \\ \epsilon_{yy} \\ \epsilon_{xy} \end{bmatrix}, \quad (6)$$

where A_{ij}, components of extensional stiffness matrix, are given by

$$A_{ij} = 2\sum_{k=1}^{m} n_k t_0 \left(\overline{Q}_{ij}\right)_k . \quad (7)$$

Principal stress components can be obtained using the following transformation [35]:

$$\begin{bmatrix} \sigma_{11} \\ \sigma_{22} \\ \tau_{12} \end{bmatrix} = \begin{bmatrix} \cos^2\theta_k & \sin^2\theta_k & 2\cos\theta_k\sin\theta_k \\ \sin^2\theta_k & \cos^2\theta_k & -2\cos\theta_k\sin\theta_k \\ -\cos\theta_k\sin\theta_k & \cos\theta_k\sin\theta_k & \cos^2\theta_k - \sin^2\theta_k \end{bmatrix}_k \times \begin{bmatrix} \sigma_{xx} \\ \sigma_{yy} \\ \tau_{xy} \end{bmatrix}. \quad (8)$$

Despite its limitations, CLPT is still a common approach utilised to determine quick and simple predictions especially for the behaviour of thinplated structures. The main simplification is that three-dimensional thick structural plates or shells are treated as two-dimensional plates or shells located through midthickness, which results in a significant reduction in the total number of variables and equations, consequently saving a lot of computational time and effort. The governing equations are easier to solve and present in closed-form solutions, which normally provides more physical or practical interpretation. This approach remains popular as it is well-known and has become the foundation for further composite plate analysis methods.

This method works relatively well for structures that are made out of a symmetric and balanced laminate, experiencing pure bending or pure tension. The error induced/introduced by neglecting the effect of transverse shear stresses becomes trivial on or close to the edges and corners of thick-sectioned configurations. The induced error increases for thick plates made of composite layers, for which the ratio of longitudinal to transverse shear elastic modulii is relatively large compared to isotropic materials [49]. It neglects transverse shear strains, underpredicts deflections and overestimates natural frequencies and buckling loads.

2.3. Out-Plane Stress. As discussed in Section 2.2, in the classical lamination theory, it was assumed that the laminate is thin compared to its lateral dimensions and that straight lines normal to the middle surface remain straight and normal to that surface after deformation. Therefore, the transverse shear stress and strain are neglected. These assumptions are not valid in the case of thicker laminates and laminates with low stiffness central plies undergoing significant transverse shear deformations. In order to overcome these limitations several theories have been proposed to analyse thicker laminated composite plates in order to consider the transfer shear effect. Most of these theories are extensions of the conventional theories developed by Reissner [47] and Mindlin [50], which are known as the shear deformation plate theories. These theories are based on the assumption that the displacement w is constant through the thickness while the displacements u and v vary linearly through the thickness of each layer (constant cross-sectional rotations w_x and w_y). Generally these theories are known as first-order shear deformation theories (FSDT) [3]. According to this theory, transverse straight lines before deformation will still be straight after deformation but they are not normal to the midplane after deformation. This theory assumes constant transverse shear stress.

In the following, referred to as first-order shear deformation laminate plate theory, the assumption of normality of straight lines is removed compared to CLPT. On the other hand straight lines normal to the middle surface remain straight but not normal to that surface after deformation [43].

For out-plane stress, (4), (5), and (6) are described as (9), (10), and (11) respectively.

$$\begin{aligned}
\overline{Q}_{11} &= Q_{11}\cos^4\theta + 2(Q_{12} + 2Q_{66}) \\
&\quad \times \sin^2\theta\cos^2\theta + Q_{22}\sin^4\theta, \\
\overline{Q}_{22} &= Q_{11}\sin^4\theta + 2(Q_{12} + 2Q_{66}) \\
&\quad \times \sin^2\theta\cos^2\theta + Q_{22}\cos^4\theta, \\
\overline{Q}_{12} &= (Q_{11} + Q_{22} - 4Q_{66})\sin^2\theta\cos^2\theta \\
&\quad + Q_{12}\left(\sin^4\theta + \cos^4\theta\right), \\
\overline{Q}_{13} &= (Q_{11} - Q_{12} - 2Q_{33})\sin\theta\cos^3\theta \\
&\quad + (Q_{12} - Q_{22} + 2Q_{33})\sin^3\theta\cos\theta, \\
\overline{Q}_{23} &= (Q_{11} - Q_{12} - 2Q_{33})\sin^3\theta\cos\theta \\
&\quad + (Q_{12} - Q_{22} + 2Q_{33})\sin\theta\cos^3\theta, \\
\overline{Q}_{33} &= (Q_{11} + Q_{22} - 2Q_{12} - 2Q_{33}) \\
&\quad \times \sin^2\theta\cos^2\theta + Q_{33}\left(\sin^4\theta + \cos^4\theta\right), \\
\overline{Q}_{55} &= Q_{55}\cos^2\theta - Q_{66}\sin^2\theta, \\
\overline{Q}_{56} &= Q_{55}\sin\theta\cos\theta - Q_{66}\sin\theta\cos\theta, \\
\overline{Q}_{66} &= Q_{55}\sin^2\theta - Q_{66}\cos^2\theta.
\end{aligned} \tag{9}$$

$$\begin{bmatrix} N_{xx} \\ N_{yy} \\ N_{xy} \end{bmatrix} = \int_{-h/2}^{h/2} \begin{bmatrix} \sigma_{xx} \\ \sigma_{yy} \\ \tau_{xy} \end{bmatrix} dz$$

$$\begin{bmatrix} M_{xx} \\ M_{yy} \\ M_{xy} \end{bmatrix} = \int_{-h/2}^{h/2} \begin{bmatrix} \sigma_{xx} \\ \sigma_{yy} \\ \tau_{xy} \end{bmatrix} z\,dz, \tag{10}$$

$$\begin{bmatrix} V_q \\ V_r \end{bmatrix} = \int_{-h/2}^{h/2} \begin{bmatrix} \tau_{xz} \\ \tau_{yz} \end{bmatrix} dz,$$

$$\begin{bmatrix} N_{xx} \\ N_{yy} \\ N_{xy} \\ M_{xx} \\ M_{yy} \\ M_{xy} \\ V_q \\ V_r \end{bmatrix} = \begin{bmatrix} N_1 \\ N_2 \\ N_{12} \\ M_1 \\ M_2 \\ M_{12} \\ V_1 \\ V_2 \end{bmatrix} = \begin{bmatrix} A_{11} & A_{12} & A_{13} & B_{11} & B_{12} & B_{13} & 0 & 0 \\ A_{12} & A_{22} & A_{23} & B_{12} & B_{22} & B_{23} & 0 & 0 \\ A_{13} & A_{23} & A_{33} & B_{13} & B_{23} & B_{33} & 0 & 0 \\ B_{11} & B_{12} & B_{13} & D_{11} & D_{12} & D_{13} & 0 & 0 \\ B_{12} & B_{22} & B_{23} & D_{12} & D_{22} & D_{23} & 0 & 0 \\ B_{13} & B_{23} & B_{33} & D_{13} & D_{23} & D_{33} & 0 & 0 \\ 0 & 0 & 0 & 0 & 0 & 0 & E_{11} & E_{12} \\ 0 & 0 & 0 & 0 & 0 & 0 & E_{12} & E_{22} \end{bmatrix} \begin{bmatrix} \epsilon_1 \\ \epsilon_2 \\ \gamma_{12} \\ k_1 \\ k_2 \\ k_{12} \\ \gamma_{13} \\ \gamma_{23} \end{bmatrix}, \tag{11}$$

where the components of this section stiffness matrix are given by

$$\begin{aligned}
\left(A_{ij}, B_{ij}, D_{ij}\right) &= \int_{-h/2}^{h/2} \overline{Q}_{ij}^{m}\left(1, z, z^2\right) dz \quad (i, j = 1, 2, 3) \\
E_{ij} &= \int_{-h/2}^{h/2} \overline{Q}_{\alpha\beta}^{m} k_i k_j\, dz \\
&\quad \times (i, j = 1, 2,\ \alpha, \beta = i + 4,\ \ j + 4).
\end{aligned} \tag{12}$$

2.4. Failure Criteria. As it is shown in previous research [35] using one failure method is not reliable enough for evaluating the results. Also in each fibre orientation the type of the failure is switched between fibre, shear, and transverse shear failure [43]. Therefore, it is logical that the results would be checked by at least two failure criteria. In this research maximum stress and Tsai-Wu criteria are chosen.

2.4.1. Maximum Stress Criterion. Maximum stress criterion is one of the simplest failure methods to apply. According to this criterion, failure is predicted whenever one of the principal stress components exceeds its corresponding strength. It is expressed in the form of the following subcriteria:

$$\sigma_1 = \begin{cases} F_{1t} & \text{when } \sigma_1 > 0, \\ -F_{1c} & \text{when } \sigma_1 < 0, \end{cases} \tag{13a}$$

$$\sigma_2 = \begin{cases} F_{2t} & \text{when } \sigma_2 > 0, \\ -F_{2c} & \text{when } \sigma_2 < 0, \end{cases} \tag{13b}$$

$$\sigma_3 = \begin{cases} F_{3t} & \text{when } \sigma_3 > 0, \\ -F_{3c} & \text{when } \sigma_3 < 0, \end{cases} \tag{13c}$$

$$\left|\tau_4\right| = F_4, \tag{13d}$$

$$\left|\tau_5\right| = F_5, \tag{13e}$$

$$\left|\tau_6\right| = F_6. \tag{13f}$$

F_{1t} and F_{1c} are the longitudinal tensile and compressive strengths, F_{2t} and F_{2c} are the transverse longitudinal tensile and compressive strengths, and F_6 is the in-plane shear strength. Four additional lamina strength parameters, which are relevant in three-dimensional analysis, are the out-plane or interlaminar tensile, compressive, and shear strengths, F_{3t}, F_{3c}, F_4, and F_5.

2.4.2. Tsai-Wu Criterion. The Tsai-Wu failure criterion is one of the most reliable static failure criteria as it provides a simple analytical expression taking components. Reddy [43] proposed a modified tensor polynomial theory by assuming the existence of a failure in the stress space. In contracted notation it takes the form:

$$f_i\sigma_i + f_{ij}\sigma_i\sigma_j = 1, \tag{14}$$

where f_i and f_{ij} are second- and fourth-order strength tensors, and $i, j = 1, 2, \ldots, 6$ [36].

By applying assumptions, some of f_i and f_{ij} are identified. Finally it is reduced to a failure envelope for constant values of shear stress $\tau_6 = kF_6$

$$f_1\sigma_1 + f_2\sigma_2 + f_{11}\sigma_1^2 + f_{22}\sigma_2^2 + 2f_{12}\sigma_1\sigma_2 = 1 - k^2 \tag{15}$$

or equivalently

$$\frac{\sigma_{11}^2}{F_{1t}|F_{1c}|} + \frac{\sigma_{22}^2}{F_{2t}|F_{2c}|} + \frac{\tau_{12}^2}{F_6} - \frac{\sigma_{11}\sigma_{22}}{\sqrt{F_{1t}F_{1c}F_{2t}F_{2c}}} + \left(\frac{1}{F_{1t}} - \frac{1}{|F_{1c}|}\right)\sigma_{11} + \left(\frac{1}{F_{2t}} - \frac{1}{|F_{2c}|}\right)\sigma_{22} < 1. \tag{16}$$

3. Optimisation

An optimised composite laminate requires finding the minimum number of layers, and the best fibre orientation and thickness for each layer. Several optimisation methods have been introduced to solve this challenging problem, which is often nonlinear, nonconvex, multimodal, and multidimensional. Nowadays usually stochastic non-linear optmisation methods are utilised for this problem as they can avoid the local minimums. One of the best algorithms in this category is simulated annealing (SA) method which is used in similar problems [7].

3.1. Simulated Annealing Algorithm. Kirkpatrick et al. [51] proposed simulated annealing as a powerful stochastic search technique in 1983. The method gets its name from the physical process whereby the temperature of a solid is raised to a melting point, where the atoms can move freely, and then slowly cooled. The method attempts to model the behaviour of the atoms in forming arrangements in solid material during annealing. Although the atoms move randomly, as their natural behaviour they tend to form lower-energy configurations [52]. However, this is a time-driven process. When a material is crystallised from its liquid phase, it must be cooled slowly if it is to assume its highly ordered, lowest-energy, perfect crystalline state. At each temperature level during this annealing process, the material should reach equilibrium. As the temperature decreases, the arrangement of the atoms gets closer and closer to the lower energy state. This process continues until the temperature finally reaches freezing point [52]. The temperature is initially assigned a higher value, which corresponds to more probability of accepting a bad move and is gradually reduced by a user-defined cooling schedule. Retaining the best solution is recommended in order to preserve the good solution [52].

At each iteration of the simulated annealing algorithm, a new point is randomly generated. The distance of the new point from the current point, or the extent of the search, is based on a probability distribution with a scale proportional to the temperature. The algorithm accepts all new points that lower the objective, but also, with a certain probability, points that raise the objective. The algorithm avoids being trapped in local minima, by accepting points that raise the objective, and is able to explore globally for more possible solutions. An annealing schedule is selected to systematically decrease the temperature as the algorithm proceeds. As the temperature decreases, the algorithm reduces the extent of its search to converge to a minimum.

If a set of configurations is considered, in each iteration the speed convergency would be increased. In this paper the SA proposed by Erdal and Sonmez [52] is applied. The number of these configurations depends on the dimension of the problem:

$$N = 7(n + 1), \tag{17}$$

where n is the dimension of the problem, that is, the number of design variables [52].

3.2. Penalty Function and Optimisation Procedure. In this step a penalty function is expressed, and then this function has to be optimized:

$$F = \sum_{k=1}^{m} n_k + w_1 P_{MS} + w_2 P_{TW} - w_1 SF_{MS} - w_2 SF_{TW} + \frac{w_{3i}}{k}\cos(\Delta\theta). \tag{18}$$

n_k is the number of plies in the kth lamina, in which the orientation angle is θ_k; m is the total number of distinct lamina; the second and third terms represent the penalty values introduced to increase the value of the objective function for designs for which failure is predicted, and thus to restrict the search to the feasible design space; P_{MS} and P_{TW} are penalty values calculated based on the maximum stress criterion and the Tsai-Wu criterion, respectively. SF_{MS} and SF_{TW} are equal to the safety factors according to the maximum stress, and Tsai-Wu criteria, respectively, if they are greater than 1.0, otherwise these terms are equal to zero; w_i are suitable coefficients [35].

This penalty function is the same as the one defined by Akbulut and Sonmez [35] in 2008 except the last term of (18). In their work, the ply angles are optimised for in-plane plate and the effect of shear stress and as a result induced twist angle of the plate was neglected. However, in this research the induced twist angles are considered. Therefore the maximum acceptable twist angle is defined, and by assuming this maximum twist angle, the appropriate coefficient for w_{3i} is obtained.

Implying several tests by finite element method (FEM) software shows that maximum twist for each material happens at the specific angles $(\theta_{\max})_i$ $i = 1, 2, 3, \ldots, i$ is the number of possible $\theta_{\max}$. The range of fibre orientations $[-90, 90]$ is divided into several areas, each $\theta_{\max}$ is the centre of the area and in each iteration the program find the θ_k to the range that it belongs to and then the program works with the appropriate $\theta_{\max}$ and corresponding w_{3i}. Except for the related w_{3i}, the other w_{3i} are equal to zero. $\Delta\theta$ is defined as

$$\Delta\theta = \theta_{\max} - \theta_k. \tag{19}$$

So, when the ply angle in each layer (θ_k) would be close to $\theta_{\max}$, the amount of $\cos(\Delta\theta)$ is bigger, and thus the last term of penalty function and as a result penalty function would be larger.

For a proper w_3, the amount of induced twist (α) always will be less than maximum acceptable twist $(\alpha_{\max})$. For example, in a composite product, the final induced twist is desired to be less than a certain amount of twist (which here it is called $\alpha_{\max}$). When fibre orientation for layers approach the $\theta_{\max}$ (where maximum twist happening), the term $\Delta\theta$ or $(\theta_{\max} - \theta_k)$ in (18) tends to zero. Then the term $\cos(\Delta\theta)$ and as a result penalty function in (18) are increased. Increasing penalty function means that this fibre orientation is not going to be selected in optimisation process and therefore the induced twist in final product stays less than $\alpha_{\max}$. This technique is what distinguished this model from others as the induced twist is ignored in them.

The reason that the objective is reduced for safe designs is that there may be many feasible designs with the same minimum thickness. Of these designs, the optimum is defined as the one with the largest failure load. Accordingly, the objective function is linearly reduced in proportion to the failure margin [53]. Similarly in another study [54], the margins to initial failure were maximized with the minimum feasible number of laminae. The safety factor of the laminate according to the maximum stress criterion, SF_{MS}, is calculated as follows [35]:

$$SF_{MS}^{k} = \min \text{ of} \begin{cases} SF_X^k = \begin{cases} \dfrac{F_{1t}}{\sigma_{11}} & \text{if } \sigma_{11} > 0 \\[2ex] \dfrac{F_{1c}}{\sigma_{11}} & \text{if } \sigma_{11} < 0 \end{cases} \\[4ex] SF_Y^k = \begin{cases} \dfrac{F_{2t}}{\sigma_{22}} & \text{if } \sigma_{22} > 0 \\[2ex] \dfrac{F_{2c}}{\sigma_{22}} & \text{if } \sigma_{22} < 0 \end{cases} \\[2ex] SF_S^k = \dfrac{S}{|\tau_{12}|} \end{cases} \tag{20}$$

$$SF_{MS} = \min \text{ of } SF_{MS}^{k} \text{ for } k = 1, 2, \ldots, m-1, m. \tag{21}$$

The safety factor for the kth lamina, SF_{TW}^k, according to the Tsai-Wu criterion is defined as the multiplier of the stress components at lamina k, σ_{ij}^k, that makes the right hand side of (16) equal to 1 then it turns into [35]

$$a\left(SF_{TW}^{k}\right)^2 + b\left(SF_{TW}^{k}\right) = 1, \tag{22}$$

where

$$a = \frac{\left(\sigma_{11}^k\right)^2}{F_{1t}|F_{1c}|} + \frac{\left(\sigma_{22}^k\right)^2}{F_{2t}|F_{2c}|} + \frac{\left(\tau_{12}^k\right)^2}{F_6^2} - \frac{\left(\sigma_{11}^k\right)\left(\sigma_{22}^k\right)}{\sqrt{(F_{1t}F_{1c}F_{2t}F_{2c})}}$$

$$b = \left(\frac{1}{F_{1t}} - \frac{1}{|F_{1c}|}\right)\sigma_{11}^k + \left(\frac{1}{F_{2t}} - \frac{1}{|F_{2c}|}\right)\sigma_{22}^k. \tag{23}$$

The root of the above equation gives the safety factor. Because a negative safety factor is not physically meaningful, the absolute value of the first root is considered as the actual safety factor

$$SF_{TW}^{k} = \left|\frac{-b \pm \sqrt{b^2 + 4a}}{2a}\right|. \tag{24}$$

Then, the minimum of SF_{TW}^k is chosen as the safety factor of the laminate:

$$SF_{TW} = \min \text{ of} SF_{TW}^{k} \quad \text{for } k = 1, 2, \ldots, m-1, m \tag{25}$$

In (24), the $|(-b + \sqrt{b^2 + 4a})/2a|$ can be considered, as b is always positive, and the aim is to find the minimum of SF_{TW}^k.

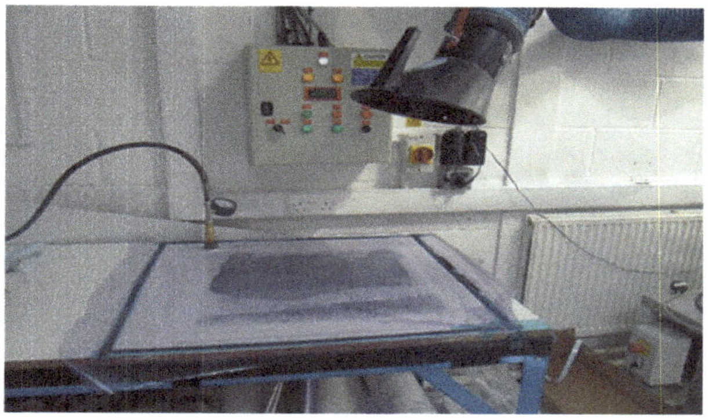

FIGURE 3: Process of making laminated plate.

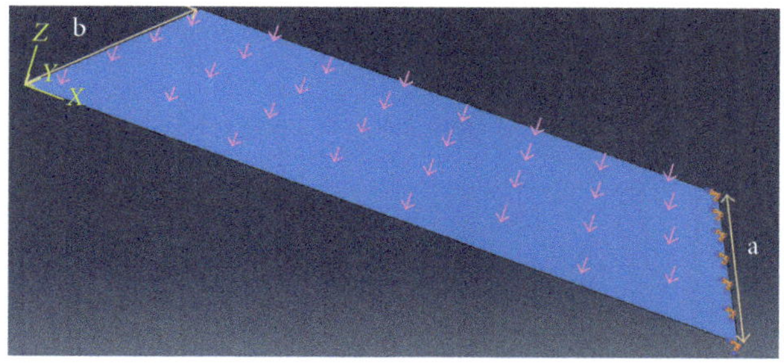

FIGURE 4: Boundary condition and loading (the plate is clamped on side a. Displacements are measured on side b).

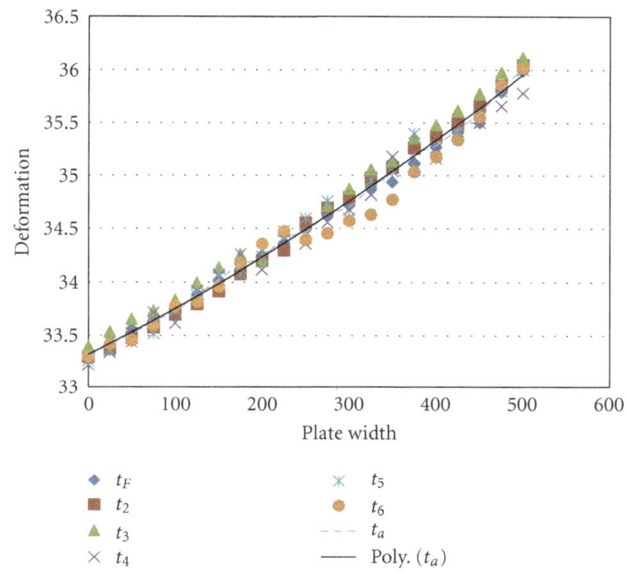

FIGURE 5: An example for one case; t_1 to t_6 are the experimental test results for six similar plates under constant load. t_a is the average of t_1 to t_6 and t_F is the FEM outputs.

The penalty value due to the violation of the maximum stress and Tsai-Wu criteria are calculated in (26) and (27), respectively:

$$P_x^k = \begin{cases} 0 & \text{if } SF_x^k \geq 1, \\ \dfrac{1}{SF_x^k} - 1 & \text{if } SF_x^k < 1, \end{cases}$$

$$P_y^k = \begin{cases} 0 & \text{if } SF_y^k \geq 1, \\ \dfrac{1}{SF_y^k} - 1 & \text{if } SF_y^k < 1, \end{cases} \qquad (26)$$

$$P_s^k = \begin{cases} 0 & \text{if } SF_s^k \geq 1, \\ \dfrac{1}{SF_s^k} - 1 & \text{if } SF_s^k < 1, \end{cases}$$

$$P_{TW}^k = \begin{cases} 0 & \text{if } SF_{TW}^k \geq 1, \\ \dfrac{1}{SF_{TW}^k} - 1 & \text{if } SF_{TW}^k < 1. \end{cases} \qquad (27)$$

The total penalty value for the laminate due to the violation of the maximum stress and Tsai-Wu criteria are then calculated by summing up the penalty values calculated for each lamina

$$P_{MS} = \sum_{k=1}^{m} P_x^k + P_y^k + P_s^k,$$
$$P_{TW} = \sum_{k=1}^{m} P_{TW}^k. \qquad (28)$$

4. Results

4.1. Experimental Results. In order to validate the FEM model some experimental tests have been performed. For each case six similar laminated plates are manufactured. One of the samples is shown in Figure 3. Carbon fibre is used for all laminated experimental tests, and the size of plate is 500*500 (mm).

Square laminated composite panels (500×500 mm) were produced by vacuum bagging. In order to minimise the error, the average results of these six plates were calculated to be compared with FEM results obtained from the design tool.

As it is shown in Figure 4, the plates are fixed on one side (a) and displacement in z direction is measured on the opposite free side of the plate (b) due to the effect of gravity only.

In Figure 5, t_1 to t_6 are the experimental test results for six similar plates under the constant load. Six plates in each case have the same layup and geometry; t_a is the average of t_1 to t_6 and t_F is the FEM results.

The experimental tests have been carried out for 15 different cases with different loads and layups. In Figure 6 the percentage difference of deformation between tests and FEM results and in Figure 7 the percentage difference of α are shown. As is shown in Figures 6 and 7 there is a very good agreement between FEM and experimental results. In Figures 6 and 7 the axis which shows error is zoomed in to 10% to distinguish the differences between each case. These Experimental tests have been done to validate the FE model which is used in this work. The good agreement, as it is shown in Figures 6 and 7, indicates the validity of this FEM. In the following case studies, FE results are compared with

Table 1: Optimum lamina orientations for material T300/5308 under different loads.

Loading: N_{xx}, N_{yy}, N_{xy} (MPa m) (Figure 2)	Optimum lamina orientations Akbulut and Sonmez [35]	Present work	Safety factor Akbulut and Sonmez [35] Max. stress	Tsai-Wu	Present work Max. stress	Tsai-Wu
10/5/0	$[37_{27}/-37_{27}]$	$[39_{29}/-39_{29}]$	1.0277	1.0068	1.1309	1.1001
20/5/0	$[31_{23}/-31_{23}]$	$[36_{27}/-36_{27}]$	1.1985	1.0208	1.3305	1.1560
40/5/0	$[26_{20}/-26_{20}]$	$[26_{22}/-26_{22}]$	1.5381	1.0190	1.6504	1.1903
80/5/0	$[21_{25}/-19_{28}]$	$[21_{25}/-21_{25}]$	1.2213	1.0113	1.2302	1.0120
120/5/0	$[17_{35}/-17_{35}]$	$[17_{35}/-17_{35}]$	1.0950	1.0030	1.0951	1.0030

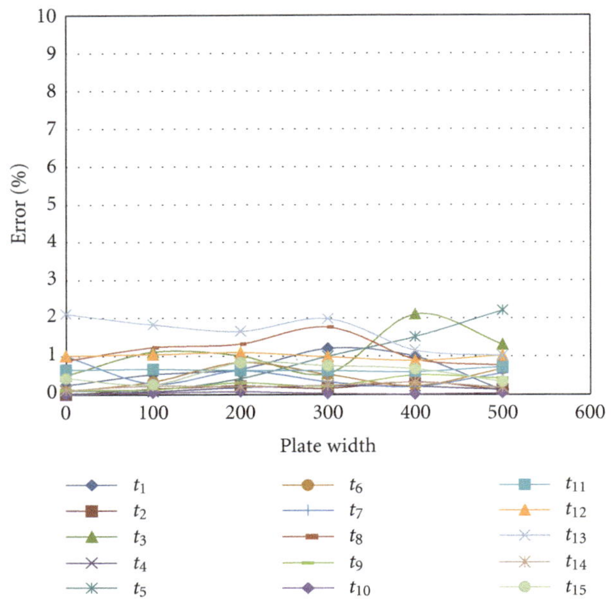

Figure 6: Percentage difference of deformation between tests and FEM results.

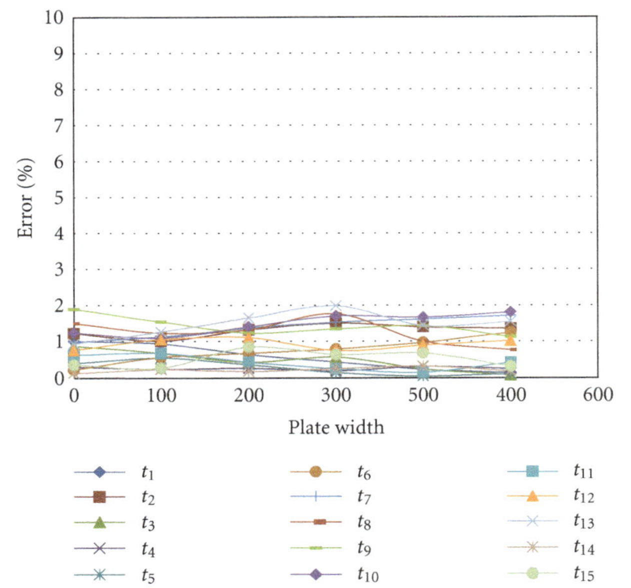

Figure 7: Percentage difference of (α) between tests and FEM results.

similar works with the same boundary condition as shown in Figure 4 with different loadings.

4.2. Case Study Results. In this Section, two case studies are considered. The first case study compares the obtained results with those which were found by Akbulut and Sonmez [35] to validate the proposed optimisation method. The graphite/epoxy materials T300/5308 with the properties of $E_{11} = 40.91$ GPa, $E_{22} = 9.88$ GPa, $G\_12 = 2.84$ GPa, $\vartheta_{12} = 0.292$, $F_{1t} = 779$ MPa, $F_{1c} = -1134$ MPa, $F_{2t} = 19$ MPa, $F_{2c} = -131$ MPa, and $S = 75$ MPa is considered for the first case study.

In this case the maximum acceptable twist angle is $\alpha_{max} = .01$. In Tables 1 and 2 the results are compared with previous work [35]. As it is shown in Table 1, the number of lay-up increase for some loads and it shows that in these cases the amount of twist angle is more than α_{max}. Clearly by increasing the lay-ups, safety factor for both Tsai-Wu and Maximum stress will be increased. In Table 2, the number of lay-ups and thickness is constant but in some cases the optimum orientation angles is different from the previous work. Although the safety factor is reduced, it avoids passing the acceptable twist angle and the safety factor is still more than 1, therefore, it is still acceptable. When the maximum twist angle is less than the α_{max}, the results are comparable with the work has been done by Akbulut and Sonmez [35].

In the second case study a highly anisotropic material is considered (material II). The elements of stiffness matrix are $D_{1111} = 138000$, $D_{1122} = 44000$, $D_{2222} = 138000$, $D_{1133} = 5000$, $D_{2233} = 6000$, $D_{3333} = 47000$, $D_{1112} = 2000$, $D_{2212} = 8000$, $D_{3312} = 3000$, $D_{1212} = 10000$, $D_{1113} = 0$, $D_{2213} = 3500$, $D_{3313} = 0$, $D_{1213} = 21000$, $D_{1313} = 58000$, $D_{1123} = 0$, $D_{2223} = 1000$, $D_{3323} = 2500$, $D_{1223} = 2000$, $D_{1323} = 4500$, and $D_{2323} = 23000$.

In Figure 8 the amount of twist angle under a constant load for different ply angles is shown. So the amount of $(\theta_{max})_i$ in (20) are (±30, ±60) which are the local maximums.

TABLE 2: Optimum lamina orientations for material T300/5308 under different loads for constant thickness.

Loading: N_{xx}, N_{yy}, N_{xy} (KPa m) (Figure 2)	Optimum lamina orientations: Akbulut and Sonmez [35]	Optimum lamina orientations: Present work	Safety factor: Akbulut and Sonmez [35]	Safety factor: Present work
200/200/0	$[50.80_4/-49.80_4/26.59_4/-49.73_4]$	$[50.80_4/-49.80_4/26.59_4/-49.73_4]$	2.14	2.14
200/0/200	$[31.72_{16}]$	$[32.40_4/-56.61_4/-7.81_4/33.87_4]$	4.84	4.59
400/200/0	$[30.98_4/-36.57_4/37.67_4/-37.20_4]$	$[-20.12_4/58.01_4/-49.90_4/20.11_4]$	1.64	1.42
200/200/200	$[45_{16}]$	$[45_{16}]$	1.11×10^{16}	1.11×10^{16}

TABLE 3: Optimum lamina orientations for second case study material under different loads for constant thickness.

Loading: $N_{xx}, N_{yy}, N_{xy}, N_{zz}$ (KPa m)	Optimum lamina orientations	Safety factor: Max. stress	Safety factor: Present work
100/100/0/20	$[-77.68_4/23.54_4/-22.78_4/49.79_4]$	1.2720	1.0931
100/0/100/20	$[53.45_8/-78.23_8]$	1.5402	1.2112
200/100/0/20	$[-22.23_4/49.26_4/-65.33_4/11.23_4]$	1.3200	1.111
100/100/100/20	$[44.8_8/-33.21_8]$	1.3401	1.0376

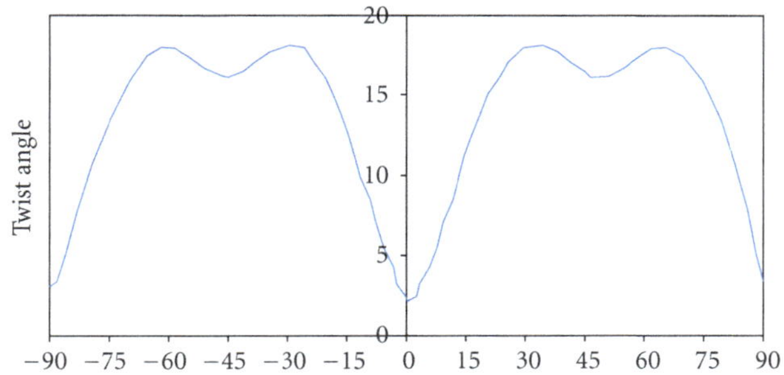

FIGURE 8: Twist angle for a first layer of material (II).

This test is for the first layer in order to find the $(\theta_{\max})_i$ and the areas which were explained in Section 3.2. FEM tests show that these $(\theta_{\max})_i$ are the same by adding the next layers. If the stiffness matrix is symmetric the curves will be symmetric about y axis. A general stiffness matrix for material II is considered, so there is no mirror about y axis. Optimum lamina orientations under different loads in this case are shown in the Table 3. In this case the pure bending load N_{zz} is also applied to the plate.

5. Conclusion

In this study, an optimisation methodology of composite plates was presented. A method was proposed in order to overcome the difficulties and shortcomings faced by the previous research. In previous work the effect of transverse shear was neglected, and therefore the induced twist angle is ignored. In some applications the twist angle, which is the direct effect of transverse shear, is undesirable. Therefore, in this research, after optimising the fibre orientations, by considering the induced twist angle as well as safety factor, the induced twist angle always stays less than the acceptable twist angle. One of the other weakness in previous work was that the plate was optimised under specific loads, such as longitudinal or in-plane loading. By the proposed method in this research the out-plane stress optimisation can be solved as well as the in-plane stresses. In order to have a reliable optimisation, simulated annealing, which is one of the stochastic optimisation methods and can escape the local minima is applied and the penalty function for this optimisation method is modified. This modified penalty function forces the induced twist to stay under a predefined induced twist. In addition, two Tsai-Wu and maximum stress failure criteria are used in the algorithm individually to avoid false optimal design.

References

[1] Y. Fu, S. Li, and Y. Jiang, "Analysis of inter-laminar stresses for composite laminated plate with interfacial damage," *Acta Mechanica Solida Sinica*, vol. 21, no. 2, pp. 127–140, 2008.

[2] M. Blanc and M. Touratier, "An efficient and simple refined model for temperature analysis in thin laminated composites," *Composite Structures*, vol. 77, no. 2, pp. 193–205, 2007.

[3] R. L. Onsy and T. Tran-Cong, "Finite strip elements for laminated composite plates with transverse shear strain discontinuities," *Composite Structures*, vol. 56, no. 3, pp. 249–258, 2002.

[4] S. C. Han, A. Tabiei, and W. T. Park, "Geometrically nonlinear analysis of laminated composite thin shells using a modified first-order shear deformable element-based Lagrangian shell element," *Composite Structures*, vol. 82, no. 3, pp. 465–474, 2008.

[5] F. Faimun, P. D. Gosling, O. Polit, and M. Touratier, "Analytical response sensitivity of laminated composite plate using trigonometric type element," in *Proceedings of the 13th Association of Computational Mechanics in Engineering Conference (ACME '05)*, Sheffield, UK, 2005.

[6] R. Khandan, S. Noroozi, P. Sewell, J. Vinney, and M. R. Ramazani, " Optimum design of fibre orientation angles in composite laminate plates for minimum thickness (ASME '10)," in *Proceedings of the International Mechanical Engineering Congress & Exposition*, British Columbia, Canada, 2010.

[7] H. Ghiasi, D. Pasini, and L. Lessard, "Optimum stacking sequence design of composite materials part I: constant stiffness design," *Composite Structures*, vol. 90, no. 1, pp. 1–11, 2009.

[8] R. S. Sandhu, "Parametric study of Tsai's strength criterion for filamentary composites," TR-68-168, Air Force Flight Dynamic Laboratory, Wright-Patterson AFB, 1969.

[9] H. E. Brandmaier, "Optimum filament orientation criteria," *Journal of Composite Materials*, vol. 4, pp. 422–425, 1970.

[10] J. Wang and B. L. Karihaloo, "Optimum in situ strength design of laminates under combined mechanical and thermal loads," *Composite Structures*, vol. 47, no. 1–4, pp. 635–641, 1999.

[11] C. C. Chao, C. T. Sun, and S. L. Kohn, "Strength optimization for cylindrical shells of laminated composites," *Journal of Composite Materials*, vol. 9, no. 1, pp. 53–66, 1975.

[12] W. J. Park, "An optimal design of simple symmetric laminates under the first ply failure criterion," *Journal of Composite Materials*, vol. 16, no. 4, pp. 341–352, 1982.

[13] H. Fukunaga and T. W. Chou, "On laminate configurations for simultaneous failure," *Journal of Composite Materials*, vol. 22, pp. 271–286, 1988.

[14] A. Miravete, "Optimization of symmetrically laminated composites rectangular plates," in *Proceedings of the 7th International Conference on Composite Materials*, Beijing, China, 1989.

[15] H. Fukunaga and G. N. Vanderplaats, "Strength optimization of laminated composites with respect to layer thickness and/or layer orientation angle," *Computers and Structures*, vol. 40, no. 6, pp. 1429–1439, 1991.

[16] Z. Gurdal and R. T. Haftka, *Elements of Structural Optimization*, Kluwer Academic, New York, NY, USA, 3rd edition, 1992.

[17] Z. Gurdal, R. T. Haftka, and P. Hajela, *Design and Optimization of Laminated Composite Materials*, Wiley-Interscience, New York, NY, USA, 1999.

[18] R. Spallino, G. Giambanco, and S. Rizzo, "A design algorithm for the optimization of laminated composite structures," *Engineering Computations*, vol. 16, no. 3, pp. 302–315, 1999.

[19] P. M. Weaver, "Designing composite structures: lay-up selection," *Proceedings of the Institution of Mechanical Engineers, Part G: Journal of Aerospace Engineering*, vol. 216, no. 2, pp. 105–116, 2002.

[20] A. Chattopadhyay, C. E. Seeley, and R. Jha, "Aeroelastic tailoring using piezoelectric actuation and hybrid optimization," *Smart Materials and Structures*, vol. 8, no. 1, pp. 83–91, 1999.

[21] M. A. Luersen and R. le Riche, "Globalized nelder-mead method for engineering optimization," *Computers and Structures*, vol. 82, no. 23–26, pp. 2251–2260, 2004.

[22] H. Ghiasi, D. Pasini, and L. Lessard, "Constrained globalized Nelder-Mead method for simultaneous structural and manufacturing optimization of a composite bracket," *Journal of Composite Materials*, vol. 42, no. 7, pp. 717–736, 2008.

[23] M. Kameyama and H. Fukunaga, "Optimum design of composite plate wings for aeroelastic characteristics using lamination parameters," *Computers and Structures*, vol. 85, no. 3, pp. 213–224, 2007.

[24] C. M. M. Soares, V. F. Correia, H. Mateus, and J. Herskovits, "A discrete model for the optimal design of thin composite plate-shell type structures using a two-level approach," *Composite Structures*, vol. 30, no. 2, pp. 147–157, 1995.

[25] J. Wang and B. L. Karihaloo, "Optimum in situ strength design of composite laminates. Part II: optimum design," *Journal of Composite Materials*, vol. 30, no. 12, pp. 1338–1358, 1996.

[26] V. M. F. Correia, C. M. M. Soares, and C. A. M. Soares, "Higher order models on the eigenfrequency analysis and optimal design of laminated composite structures," *Composite Structures*, vol. 39, no. 3-4, pp. 237–253, 1997.

[27] C. M. Mota Soares, C. A. Mota Soares, and V. M. Franco Correia, "Optimization of multilaminated structures using higher-order deformation models," *Computer Methods in Applied Mechanics and Engineering*, vol. 149, no. 1–4, pp. 133–152, 1997.

[28] A. Y. Abu-Odeh and H. L. Jones, "Optimum design of composite plates using response surface method," *Composite Structures*, vol. 43, no. 3, pp. 233–242, 1998.

[29] S. A. Barakat and G. A. Abu-Farsakh, "The use of an energy-based criterion to determine optimum configurations of fibrous composites," *Composites Science and Technology*, vol. 59, no. 12, pp. 1891–1899, 1999.

[30] J. S. Moita, J. I. Barbosa, C. M. M. Soares, and C. A. M. Soares, "Sensitivity analysis and optimal design of geometrically non-linear laminated plates and shells," *Computers and Structures*, vol. 76, no. 1, pp. 407–420, 2000.

[31] J. H. Oh, Y. G. Kim, and D. G. Lee, "Optimum bolted joints for hybrid composite materials," *Composite Structures*, vol. 38, pp. 329–341, 1997.

[32] S. Adali, A. Richter, and V. E. Verijenko, "Optimization of shear-deformable laminated plates under buckling and strength criteria," *Composite Structures*, vol. 39, no. 3-4, pp. 167–178, 1997.

[33] C. W. Kim, W. Hwang, H. C. Park, and K. S. Han, "Stacking sequence optimization of laminated plates," *Composite Structures*, vol. 39, no. 3-4, pp. 283–288, 1997.

[34] P. Y. Tabakov and M. Walker, "A technique for optimally designing engineering structures with manufacturing tolerances accounted for," *Engineering Optimization*, vol. 39, no. 1, pp. 1–15, 2007.

[35] M. Akbulut and F. O. Sonmez, "Optimum design of composite laminates for minimum thickness," *Computers and Structures*, vol. 86, no. 21-22, pp. 1974–1982, 2008.

[36] I. M. Daniel and O. Ishai, *Engineering Mechanics of Composite Materials*, Oxford University Press, Oxford, UK, 2nd edition, 2006.

[37] E. Reissner and Y. Stavsky, "Bending and stretching of certain types of aeolotropic elastic plates," *Journal of Applied Mechanics*, vol. 28, pp. 402–408, 1961.

[38] Y. Stavsky, "Bending and stretching of laminated aeolotropic plates," *Journal of Engineering Mechanics*, vol. 87, no. 6, pp. 31–56, 1961.

[39] S. B. Dong, K. S. Pister, and R. L. Taylor, "On the theory of laminated anisotropic shells and plates," *Journal of Aeronautical Science*, vol. 29, no. 8, pp. 969–975, 1962.

[40] P. C. Yang, C. H. Norris, and Y. Stavsky, "Elastic wave propagation in heterogeneous plates," *International Journal of Solids and Structures*, vol. 2, no. 4, pp. 665–684, 1966.

[41] S. A. Ambartsumyan, *Theory of Anisotropic Plates*, Technomic, Stamford, Conn, USA, 1969, Translated from Russian by T. Cheron.

[42] J. M. Whitney and A. W. Leissa, "Analysis of heterogeneous anisotropic plates," *Journal of Applied Mechanics*, vol. 36, no. 2, pp. 261–266, 1969.

[43] J. N. Reddy, *Mechanics of Laminated Composite Plates and Shells*, CRC Press, New York, NY, USA, 2nd edition, 2004.

[44] G. Kirchhoff, "Uber das Gleichgewicht und die Bewegung einer elestischen Scheibe," *Journal Fur Reine Und Angewandte Mathematik*, vol. 40, pp. 51–88, 1850.

[45] A. E. H. Love, *The Mathematical Theory of Elasticity*, Cambridge University Press, Cambridge, UK, 1934.

[46] S. Timoshenko, *Theory of Elasticity*, McGraw-Hill, NewYork, NY, USA, 1934.

[47] E. Reissner, "The effect of transverse shear deformation on the bending of elastic plates," *Journal of Applied Mechanics*, vol. 12, pp. 69–77, 1945.

[48] R. M. Jones, *Mechanics of Composite Materials*, Taylor & Francis, Boca Raton, Fla, USA, 2nd edition, 1999.

[49] J. N. Reddy, *Energy and Variational Methods in Applied Mechanics*, John Wiley & Sons, New York, NY, USA, 1984.

[50] R. D. Mindlin, "Influence of rotary inertia and shear on flexural motions of isotropic elastic plates," *Journal of Applied Mechanics*, vol. 18, pp. 31–38, 1951.

[51] S. Kirkpatrick, C. D. Gelatt, and M. P. Vecchi, "Optimization by simulated annealing," *Science*, vol. 220, pp. 671–680, 1983.

[52] O. Erdal and F. O. Sonmez, "Optimum design of composite laminates for maximum buckling load capacity using simulated annealing," *Composite Structures*, vol. 71, no. 1, pp. 45–52, 2005.

[53] R. le Riche and R. T. Haftka, "Improved genetic algorithm for minimum thickness composite laminate design," *Composites Engineering*, vol. 5, no. 2, pp. 143–161, 1995.

[54] P. Kere, M. Lyly, and J. Koski, "Using multicriterion optimization for strength design of composite laminates," *Composite Structures*, vol. 62, no. 3-4, pp. 329–333, 2003.

6

Lanthanum Influence on $EuAlO_3$ Perovskite Structural Properties: Experimental and Molecular Dynamics Studies

Enrique Lima,[1] María Elena Villafuerte-Castrejón,[1] José Saniger,[2] Victor Lara,[3] Jorge E. Sánchez-Sánchez,[4] and Luis Javier Álvarez[4]

[1] *Instituto de Investigaciones en Materiales, Universidad Nacional Autónoma de México, Circuito Exterior, A. P. 70-360, 04510 México, DF, Mexico*
[2] *CCADET, Universidad Nacional Autónoma de México, Circuito Exterior, A. P. 70-188, 04510 México, DF, Mexico*
[3] *Departamento de Química, Universidad Autónoma Metropolitana, Iztapalapa, Avenida San Rafael Atlixco No. 186, 09340 México, DF, Mexico*
[4] *Laboratorio de Simulación, Instituto de Matemáticas, Unidad Cuernavaca, Universidad Nacional Autónoma de México. AP-273-3 Admon. 3, Cuernavaca, Morelos 62251, Mexico*

Correspondence should be addressed to Enrique Lima, lima@iim.unam.mx

Academic Editor: Peter Majewski

X-ray diffraction, ^{27}Al MAS NMR, and FTIR spectra along with results of molecular dynamics simulations were used to characterise $La_xEu_{1-x}AlO_3$ perovskites for $x = 0.3$, 0.1. Experimental and simulation results show that local changes in the perovskite-like structure can be achieved as lanthanum ions substitute europium ones. The introduction of La^{3+} ions in the $EuAlO_3$ parent causes an increase in the mobility of oxygen network.

1. Introduction

Various physical properties of $LaAlO_3$ as well as $EuAlO_3$ perovskites have both scientific and technological interest when doped with lanthanides [1, 2]. The 4f electrons of doping species are responsible for changes in their magnetic, electric, and optical properties. Rare earths containing perovskites have attracted attention mainly as semiconductor and/or optical materials [3–5]. A particularly interesting system that has not been thoroughly studied is the $La_{1-x}Eu_xAlO_3$ perovskite. In both parent perovskites, $LaAlO_3$, and $EuAlO_3$, aluminium cations are octahedrally coordinated and lanthanum and europium cations are located in dodecahedral sites [6, 7].

$La_{1-x}Eu_xAlO_3$ perovskites have been obtained through substitution of La^{3+} by Eu^{3+} in $LaAlO_3$ parent solid. These studies have been carried out previously and lattice parameters variations have been reported. However, these studies are limited to low percentages of substitution of La^{3+} by Eu^{3+} and were focused on the optical applications of these materials [8]. For instance, with the goal to maximize the fluorescence intensity on $La_{1-x}Eu_xAlO_3$, Klimke and Wulff studied the introduction of Eu^{3+} in a $LaAlO_3$ perovskite by solid-state reactions. They reported an optimal Eu^{3+} concentration of 1 mol% which is within the range of doping level for fluorescent materials [9]. However, diffusion of europium into perovskite lattice, induced by thermal treatment, is often restricted because microdomains of $EuAlO_3$ could appear. Then, samples with larger content of europium are prepared by chemical methods such as coprecipitation. Lanthanum is often used to stabilize perovskite-like catalysts [10] and europium due to its paramagnetic properties is used to modify the ionic conductivity of many perovskite-like materials [11]. Then, it is interesting to study lanthanum doping into $EuAlO_3$ because a more rigid structure could be obtained.

The crystalline structure of perovskite is expected to be preserved upon replacement of La^{3+} by Eu^{3+} given that Eu^{3+} can occupy the dodecahedral sites of La^{3+}. At low content of europium this assumption probably holds but at higher europium loadings the structure should be closer to

$EuAlO_3$ than $LaAlO_3$ structure. Besides, the differences of La^{3+} and Eu^{3+} regarding quadrupole and size could be important and cause alterations in the perovskite-like structure. Furthermore, La^{3+} and Eu^{3+} have very different electronic configurations, the first one has a configuration of noble gas (Xe) and the second one has a configuration $[Xe]4f^6$. These properties could be determinant to produce structural changes in the perovskite structure and of course can be used conveniently to tune the physicochemical properties of these materials. In order to investigate what happens upon a high percentage replacement of La^{3+} by Eu^{3+} in $LaAlO_3$, which should be equivalent to a low percentage replacement of Eu^{3+} by La^{3+} in $EuAlO_3$, we have studied the system $La_xEu_{1-x}AlO_3$ by applying a combination of experimental techniques of FTIR, XRD, and NMR and molecular dynamics simulations. Our results show how the introduction of La^{3+} produces strong modifications in the physical properties of the material, without significant changes in the structure as referred to that of $EuAlO_3$ perovskite.

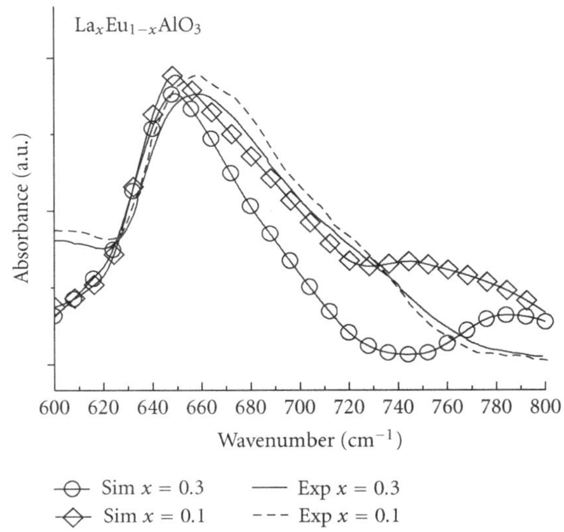

FIGURE 1: Infrared spectra, simulated and experimental, of perovskite samples with variable content of europium.

2. Experiments and Simulations

A series of $La_xEu_{1-x}AlO_3$ perovskites were synthesized by the coprecipitation method. Aqueous solutions of $La(NO_3)_2 \cdot 6H_2O$, $Al(NO_3)_3 \cdot 6H_2O$, and $Eu(NO_3)_3 \cdot 9H_2O$ with the desired La/Eu ratio were mixed and precipitated with NH_4OH. The precipitated solid was filtered in vacuum, washed with deionised water, and dried at 383 K. Finally, the solid was thermally treated twice, the first time at 973 K and the second one at 1673 K for 2 h and 6 h, respectively. Eu-containing perovskites were prepared with the following nominal composition: $LaAlO_3$, $La_{0.30}Eu_{0.70}AlO_3$, and $La_{0.10}Eu_{0.90}AlO_3$. Since there are not any sublimation phenomena to be expected we assumed the synthesis composition is the final one.

Powder X-ray diffraction (XRD) patterns were obtained using a D8 Advance Bruker AXS diffractometer, where a Cu target K_α ray ($\lambda = 0.154$ nm) was used as an X-ray source. Conventional identification of crystalline compounds was performed by comparing the diffractograms with JCPDS files (PDF card number 031-0022) [12]. Radial distribution functions, $g(r)$, were calculated from the full diffraction patterns as shown by Magini and Cabrini [13] which essentially consists of the Fourier transform of the structure factor S(k). A molybdenum anode X-ray tube was used to reach the required high values of the angular parameter $h = (4\pi \sin\theta)/\lambda$ and the X-ray pattern was measured by step scanning at angular intervals of 0.08° in a Siemens D5000 diffractometer. The cell parameters were obtained using a quartz standard to correct the shift of the (110) peak of the perovskite structure.

Solid-state ^{27}Al MAS NMR spectra were acquired under MAS conditions using an Avance 300 Bruker spectrometer with a magnetic field strength of 7.05 T, corresponding to a ^{27}Al Larmor frequency of 78.3 MHz. Short single pulses ($\pi/12$) and a recycle time of 0.5 s were used. The ^{27}Al chemical shift was referenced using an 1 M aqueous solution $Al(NO_3)_3$ as external standard.

For the infrared spectra (FTIR) all samples were diluted in CsI and then pressed to form pellets. Experiments were carried out at 2 cm^{-1} resolution and at room temperature using a Perkin-Elmer FTIR 2000 IR spectrometer and a DTGS Mid-IR detector.

Molecular dynamics simulations in the microcanonical ensemble were performed on two perovskite samples of composition $La_xEu_{1-x}AlO_3$, $x = 0.3, 0.1$. The interaction potential was of Pauling type given by

$$V\left(\vec{r_{ij}}\right) = \frac{q_i q_j}{r_{ij}} + \frac{e^2}{n\left(\sigma_i + \sigma_j\right)}\left(\frac{\sigma_i + \sigma_j}{r_{ij}}\right)^n, \qquad (1)$$

where q_i are the effective charges and σ_i are effective ionic radii. Charges were taken from reference [14] and long-range interactions were handled with Ewald sums. Since the local environment of Eu was not known at the beginning of the work the effective radius involved in the interaction potential had to be determined. In order to do so, we performed a series of MD simulations assigning different radii to particles representing Eu ions. One physical characteristic which depends on the ionic radius is the vibrational spectrum. The effective radius yielding the best fit to the IR spectrum for the concentration of Eu in the samples was $\sigma = 1.09$ Å which would correspond to a 6 coordinated Eu in the standard ionic radii data. Potential parameters used in the simulations are summarized in Table 1. Figure 1 shows the experimental and simulated IR spectra for $x = 0.3, 0.1$.

Simulation results were very similar for the two samples and therefore we constrain the report to the sample with $x = 0.7$. This simulated sample consisted of 2835 particles of which 567 represented Al ions, 1701 oxygen ions, 171 lanthanum ions, and 396 europium ions in an orthorhombic box of dimensions $a = 34.092$ Å, $b = 34.092$ Å, and $c = 26.516$ Å. The system was equilibrated at 300 K for 10 ps with temperature control previous to relaxation and production

TABLE 1: Potential parameters.

Atom	q, e	σ, Å
La	2.23	1.4
Eu	2.23	1.09
Al	1.92	0.67
O	−1.383	1.2

runs of 10 ps. In all runs, variations of total energy were of less than 0.01 per cent around the mean value, and temperature variations were of less than 2 K.

3. Results and Discussion

IR spectra were obtained to fit the interaction potential parameters. They presented only a single absorption band centered close to 655 cm^{-1}. The position of the IR absorption band moved to higher frequencies as the europium content increased. This shift to higher frequencies could be explained by the alteration of electrostatic interactions between ions. Europium and lanthanum cations have the same charge, however their masses and sizes are different and therefore there is a higher force constant for the system containing europium (La is ~11% heavier than Eu, see Table 1 for radii).

Figure 2 shows the ^{27}Al MAS NMR spectra of samples $EuAlO_3$, $La_{0.3}Eu_{0.7}AlO_3$, and $LaAlO_3$. The spectrum of the europium-free sample, Figure 2(c), presents an isotropic narrow peak at 14 ppm which is assigned to sixfold coordinated aluminium in an oxygen environment [15], in agreement with the perovskite structure. The spectra of the europium content samples were very different to those of $LaAlO_3$ sample. Then, in order to find the position of the isotropic peak, the spectra acquired at variable spinning rate were performed (not shown); they have shown that in two europium-containing samples the isotropic peak is positioned at 213 ppm, Figure 2(a), 2(b). The NMR peak was significantly broader in $EuAlO_3$ than in $LaAlO_3$; of course this modification is mainly attributed to the electronic local environment created by europium introduction. Note that Eu^{3+} is a paramagnetic lanthanide ion that broads and shifts the NMR peak to down field. In spectrum of sample containing both europium and lanthanum, a shoulder close to 167 ppm and the small peak at 30 ppm suggest that lanthanum and europium are not homogeneously distributed into the perovskite structure, but some europium and lanthanum-rich domains could be formed. Interestingly, the NMR peak for sample $La_{0.3}Eu_{0.7}AlO_3$ is broader than in $EuAlO_3$ parent solid, suggesting that introduction of lanthanum modifies the distribution of charge and possibly a field electric gradient in this sample was created. This effect is discussed below in terms of the results of molecular dynamics simulations.

X-ray powder diffraction patterns confirm that all samples have a perovskite-like structure as can be observed from Figure 3. As lanthanum is incorporated to perovskite the cell parameter increases as can be seen from X-ray diffraction data in Figure 4. Table 2 shows the calculated cell parameters as a function of sample composition. This is attributed to the difference of La^{3+} and Eu^{3+} ionic radii. Note that even though

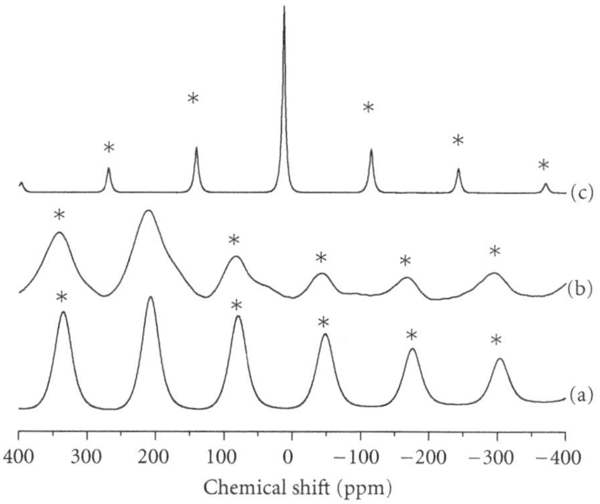

FIGURE 2: ^{27}Al MAS NMR spectra of $EuAlO_3$, $La_{0.30}Eu_{0.70}AlO_3$, and $LaAlO_3$; (a), (b), and (c), respectively. *indicates spinning side bands.

TABLE 2: Cell parameter of europium containing perovskites.

Sample	Cell parameter, a, Å
$LaAlO_3$	5.36
$La_{0.30}Eu_{0.70}AlO_3$	5.31
$La_{0.10}Eu_{0.90}AlO_3$	5.29
$EuAlO_3$	5.27

electrons in Eu^{3+} occupy f atomic orbitals, the cell parameter of $EuAlO_3$ is the lowest because the interstitial sites are determined by the oxygen packing and the ion-ion interactions.

Up to this point, experimental results show that lanthanum cations can replace europium cations in the parent $EuAlO_3$ perovskite. Even replacing as high as 30% Eu by La, XRD patterns show that the perovskite structure is preserved. NMR experiments, in addition, confirm that some local changes could have taken place when both lanthanum and europium are simultaneously present in perovskite structure. FTIR experiments also suggest some local changes as can be seen from the experimental curves, Figure 1. ^{27}Al MAS NMR results reveal that the local environment of aluminium is strongly altered. This modification could be attributed to electronic differences between europium and lanthanum. In fact La^{3+} is a diamagnetic species while, Eu^{3+} has non paired electrons promoting relaxation of NMR signal by paramagnetic effects which means that distances between aluminium and europium ions are short enough (in the second coordination sphere). It is not clear, however, what the consequences of this modification imply. Actually, short distances between aluminum and europium appear because of the subtle distortion of the lattice produced by heavy doping with Eu. The lattice distortions, however, are not so great that coordination of cations could occur in other complex perovskites [16]. Molecular dynamics (MD) allows for the elucidation of very subtle changes in some characteristics of

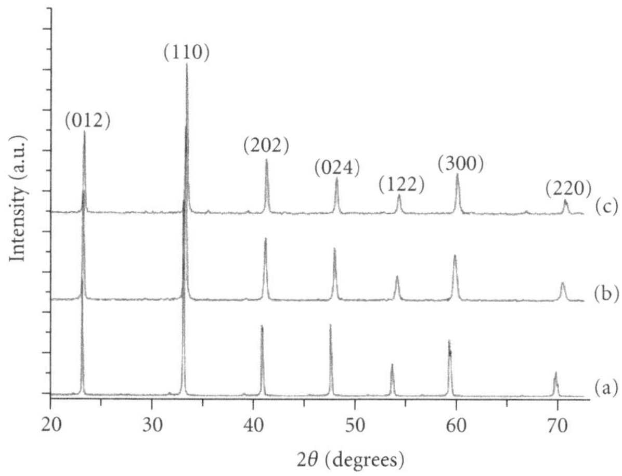

FIGURE 3: XRD patterns of (a) $LaAlO_3$, (b) $La_{0.3}Eu_{0.7}AlO_3$, and (c) $La_{0.1}Eu_{0.9}AlO_3$. Perovskite-like structure is maintained although a small change in cell parameter a is observed. Peaks are indexed with PDF card number 031-0022.

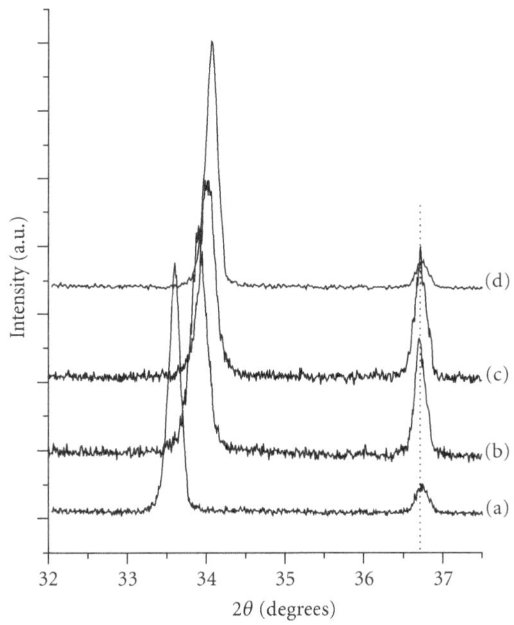

FIGURE 4: (110) XRD peak of perovskite-like structure used to calculate the cell parameters of (a) $LaAlO_3$, (b) $La_{0.3}Eu_{0.7}AlO_3$, (c) $La_{0.1}Eu_{0.9}AlO_3$, and (d) $EuAlO_3$. As standard, the peak of quartz was used (marked with the dotted line).

microscopic systems. Figure 5 shows the total radial distribution functions, $g(r)$, of europium-containing perovskites calculated from MD simulations. Even though some peaks are difficult to resolve experimentally, the rough agreement between simulated and experimental data is fairly good as far as the position of the peaks and the general shape of the functions is concerned. Some features of the radial distribution functions are (1) the peak at 2.6 Å increases as europium content increases; (2) peaks between 5 and 8 Å go to longer distances as lanthanum content increases; (3) the

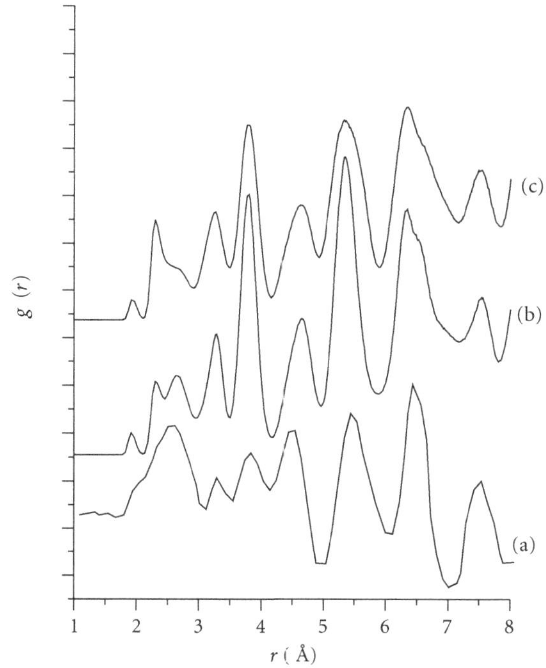

FIGURE 5: Radial distribution functions from experimental (a) and molecular dynamics simulations (b and c); $La_{0.3}Eu_{0.7}AlO_3$ (a), $La_{0.3}Eu_{0.7}AlO_3$ (b), and $La_{0.1}Eu_{0.9}AlO_3$ (c).

peak at 1.9 Å, due to Al–O distances, remains constant in intensity and position. This last observation implies that octahedral coordination of aluminium ions is not altered or its alteration is very subtle as a consequence of replacement of europium by lanthanum. Therefore, the observed changes in ^{27}Al NMR spectra are mainly due to differences in electronic properties of La^{3+} and Eu^{3+} ions. In addition, as mentioned above, lanthanide ions are heterogeneously distributed into the perovskite structure.

The structural effects of the presence of La^{3+} and Eu^{3+} in the perovskite lattice can be described in some detail from the partial radial distribution functions, $g(r)$. Figure 6 shows the Al–O, La–O, Al–Al, and O–O pair functions for the simulated $LaAlO_3$, $La_{0.3}Eu_{0.7}AlO_3$, and $EuAlO_3$ samples. The first neighbour distances Al–O and La–O are very well defined and are unchanged upon substitution as can be seen from the figure. The three La–O first neighbour bond distances are a bit more differentiated in the sample with europium as can be noted from the asymmetry of the first peak. A coordination analysis performed after the molecular dynamics runs yielded the same coordination numbers for both samples Al (6) and La or Eu (12). It can be noted that all other peaks broaden implying that long-range order is not totally preserved. Note that the second peak of $g(r_{AlO})$ splits in two peaks upon the presence of europium in perovskites. This may be interpreted in terms of the AlO_6 octahedron tilting known to appear in $ABXO_3$ perovskites. Upon heavy doping with Eu a more complex tilting may occur that produces a lowering of symmetry and is reflected in this second peak of $g(r_{AlO})$. On the other hand,

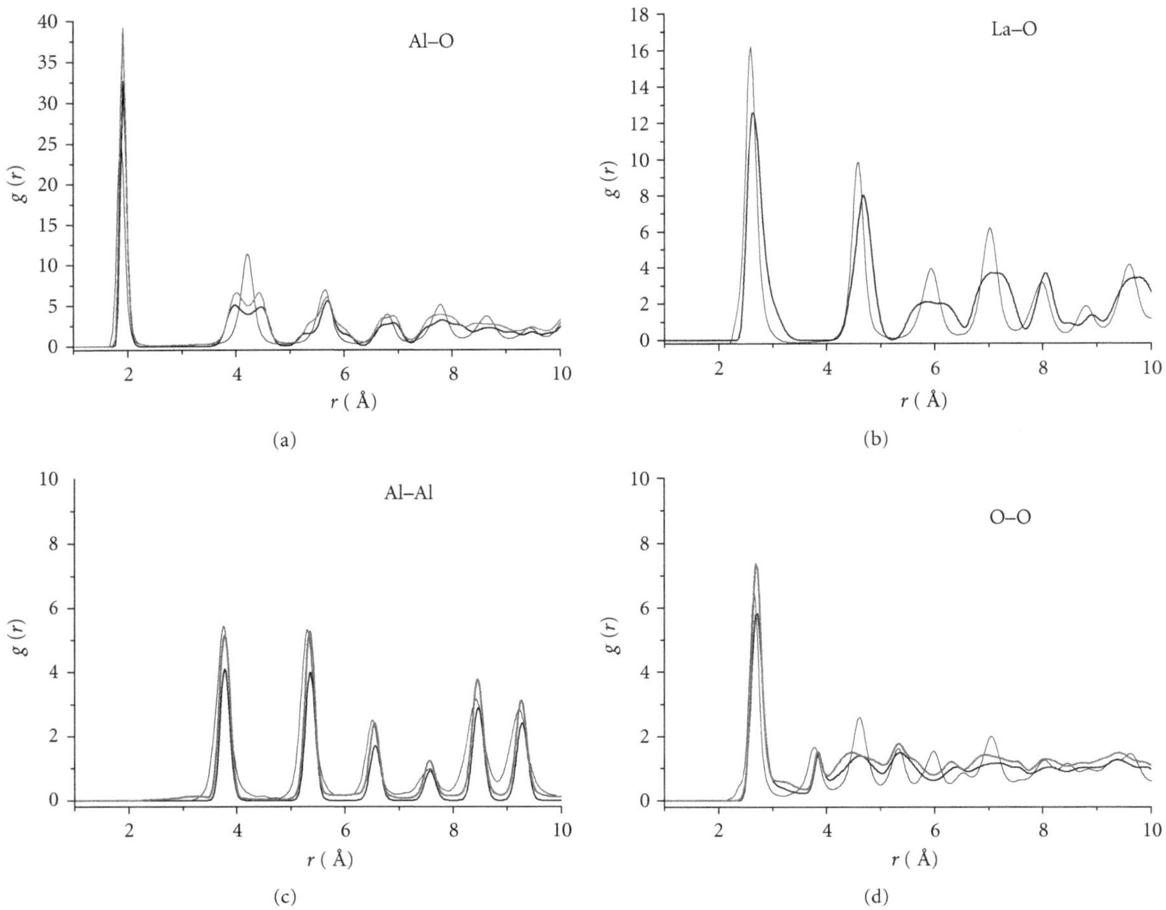

FIGURE 6: Radial distribution functions from molecular dynamics simulations for the pairs shown for samples $LaAlO_3$ (black thin line), $La_{0.3}Eu_{0.7}AlO_3$ (black line), and $EuAlO_3$ (gray line).

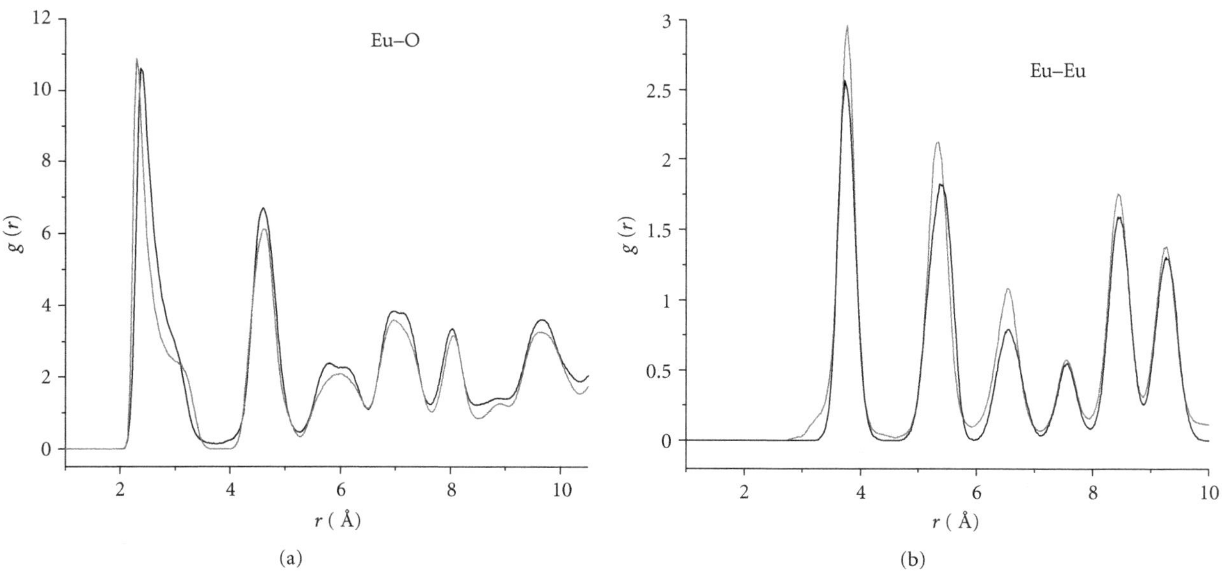

FIGURE 7: Radial distribution functions from molecular dynamics simulations for the pairs shown for samples $La_{0.3}Eu_{0.7}AlO_3$ (black line) and $EuAlO_3$ (gray line).

the aluminium sublattice remains practically unaltered upon substitution, as it can be seen from $g(r_{AlAl})$. Finally, an interesting conclusion is derived from the O–O partial radial distribution function, $g(r_{OO})$, since a remarkable broadening of peaks is presented from 4 Å on samples containing Eu, shown in Figure 6. This implies a constant rearrangement of oxygen atoms in such a way that in the long range they appear as disordered. This is shown by the O–O $g(r)$ asymptotic trend to unity. This means that in the real world the presence of Eu^{3+} significantly increases the entropy. From a purely classical mechanics point of view, one can interpret these effects as due to the fact that europium is about 11% heavier than lanthanum; therefore for a given temperature, oxygen atoms in the vicinity of europium increase the amplitude of their vibrations. If oxygen vibrations increase in amplitude and therefore if average oxygen atoms are not at the crystalline positions, the energy activation barriers for diffusion processes either disappear or else are lowered in a significant way.

Figure 7 displays the $g(r_{EuO})$ and $g(r_{EuEu})$ functions, confirming that europium sublattice is also preserved and only the position of oxygen ions changes, modifying then the distances at first neighboring Eu–O.

Factors governing the electrical ionic conductivity of perovskites are concentration of oxygen vacancies, oxygen partial pressure, crystal structure, and defects, among others [17–19]. Our findings indicate that even in the absence of defects, such as oxygen vacancies, one important factor that may substantially change the ionic conductivity is the increase of oxygen mobility if the proper cation is introduced in the perovskite network.

4. Conclusions

We have performed a structural analysis of La^{3+}-doped $EuAlO_3$. Molecular dynamic simulations, as well as ^{27}Al MAS NMR spectra analysis, indicates that coordination of cations in the doped sample remains the same as in the parent perovskite. Since substitution of La by Eu is isomorphic, no O vacancies are generated. The only structural change observed is that the oxygen sublattice gains rigidity because of the decrease of oxygen vibration amplitude.

Acknowledgments

Thanks are due to G. Cedillo and A. Tejeda (IIM-UNAM) for the technical help. The authors thank PAPIIT-UNAM (IN107110) for financial support. All calculations were carried out on an IBM SP3 supercomputer kindly donated to Laboratorio de Simulación of Instituto de Matemáticas, Unidad Cuernavaca, UNAM by Fundación Clínica Médica Sur, A. C. from Mexico City.

References

[1] J. B. A. A. Elemans, B. Van Laar, K. R. Van Der Veen, and B. O. Loopstra, "The crystallographic and magnetic structures of $La_{1-x}Ba_xMn_{1-x}Me_xO_3$ (Me = Mn or Ti)," *Journal of Solid State Chemistry*, vol. 3, no. 2, pp. 238–242, 1971.

[2] J. Mizusaki, I. Yasuda, J. I. Shimoyama, S. Yamauchi, and K. Fueki, "Electrical conductivity, defect equilibrium and oxygen vacancy diffusion coefficient of $La_{1-x}Ca_xAlO_{3-\delta}$ single crystals," *Journal of the Electrochemical Society*, vol. 140, no. 2, pp. 467–471, 1993.

[3] Y. H. Huang, S. Wang, F. Luo, S. Jiang, and C. H. Yan, "Enhanced magnetoresistance at high temperatures in $La_{0.7}Sr_{0.3}MnO_3$/silica nanocomposites," *Chemical Physics Letters*, vol. 362, no. 1-2, pp. 114–118, 2002.

[4] G. Thornton, B. C. Tofield, and A. W. Hewat, "A neutron diffraction study of $LaCoO_3$ in the temperature range 4.2 < T < 1248 K," *Journal of Solid State Chemistry*, vol. 61, no. 3, pp. 301–307, 1986.

[5] M. Yashima, M. Mori, T. Kamiyama et al., "High-temperature phase transition in lanthanum titanate perovskite $La_{0.64}(Ti_{0.92},Nb_{0.08})O_3$," *Chemical Physics Letters*, vol. 375, no. 1-2, pp. 240–246, 2003.

[6] R. M. Hazen, "A useful fiction: Polyhedral modeling of mineral properties," *American Journal of Science*, vol. 74, pp. 242–269, 1988.

[7] A. R. West, *Solid State Chemistry and Its Application*, John Wiley & Sons, New York, NY, USA, 1990.

[8] D. Hreniak, W. Stręk, P. Dereń, A. Bednarkiewicz, and A. Łukowiak, "Synthesis and luminescence properties of Eu^{3+}-doped $LaAlO_3$ nanocrystals," *Journal of Alloys and Compounds*, vol. 408–412, pp. 828–830, 2006.

[9] J. Klimke and H. Wulff, "Determination of the optimum Eu^{3+}-concentration in $LaAlO_3$: euphosphors by X-ray diffraction and fluorescence measurements," *Fresenius' Journal of Analytical Chemistry*, vol. 349, no. 1–3, pp. 245–246, 1994.

[10] J. L. Hueso, A. Caballero, J. Cotrino, and A. R. González-Elipe, "Plasma catalysis over lanthanum substituted perovskites," *Catalysis Communications*, vol. 8, no. 11, pp. 1739–1742, 2007.

[11] M. M. Seikh, V. Caignaert, V. Pralong, C. Simon, and B. Raveau, "Expansion of ferromagnetism by calcium doping in the ordered oxygen deficient perovskite $EuBaCo_2O_{5.50\pm\delta}$," *Journal of Physics Condensed Matter*, vol. 20, no. 1, Article ID 015212, 2008.

[12] Join Committee on Powder Diffraction Standards JCPDS Files, International Center for Diffraction Data, 2000.

[13] M. Magini and A. Cabrini, "Programme en FORTRAN IV pour l'analyse des données expérimentales relatives à la diffusion des rayons X par des substances liquides, amorphes et microcristallisées," *Journal of Applied Crystallography*, vol. 5, pp. 14–18, 1972.

[14] L. J. Alvarez, J. F. Sanz, M. J. Capitán, and J. A. Odriozola, "Onset of perovskite formation in the catalytic system $La_2O_{3/y}$-Al_2O_3," *Catalysis Letters*, vol. 21, no. 1-2, pp. 89–97, 1993.

[15] E. Lippmaa, A. Samoson, and M. Mägi, "High-resolution ^{27}Al NMR of aluminosilicates," *Journal of the American Chemical Society*, vol. 108, no. 8, pp. 1730–1735, 1986.

[16] V. I. Torgashev, V. B. Shirokov, A. S. Prokhorov, and L. A. Shuvalov, "Competition between rotational and polar structural distortions in perovskites," *Crystallography Reports*, vol. 50, no. 4, pp. 637–645, 2005.

[17] B. P. Burton, "Long-range versus short-range interactions and the configurational energies of $Ba(B,B')O_3$ and $Pb(B,B')O_3$ perovskites," *Modelling and Simulation in Materials Science and Engineering*, vol. 8, no. 3, pp. 211–219, 2000.

[18] T. L. Nguyen, M. Dokiya, S. Wang, H. Tagawa, and T. Hashimoto, "The effect of oxygen vacancy on the oxide ion

mobility in $LaAlO_3$-based oxides," *Solid State Ionics*, vol. 130, no. 3-4, pp. 229–241, 2000.

[19] E. Lima, M. E. Villafuerte-Castrejon, J. M. Saniger, A. Ibarra-Palos, J. E. Sanchez-Sanchez, and L. J. Álvarez, "Experimental XRD and NMR, and molecular dynamics study of Sr containing $LaAlO_3$ perovskite," *Solid State Ionics*, vol. 178, no. 39-40, pp. 1944–1949, 2008.

7

Stress-Softening and Residual Strain Effects in Suture Materials

Alex Elías-Zúñiga,[1] Beatriz Montoya,[1] Wendy Ortega-Lara,[1] Eduardo Flores-Villalba,[1,2] Ciro A. Rodríguez,[1] Hector R. Siller,[1] José A. Díaz-Elizondo,[2] and Oscar Martínez-Romero[1]

[1] *Centro de Innovación en Diseño y Tecnología, Tecnológico de Monterrey—Campus Monterrey, Avenida E. Garza Sada 2501 Sur, 64849 Monterrey, NL, Mexico*
[2] *Escuela de Medicina y Ciencias de la Salud, Tecnológico de Monterrey—Campus Monterrey, Avenida E. Garza Sada 2501 Sur, 64849 Monterrey, NL, Mexico*

Correspondence should be addressed to Alex Elías-Zúñiga; aelias@itesm.mx

Academic Editor: Pavel Lejcek

This work focuses on the experimental characterization of suture material samples of MonoPlus, Monosyn, polyglycolic acid, polydioxanone 2–0, polydioxanone 4–0, poly(glycolide-co-epsilon-caprolactone), nylon, and polypropylene when subjected to cyclic loading and unloading conditions. It is found that all tested suture materials exhibit stress-softening and residual strain effects related to the microstructural material damage upon deformation from the natural, undistorted state of the virgin suture material. To predict experimental observations, a new constitutive material model that takes into account stress-softening and residual strain effects is developed. The basis of this model is the inclusion of a phenomenological nonmonotonous softening function that depends on the strain intensity between loading and unloading cycles. The theory is illustrated by modifying the non-Gaussian average-stretch, full-network model to capture stress-softening and residual strains by using pseudoelasticity concepts. It is shown that results obtained from theoretical simulations compare well with suture material experimental data.

1. Introduction

Technical advances in the field of surgery have exponentially grown during the last few years. Progressive research, better understanding of the physiopathological processes behind every procedure, more experienced surgeons, and the increasing interest of biomedical companies in developing new products have made possible the excellent results seen in surgery nowadays.

Sutures have remained for many decades a cornerstone for most surgical procedures. Wound closure, vascular and intestinal anastomosis, structures fixation, bleeding control, and tissue approximation are only a few examples from the many uses sutures are given; however, there is no perfect suture for all purposes, and the enormous variables involved in the cicatrization process make this task a difficult one to achieve.

The ideal suture should have certain characteristics that will abet the therapeutic course; it is supposed to have an adequate tensile strength in each phase of the healing process, should be surgeon friendly, induce minimal or no tissue reaction, and must not stimulate infection. Also it should be biologically inert and should be able to have a suitable response to edema and tolerate the different environments within the human body. Since there is no such a product available at present, it is essential to comprehend not only the biological responses to the materials but also understand the precise behavior of each suture in order to decide on the best and most efficient way to take advantage of each particular property of any given suture [1]. Nichols et al. cautioned surgeons about the handling of sutures by surgical instruments since this could result in premature suture failure [2]. They indicated that their rough handling and the usage of clamps and forceps could damage and weaken these.

Therefore, learning the biomechanical performance of sutures will help surgeons not only to determine the appropriate clinical application for each type but also to improve surgical techniques to take advantage of each suture properties [3].

Some of the most important characteristics of a suture regarding its mechanical material properties are related to softening and permanent set effects which appear when sutures are subjected to cyclic load, and, thus, the normal stress versus stretch curve shows a reduction on the stress magnitude during the unloading process [4–13]. This stress-softening effect known as the Mullins effect becomes clinically relevant because initial characteristics exhibited by a suture material immediately after it has been manufactured could change dramatically when stress is applied during the suturing and healing processes [2]. Understanding the material response behavior of suture materials could help surgeons in the selection of the most appropriate suture material.

The aim of this work is to characterize the stress-softening and residual strain effects in suture materials such as polyglycolic acid, polydioxanone, nylon, and polypropylene commonly used in surgical procedures when subjected to loading and unloading cycles by developing a phenomenological nonmonotonous stress-softening hyperelastic material model that depends on the amount of strain [14] and permanent set effects [15].

We have organized this paper as follows. In Section 2, we provide detail of the uniaxial experimental tests performed on suture materials. A brief review of the required equations to describe finite deformations of an incompressible elastic material is introduced in Section 3. In Section 4, we have characterized the experimental data by assuming a nonmonotonous damage softening function and by modifying the Holzapfel et al. constitutive equation to include residual strain effects [15]. Also, we have developed the corresponding stress-stretch constitutive equations by using the non-Gaussian average-stretch, full-network model of arbitrarily oriented molecular chains [16]. A comparison of the results corresponding to simulated and experimental data is done in Section 5. Finally, in Section 6, we address some conclusions related to our experimental observations and theoretical predictions of suture materials.

2. Experimental Work

2.1. Suture Materials. Eight batches of two suture commercial manufactures were selected to be tested in uniaxial deformation. The absorbable sutures materials tested were MonoPlus, Monosyn, polyglycolic acid, polydioxanone, polydioxanone 4–0, poly(glycolide-co-epsilon-caprolactone) (PGC25 3–0); the nonabsorbable suture materials were nylon and polypropylene. The mean diameter values used to characterize the suture samples were taken from suppliers specifications. The mean values considered here were 0.397 mm for MonoPlus and Monosyn sutures, 0.334 mm for polyglycolic acid 2–0, 0.377 mm for polydioxanone 2–0, 0.23 mm for polydioxanone 4–0, 0.29 mm for PGC25 3–0, 0.247 mm for nylon, and 0.241 mm for polypropylene.

2.2. Uniaxial Tensile Tests. The experimental tests were performed in two electromechanical universal testing machines. The suture materials identified as MonoPlus, Monosyn, polyglycolic acid 2–0, polydioxanone 2–0, nylon, and polypropylene were tested in an MTS Insight 2 tensile machine with a maximum cell load capacity of 2.5 kN while the suture materials of polydioxanone 4–0 and PGC25 3–0 were tested

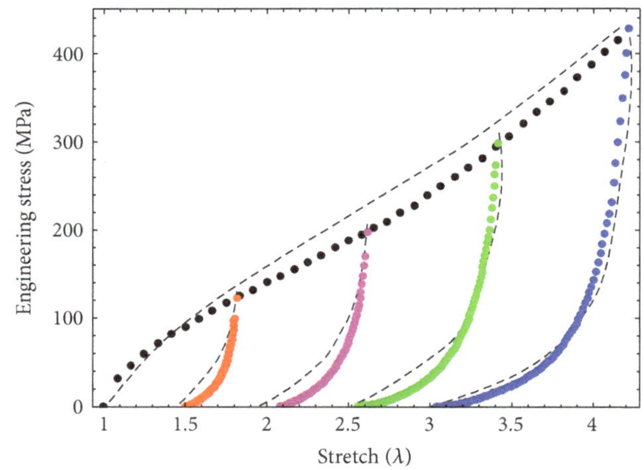

FIGURE 1: Experimental data collected from uniaxial tension cyclic loading-unloading tests for Monosyn sutures.

in an Instron tensile machine Model 3365 with a maximum cell load capacity of 1.6 kN. The selected samples length between machine grips was 50 mm. All tests were run at the machine speed of 500 mm/min at the average room temperature value of 24°C. All samples were subjected to cyclic loading-unloading conditions to obtain softening and permanent set effects as shown in Figure 1. Figure 1 illustrates that when the suture material is loaded from its virgin state, unloaded, and then reloaded again, its stress magnitude becomes smaller than the stress magnitude at the same amount of stretch during virgin loading. This reduction in the stress magnitude is known as the Mullins effect [4, 5]. This softening effect becomes associated with residual strain or permanent set effects which implies that the initial length of the suture sample has increased during the application of the tensile load. Table 1 illustrates the experimental average value of residual strain measured in each suture batch. Notice that suture materials response behavior agrees with Nichols et al. qualitative observation on sutures materials [2]. However, to describe quantitatively the stress-softened and permanent set effects observed on suture materials, a material model must be used. Therefore, to understand the physical relationships behind a material model, we first briefly review some basic knowledge on finite deformations, and, then, we shall derive a material model that is based on non-Gaussian statistical mechanics.

3. Basic Concepts on Finite Deformations

We consider the deformation of an incompressible elastic body which in its natural configuration occupies the region Ω. A material particle is considered to be in its undeformed reference configuration of a body at the place $\mathbf{X} = X_k \mathbf{e}_k$. After a prescribed deformation the body occupies the region Ω_c, the current configuration, and the particle at $\mathbf{X}$ moves to the place $\mathbf{x} = \mathbf{x}_k \mathbf{e}_k$ in a common rectangular Cartesian frame $\varphi = \{O; \mathbf{e}_k\}$ with origin O and orthonormal basis $\mathbf{e}_k$. Thus, the Cauchy-Green deformation tensor $\mathbf{B} \equiv \mathbf{F}\mathbf{F}^T$ has the form

$$\mathbf{B} = \lambda_1^2 \mathbf{e}_{11} + \lambda_2^2 \mathbf{e}_{22} + \lambda_3^2 \mathbf{e}_{33}, \tag{1}$$

TABLE 1: Comparison between experimental and predicted residual strain deformations of the selected suture materials.

Suture material	Maximum previous stretch λ_{max}	Experimental residual strain	Predicted residual strain	Error (%)
MonoPlus	2.09363	1.6027	1.498	6.5327
	3.1709	2.2103	2.056	6.9809
	4.2483	2.7361	2.646	3.2930
	5.3309	3.3038	3.234	2.1127
Monosyn	1.8165	1.4996	1.464	2.3739
	2.6165	2.0828	1.962	5.7998
	3.4165	2.5559	2.484	2.8131
	4.2165	3.0550	3.007	1.5711
Polyglycolic acid 2-0	1.22	1.100	1.125	2.2727
	1.44	1.230	1.251	1.7070
	1.66	1.380	1.381	0.0724
	1.88	1.537	1.515	1.4313
Polydioxanone 2-0	1.5	1.200	1.277	6.4166
	2.0	1.512	1.566	3.5714
	2.5	1.815	1.862	3.5895
	3	2.167	2.154	0.5999
Polydioxanone 4-0	1.2	1.051	1.051	0.7347
	1.4	1.101	1.099	1.3187
	1.6	1.149	1.147	0.7879
	1.8	1.194	1.198	0.1115
Poly(glycolide-co-epsilon-caprolactone) (PGC25 3-0)	1.3	1.123	1.104	1.7270
	1.6	1.216	1.206	0.8818
	1.9	1.307	1.306	0.0238
	2.20	1.403	1.407	0.2851
Nylon	1.48	1.109	1.235	11.36
	1.96	1.366	1.481	8.4187
	2.44	1.67	1.744	4.4311
	2.92	2.011	2.019	0.3978
Polypropylene	1.4	1.154	1.197	3.72
	1.8	1.361	1.4	2.8655
	2.2	1.564	1.615	3.2608
	2.6	1.855	1.842	0.7008

where $\mathbf{e}_{jk} \equiv \mathbf{e}_j \otimes \mathbf{e}_k$, $\mathbf{e}_i$ are the associated orthonormal principal directions, $\mathbf{F}$ is the deformation gradient, and λ_i denote the principal stretches in φ. Note that the magnitude of the strain intensity at a material point $\mathbf{X}$ denoted by m is defined by $m \equiv \sqrt{\mathbf{B}\cdot\mathbf{B}} = \sqrt{\operatorname{tr}\mathbf{B}^2}$, where tr is the trace operation. In the undeformed state $\mathbf{B} = \mathbf{1}$, the identity tensor and $m = \sqrt{3}$; otherwise, $m > \sqrt{3}$ for all isochoric deformations [8]. Also $m \geq \sqrt{3}$ for all λ, the equality holding when and only when $\lambda = 1$, the undeformed state. Recalling that the principal invariants I_k of $\mathbf{B}$ are defined by

$$I_1 = \operatorname{tr}\mathbf{B}, \qquad I_2 = \frac{1}{2}\left[I_1^2 - \operatorname{tr}\left(\mathbf{B}^2\right)\right], \qquad I_3 = \det\mathbf{B}, \quad (2)$$

thus, the magnitude of the strain intensity m is given as

$$m = \sqrt{I_1^2 - 2I_2}. \quad (3)$$

4. A Nonmonotonous Stress-Softening Material Model

To characterize stress-softening effects, there exist in the literature many different micromechanical models have been developed to explain material damage mechanisms. See, for instance, the papers of Govindjee and Simo [6], Ogden and Roxburgh [7], Beatty and Krishnaswamy [8], Elías-Zúñiga and Beatty [9], Elías-Zúñiga [11], Diani et al. [12], de Tommasi et al. [13], Holzapfel et al. [15], Johnson and Beatty [17], de Souza Neto et al. [18], Marckmann et al. [19], Dorfmann and Ogden [20], Kazakevičiute-Makovska and Kačianauskas [21], and references cited therein for an overview of the main features of these models.

In this section, we derive the corresponding equations that describe the non-monotonic behavior of suture biocompatible materials subjected to loading and unloading cycles by assuming that the stress-softened material behavior can be obtained from the virgin material response constitutive equation. Here, we assume an incompressible and isotropic virgin elastic material whose corresponding time independent Cauchy stress constitutive equation has the form

$$\mathbf{T} = -p\mathbf{1} + \aleph_1 (I_1, I_2)\,\mathbf{B} + \aleph_{-1} (I_1, I_2)\,\mathbf{B}^{-1}, \quad (4)$$

in which $\mathbf{T}$ is the Cauchy stress, p is an undetermined pressure, and $\aleph_\Gamma = \aleph_\Gamma(I_1, I_2)$, $\Gamma = 1, -1$, denote the virgin material response functions related to the strain energy function $W = \widehat{W}(I_1, I_2)$, per unit reference volume, in accordance with

$$\aleph_1 = 2W_1, \qquad \aleph_{-1} = -2W_2, \quad (5)$$

wherein $W_\alpha \equiv \partial\widehat{W}/\partial I_\alpha$ [16]. By using (4), Elías-Zúñiga and Beatty in [9] proposed a damage type model to describe the stress-softened material behavior of the form

$$\boldsymbol{\tau} = F(m; M)\,\mathbf{T}, \quad (6)$$

in which $\boldsymbol{\tau}$ denotes the Cauchy stress in the stress-softened material, M denotes the maximum previous strain at which the material is unloaded from the primary path, and $F(m; M)$ is an isotropic softening function at the damage level $m_{\max} = M$ on the interval $m \in [\sqrt{3}, M]$. They assumed that this softening function $F(m; M)$ is a monotone increasing function of the strain intensity that satisfies the conditions

$$0 < F(m; M) < 1, \qquad F(M; M) = 1. \quad (7)$$

Based on this assumption, Elías-Zúñiga and Beatty proposed the following softening function:

$$F(m; M) = e^{-b\sqrt{(M-m)}}, \quad (8)$$

where b is a dimensionless positive material-softening parameter. After substituting (8) into (6), Elías-Zúñiga and Beatty obtained the following stress-softened phenomenological material model:

$$\boldsymbol{\tau} = e^{-b\sqrt{(M-m)}}\mathbf{T}. \quad (9)$$

Theoretical predictions provided by (9) were computed in Elías-Zúñiga and Beatty [9] and Elías-Zúñiga and Beatty [10] and compared to experimental data for uniaxial extension, pure shear, and equibiaxial deformation states. There, the theoretical predictions showed reasonably good agreement with experimental data not only for the virgin loading path but also for the reloading paths. However, Kazakevičiūtė-Makovska [22] observed that the experimental data when plotted as the normalized stress τ/T versus the stretch ratio $\lambda/\lambda_{\max}$ showed nonmonotonous behavior with the characteristic S-shaped form, and $\lambda_{\max}$ defines the maximum amount of stretch on the primary loading path corresponding to the value at which the unloading starts in a particular deformation cycle. Kazakevičiūtė-Makovska concluded that, because of the variations in shapes of the curves for different deformation cycles, different values of the softening parameters were needed to fit experimental data for a particular choice of the softening function. Moreover, Kazakevičiūtė-Makovska showed that the softening function given by (8) fails in predicting the nonmonotonous behavior exhibited by experimental data collected by Mullins and Tobin [5], Cheng and Chen [23], and Mars and Fatemi [24] at higher stretch values.

On the other hand, de Tommasi and coworkers showed the importance of microscopic inhomogeneity to describe known experimental effects observed in amorphous materials such as the transition from diffuse to localized damage as the distribution properties are varied [25]. In fact, they showed that the monotone stress-stretch loading curve behavior is mainly due to a diffuse damage mechanism. They also considered that amorphous materials may be characterized by unstable strain domain, which gives the possibility of having homogeneous or localized damage with nonmonotone primary loading curve.

To confirm these observations, Elías-Zúñiga and Rodrguez in [14] used the nonmonotonous stress-softening function

$$F(m; M) = e^{-b[(M-m)(m/M)^\gamma]^\alpha}, \quad (10)$$

where b is a positive softening material parameter and α and γ are positive scaling constants chosen to best fit experimental data. They chose the values of $\alpha = 1/2$ and $\gamma = 1$ for the scaling constants and fit the value of the softening parameter b according to the unloading experimental data of the unloading path at which the amount of stretch has the maximum value. Then, they used the equation

$$\boldsymbol{\tau} = \mathbf{T}e^{-b[(M-m)(m/M)]^{1/2}} \quad (11)$$

to predict the corresponding stress-softened values during the inflation and deflation of rubber balloons.

To account permanent set effects during the material unloading processes, Elías-Zúñiga and Rodrguez [14] modified Holzapfel et al. model [15] and proposed an energy model based on pseudoelastic theory of the form:

$$W_s = W(\lambda_1, \lambda_2, \lambda_3) + \frac{\mu}{C}\sum_{a=1}^{3}\left[\frac{1}{2}(\lambda_{\max a}^n - \lambda_a^n)^2 - Cd_0\right], \quad (12)$$

where $W(\lambda_1, \lambda_2, \lambda_3)$ represents the strain energy function associated with the primary loading path, μ is the material shear modulus, C is a positive dimensionless material constant, d_0 is an integration constant, n is a fitting parameter that in general takes the value of 1/2, λ_a represents the principal stretches, and $\lambda_{\max a}$, $a = 1, 2, 3$, are the maximum values of the principal stretches at which unloading begins on the primary loading path.

Here, we use the softening function given by (10) and assume that $W(\lambda_1, \lambda_2, \lambda_3)$ is provided by the non-Gaussian Arruda-Boyce constitutive equation for an average-stretch, full-network of arbitrarily oriented molecular chains to predict softening and residual strain effects on suture materials.

4.1. A Nonmonotonous Amended Averaged Stretch Material Model. For this material model, it is well known that the strain energy per unit volume for the loading path is given by

$$W(\lambda_1, \lambda_2, \lambda_3) = \mu\left[N_8\left(\beta\lambda_r + \ln\left(\frac{\beta}{\sinh\beta}\right)\right) - \ln\left(\frac{\beta}{\lambda_r}\right)\right] - c_8, \quad (13)$$

where λ_r is the relative chain stretch defined by

$$\lambda_r = \frac{\lambda_{\text{chain}}}{\lambda_L}, \quad (14)$$

$\lambda_L = \sqrt{N_8}$ represents the fully extended chain stretch, N_8 is the chain number of rigid links, each of length l, λ_{chain} is the chain deformation that in the affine deformation is determined by

$$\lambda_{\text{chain}} \equiv \sqrt{\frac{I_1}{3}}, \quad (15)$$

β defined by $\beta \equiv \mathscr{L}^{-1}(\lambda_r)$ is the inverse of the Langevin function $\mathscr{L}(\beta)$ which is defined as

$$\lambda_r = \mathscr{L}(\beta) \equiv \coth\beta - \frac{1}{\beta}, \quad (16)$$

and c_8 is a constant that ensures that the strain energy density vanishes in the undeformed state [16, 26]. Substitution of (13) into (12) provides the modified non-Gaussian pseudo strain energy per unit volume that accounts for residual strains on the unloading path; that is,

$$W_s = \mu\left[N_8\left(\beta\lambda_r + \ln\left(\frac{\beta}{\sinh\beta}\right)\right) - \ln\left(\frac{\beta}{\lambda_r}\right)\right] + \frac{\mu}{C}\sum_{a=1}^{3}\left[\frac{1}{2}\left(\lambda_{\max a}^n - \lambda_a^n\right)^2\right] + D, \quad (17)$$

where D is an energy constant.

The Cauchy stress-stretch averaging network model components for the virgin material are obtained by substituting (13) into (4):

$$T_k = -p + \aleph(I_1)\lambda_k^2, \quad (18)$$

where $\aleph(I_1)$ is a material response function given as

$$\aleph(I_1) \equiv \frac{\mu}{3\lambda_r}\left[\beta + \frac{1}{N_8}\left(\frac{1}{\lambda_r} - \frac{1}{\beta(1 - \lambda_r^2 - 2\lambda_r/\beta)}\right)\right]. \quad (19)$$

Eliminating the pressure from (18) gives

$$T_j - T_k = \aleph(I_1)\left(\lambda_j^2 - \lambda_k^2\right), \quad (20)$$

where $j \neq k = 1, 2, 3$ (no sum). Similarly, the Cauchy stress-stretch constitutive equation for a stress-softened material can be obtained by substituting (17) into (4) and by using (5) and (11) yields the following stress-softened components:

$$\tau_k = \left[-p + \aleph(I_1)\lambda_k^2 + \frac{\mu\lambda_k}{2C}f_k(\lambda_1, \lambda_2, \lambda_3)\right] \times e^{-b\sqrt{(M-m)(m/M)}}, \quad k = 1, 2, 3\ (\text{nosum}), \quad (21)$$

where

$$f_k(\lambda_1, \lambda_2, \lambda_3) = \frac{\partial\sum_{a=1}^{3}\left(\lambda_{\max a}^n - \lambda_a^n\right)^2}{\partial\lambda_k}. \quad (22)$$

Then, on elimination of p from (21), it yields

$$\tau_j - \tau_k = \left[\aleph(I_1)\left(\lambda_j^2 - \lambda_k^2\right) + \frac{\mu}{2C}\left(\lambda_j f_j(\lambda_1, \lambda_2, \lambda_3) - \lambda_k f_k(\lambda_1, \lambda_2, \lambda_3)\right)\right] \times e^{-b\sqrt{(M-m)(m/M)}}, \quad (23)$$

where, in general, $j \neq k = 1, 2, 3$ (no sum).

Recalling that, for an incompressible material, the engineering stress $\boldsymbol{\sigma}$ is related to the Cauchy stress by

$$\boldsymbol{\sigma} = \mathbf{T}\mathbf{F}^{-1}, \quad (24)$$

then, the uniaxial engineering stress-stretch relation for an average-stretch, full-network stress-softened material model is obtained by using (22), (23), and (24):

$$\sigma_s = \left[\aleph(I_1)\left(\lambda - \lambda^{-2}\right) + \frac{\mu}{C}\left(-n\lambda^{(n-1)}\left(\lambda_{\max}^n - \lambda^n\right) + n\lambda^{-(1+n/2)}\left(\lambda_{\max}^{-n/2} - \lambda^{-n/2}\right)\right)\right] \times e^{-b\sqrt{(M-m)(m/M)}}. \quad (25)$$

Here,

$$m = \sqrt{\lambda^4 + 2\lambda^{-2}}, \quad (26)$$

and the relative chain stretch which can be obtained from (14) and (15) is given as

$$\lambda_r = \sqrt{\frac{1}{3N_8}\left(\lambda^2 + 2\lambda^{-1}\right)}. \quad (27)$$

Of course, other material models may be modified by using (10) and our derived pseudo strain energy per unit volume given by (12) to account for a nonmonotonous stress-softened behavior as well as permanent set effects, respectively.

We next examine the degree of accuracy attained by our proposed material model in predicting experimental data of biocompatible suture materials.

5. Comparison with Suture Experimental Data

To assess the accuracy of the derived constitutive (25) which includes residual strains and has a nonmonotonous stress softening function that describes Mullins effect, we used the experimental data collected during uniaxial extension test of the aforementioned suture material samples. Notice from

(25) that only four constitutive material constants and one fitting parameter need to be computed, that is, the shear modulus μ, the chain number of links N, the stress softening parameter b, the residual strain material constant C, and the fitting parameter n. However, we have found that in general the value of $n = 1$ for the uniaxial stress-softened material model described by (25) provides good fit to the collected experimental data.

We begin with the stress-stretch data for suture samples of MonoPlus and Monosyn materials. Figures 2 and 3 illustrate the predicted engineering stress response curves obtained from (24) and (25). We can see from Figures 2 and 3 that theoretical results are in good agreement with experimental data for the several loading and unloading cycles that exhibit residual strains. The constitutive material constants used to best fit experimental data are $\mu = 100$ MPa, $N = 20$, $b = 0.45$, and $C = 0.0065$ MPa for MonoPlus sutures and $\mu = 92$ MPa, $N = 20$, $b = 0.85$, and $C = 0.0045$ MPa for Monosyn material. The amount of error attained between experimental and predicted residual strains is shown in Table 1. From Figures 2 and 3 and Table 1, it is concluded that Monosyn sutures tend to soften and have bigger residual strains than MonoPlus sutures. In all figures, the dashed black lines represent theoretical predictions, and the blue solid lines describe experimental data.

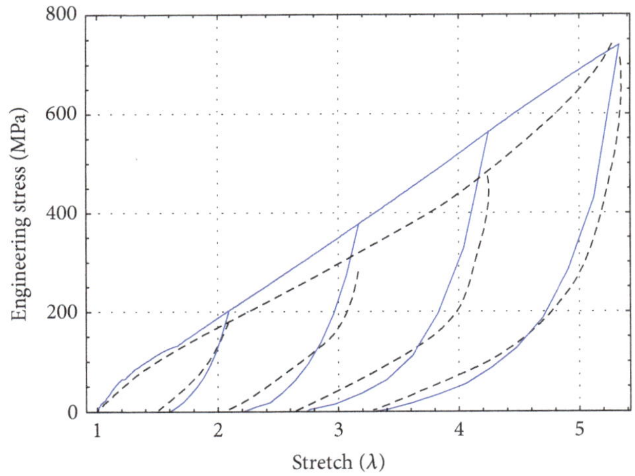

FIGURE 2: Engineering stress-stretch data for MonoPlus sutures compared with theoretical predictions of the nonmonotonous amended average-stretch, full-network model for which $\mu = 100$ MPa, $N = 20$, $b = 0.45$, and $C = 0.0065$ MPa. The dashed black lines represent theoretical predictions, and the blue solid lines describe experimental data.

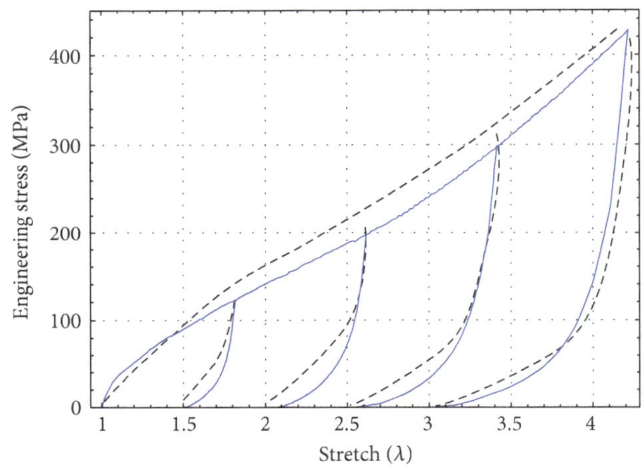

FIGURE 3: Engineering stress-stretch data for Monosyn sutures compared with theoretical predictions of the nonmonotonous amended average-stretch, full-network model for which $\mu = 92$ MPa, $N = 20$, $b = 0.85$, and $C = 0.0045$ MPa. The dashed black lines represent theoretical predictions, and the blue solid lines describe experimental data.

Figures 4 through 7 illustrate the stress-stretch curves collected from polyglycolic acid, polydioxanone, polydioxanone 4.0, and poly(glycolide-co-epsilon-caprolactone) (PGC25 3–0) suture material samples, respectively. We can see from Figures 4, 5, 6, and 7 that the predicted response stress-stretch curves computed from (25) stand in good agreement with experimental data for the several loading and unloading cycles. In these figures, the blue lines represent the experimental collected data, and the dashed black lines represent theoretical results obtained from (24) and (25). The material constants used in the material model are provided by a best fit analysis and these are listed in the figure captions. It is clear from Figures 4–7 that each suture material exhibits different qualitative and quantitative response material behavior. In fact, the amount of softening on the polyglycolic acid 2–0 and polydioxanone 2–0 suture materials is bigger than those of polydioxanone 4.0 and PGC25 3–0.

Finally, Figures 8 and 9 show the stress-stretch curves of the nonabsorbable nylon and polypropylene suture materials. Although the polypropylene sutures are stiffer than the nylon ones, the amount of strength and residual strain are quite similar. Both sutures material experienced stress-softened and permanent set that must be taken into account during suture manipulation to prevent damaging and weakening undesirable effects.

6. Conclusions

In this paper, we have examined the material behavior of eight different types of suture materials and found that, when these are subjected to loading and unloading cycles, its stress magnitude becomes lower than that of the virgin material. Furthermore, all tested sutures exhibit residual strains which is related to microstructural material damage upon deformation from the natural, undistorted state of the virgin material. To predict the suture materials response behavior observed during uniaxial tension test, we have introduced a new nonmonotonous stress-softened material model that takes into account permanent set effects for the unloading paths as described by the simple constitutive relation (25).

For each suture material, we have compared experimental data with theoretical predictions obtained from (25). In each case, we have determined the corresponding four material constants: the material shear modulus μ, the chain number of rigid links N, the material softening parameter b, and a positive material constant C that is related to the pseudoelastic

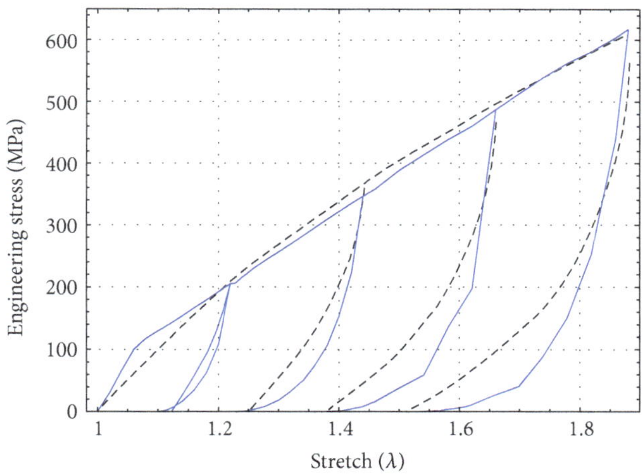

FIGURE 4: Engineering stress-stretch data for polyglycolic acid sutures compared with theoretical predictions of the nonmonotonous amended average-stretch, full-network model for which μ = 385 MPa, N = 70.5, b = 1.3, and C = 0.001 MPa. The dashed black lines represent theoretical predictions, and the blue solid lines describe experimental data.

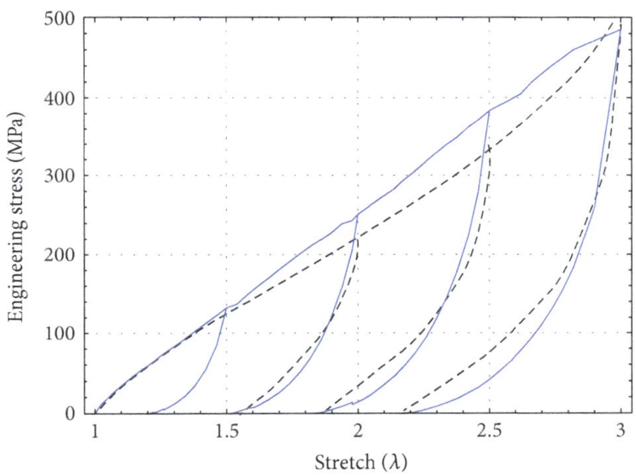

FIGURE 5: Engineering stress-stretch data for polydioxanone 2–0 compared with theoretical predictions of the nonmonotonous amended average-stretch, full-network model for which μ = 126 MPa, N = 6, b = 0.65, and C = 0.0035 MPa. The dashed black lines represent theoretical predictions, and the blue solid lines describe experimental data.

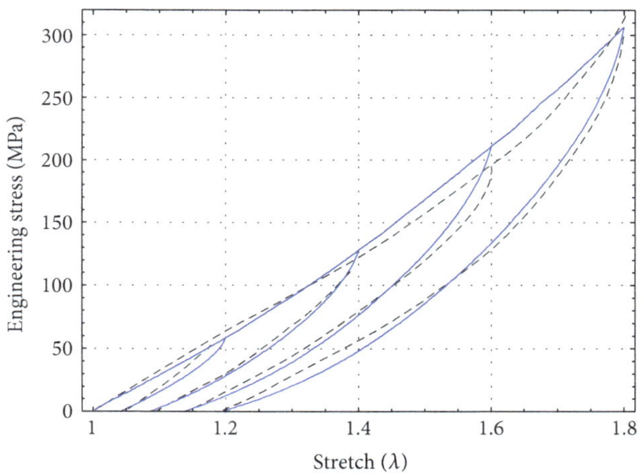

FIGURE 6: Engineering stress-stretch data for polydioxanone 4.0 compared with theoretical predictions of the nonmonotonous amended average-stretch, full-network model for which μ = 148 MPa, N = 1.95, b = 0.445, and C = 0.0115 MPa. The dashed black lines represent theoretical predictions, and the blue solid lines describe experimental data.

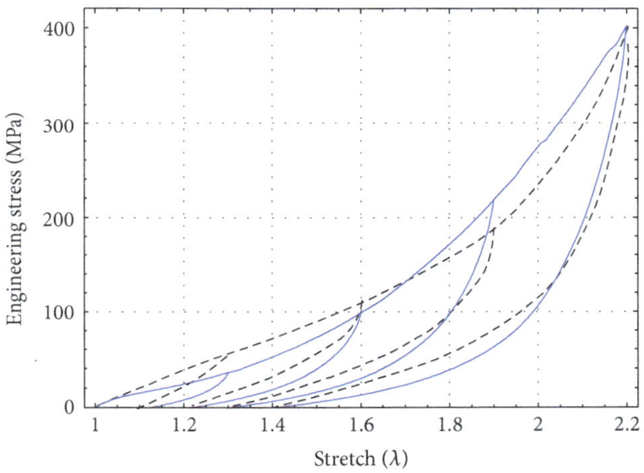

FIGURE 7: Engineering stress-stretch data for PGC25 3–0 sutures compared with theoretical predictions of the nonmonotonous amended average-stretch, full-network model for which μ = 90 MPa, N = 2.35, b = 0.75, and C = 0.012 MPa. The dashed black lines represent theoretical predictions, and the blue solid lines describe experimental data.

residual strain energy. Based on the accuracy of our proposed nonmonotonous model to predict experimental data, we can conclude that the extent of damage of suture biocompatible materials can be conveniently determined by considering its softening behavior observed during experimental tests. Nevertheless, there is a variation in the theoretical predictions, as shown in Table 1, that we believe is due to some viscoelastic effects that were not considered in the proposed material model.

Finally, the present study confirms that stress-softening and residual strain effects appear in the suture materials tested here. The experimental work of this paper proves that suture materials change dramatically when tensile loads are applied during the suturing and healing processes [2]. We have also found that the aforementioned effects are more evident when *in vitro* sutures are subjected to cyclic loading conditions. However, the results of this new experimental work will be reported in a subsequent paper.

Conflict of Interests

The authors declare that they have no conflict of interests.

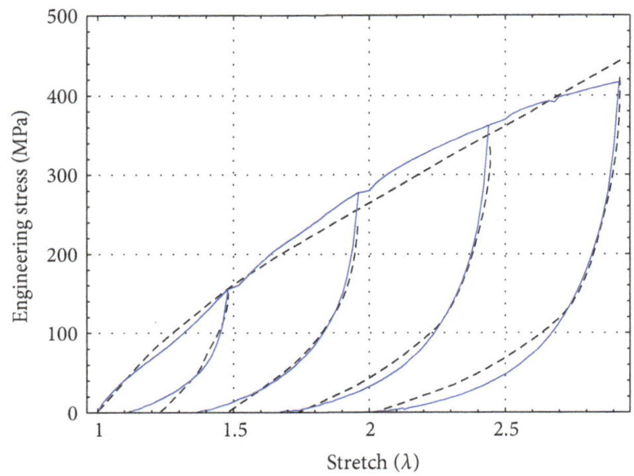

FIGURE 8: Engineering stress-stretch data for nylon sutures compared with theoretical predictions of the nonmonotonous amended average-stretch, full-network model for which $\mu = 155$ MPa, $N = 20.5$, $b = 1$, and $C = 0.0035$ MPa. The dashed black lines represent theoretical predictions, and the blue solid lines describe experimental data.

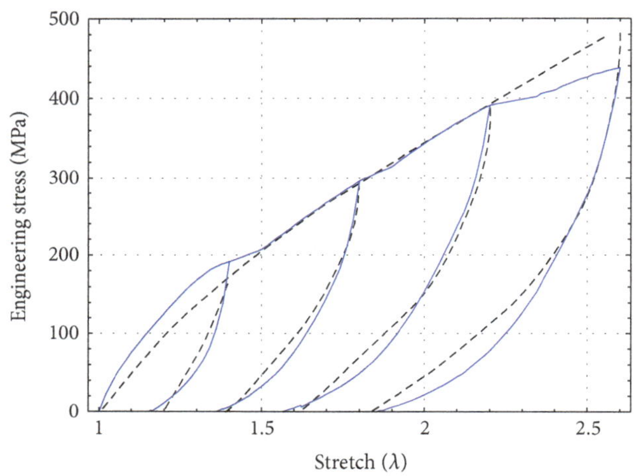

FIGURE 9: Engineering stress-stretch data for polypropylene sutures compared with theoretical predictions of the nonmonotonous amended average-stretch, full-network model for which $\mu = 200$ MPa, $N = 30.5$, $b = 0.65$, and $C = 0.00265$ MPa. The dashed black lines represent theoretical predictions, and the blue solid lines describe experimental data.

Acknowledgments

This work was funded by Tecnológico de Monterrey—Campus Monterrey, through the Research Chair in Nanomaterials for Medical Devices and Research Chair in Intelligent Machines. Additional support was provided by the project FOMIX Nuevo León M0014-2010-30 #145045 and from the European Union Seventh Framework Programme (FP7-PEOPLE-2009) under the grant agreement IRSES no. 247476.

References

[1] N. P. Ingle and M. W. King, "Optimizing the tissue anchoring performance of barbed sutures in skin and tendon tissues," *Journal of Biomechanics*, vol. 43, no. 2, pp. 302–309, 2010.

[2] W. K. Nichols, M. Stanton, D. Silver, and W. F. Keitzer, "Anastomotic aneurysms following lower extremity revascularization," *Surgery*, vol. 88, no. 3, pp. 366–374, 1980.

[3] M. Poukalova, C. M. Yakacki, R. E. Guldberg et al., "Pullout strength of suture anchors: effect of mechanical properties of trabecular bone," *Journal of Biomechanics*, vol. 43, no. 6, pp. 1138–1145, 2010.

[4] L. J. Mullins, "Effect of stretching on the properties of rubber," *Journal of Rubber Research*, vol. 16, pp. 275–289, 1947.

[5] L. Mullins and N. R. Tobin, "Theoretical model for the elastic behavior of filled-reinforced vulcanized rubbers," *Journal of Rubber Chemestry and Technology*, vol. 30, pp. 555–571, 1957.

[6] S. Govindjee and J. Simo, "A micro-mechanically based continuum damage model for carbon black-filled rubbers incorporating Mullins' effect," *Journal of the Mechanics and Physics of Solids*, vol. 39, no. 1, pp. 87–112, 1991.

[7] R. W. Ogden and D. G. Roxburgh, "A pseudo-elastic model for the Mullins effect in filled rubber," *Proceedings of the Royal Society A: Mathematical, Physical and Engineering Sciences*, vol. 455, no. 1988, pp. 2861–2877, 1999.

[8] M. F. Beatty and S. Krishnaswamy, "Theory of stress-softening in incompressible isotropic materials," *Journal of the Mechanics and Physics of Solids*, vol. 48, no. 9, pp. 1931–1965, 2000.

[9] A. Elías-Zúñiga and M. F. Beatty, "A new phenomenological model for stress-softening in elastomers," *Zeitschrift fur Angewandte Mathematik und Physik*, vol. 53, no. 5, pp. 794–814, 2002.

[10] A. Elías-Zúñiga and M. F. Beatty, "Stress-softening effects in the transverse vibration of a non-Gaussian rubber string," *Meccanica*, vol. 38, no. 4, pp. 419–433, 2003.

[11] A. Elías-Zúñiga, "A phenomenological energy-based model to characterize stress-softening effect in elastomers," *Polymer*, vol. 46, no. 10, pp. 3496–3506, 2005.

[12] J. Diani, M. Brieu, and J. M. Vacherand, "A damage directional constitutive model for Mullins effect with permanent set and induced anisotropy," *European Journal of Mechanics, A/Solids*, vol. 25, no. 3, pp. 483–496, 2006.

[13] D. de Tommasi, G. Puglisi, and G. Saccomandi, "A micromechanics-based model for the Mullins effect," *Journal of Rheology*, vol. 50, no. 4, pp. 495–512, 2006.

[14] A. Elías-Zúñiga and C. A. Rodríguez, "A non-monotonous damage function to characterize stress-softening effects with permanent set during inflation and deflation of rubber balloons," *International Journal of Engineering Science*, vol. 48, no. 12, pp. 1937–1943, 2010.

[15] G. A. Holzapfel, M. Stadler, and R. W. Ogden, "Aspects of stress softening in filled rubbers incorporating residual strains," in *Proceedings of the first European Conference on Constitutive Models for Rubber*, A. Dorfmann and A. Muhr, Eds., pp. 189–193, Rotterdam, The Netherlands, 1999.

[16] M. F. Beatty, "An average-stretch full-network model for rubber elasticity," *Journal of Elasticity*, vol. 70, no. 1–3, pp. 65–86, 2003.

[17] M. A. Johnson and M. F. Beatty, "The mullins effect in uniaxial extension and its influence on the transverse vibration of a rubber string," *Continuum Mechanics and Thermodynamics*, vol. 5, no. 2, pp. 83–115, 1993.

[18] E. A. de Souza Neto, D. Perić, and D. R. J. Owen, "A phenomenological three-dimensional rate-idependent continuum damage model for highly filled polymers: formulation and computational aspects," *Journal of the Mechanics and Physics of Solids*, vol. 42, no. 10, pp. 1533–1550, 1994.

[19] G. Marckmann, E. Verron, L. Gornet, G. Chagnon, P. Charrier, and P. Fort, "A theory of network alteration for the Mullins effect," *Journal of the Mechanics and Physics of Solids*, vol. 50, no. 9, pp. 2011–2028, 2002.

[20] A. Dorfmann and R. W. Ogden, "A constitutive model for the Mullins effect with permanent set in particle-reinforced rubber," *International Journal of Solids and Structures*, vol. 41, no. 7, pp. 1855–1878, 2004.

[21] R. Kazakevičiute-Makovska and R. Kačianauskas, "Modelling of stress softening in elastomeric materials: foundations of simple theories," *Mechanics Research Communications*, vol. 31, no. 4, pp. 395–403, 2004.

[22] R. Kazakevičiūtė-Makovska, "Experimentally determined properties of softening functions in pseudo-elastic models of the Mullins effect," *International Journal of Solids and Structures*, vol. 44, no. 11-12, pp. 4145–4157, 2007.

[23] M. Cheng and W. Chen, "Experimental investigation of the stress-stretch behavior of EPDM rubber with loading rate effects," *International Journal of Solids and Structures*, vol. 40, no. 18, pp. 4749–4768, 2003.

[24] W. V. Mars and A. Fatemi, "Observations of the constitutive response and characterization of filled natural rubber under monotonic and cyclic multiaxial stress states," *Journal of Engineering Materials and Technology*, vol. 126, no. 1, pp. 19–28, 2004.

[25] D. de Tommasi, G. Puglisi, and G. Saccomandi, "Localized versus diffuse damage in amorphous materials," *Physical Review Letters*, vol. 100, no. 8, Article ID 085502, pp. 1–4, 2008.

[26] A. Elías-Zúñiga and M. F. Beatty, "Constitutive equations for amended non-Gaussian network models of rubber elasticity," *International Journal of Engineering Science*, vol. 40, no. 20, pp. 2265–2294, 2002.

Simple Method to Synthesize Functionalized Carbon Nanotubes Employing Cobalt Nitrate and Acetone by Using Spray Pyrolysis Deposition Technique

Jorge A. Gómez,[1] Alfredo Marquez,[2] Antonino Pérez,[2] and A. Duarte-Moller[2]

[1] *Instituto Potosino de Investigación Científica y Tecnológica (IPICyT), Camino a la Presa San José 2055, Colonia Lomas 4 Sección, 78216 San Luis Potosí, SLP, Mexico*
[2] *Centro de Investigación en Materiales Avanzados (CIMAV), Miguel de Cervantes 120, Complejo Industrial Chihuahua, 31109 Chihuahua, CHIH, Mexico*

Correspondence should be addressed to A. Duarte-Moller, alberto.duarte@cimav.edu.mx

Academic Editor: Wen-Hua Sun

Recently alcohols and ketones have been employed to sensitize CNT by CVD. A study has shown the importance of the chemical nature of those carbon precursors on the characteristics of the CNT (carbon nanotubes) obtained. In the present work we show the influence of the catalyst employed on the synthesis of functionalized multiwall carbon nanotubes (MWCNTs) utilizing acetone as carbon source and cobalt nitrate $Co(NO_3)_2$ as catalyst.

1. Introduction

For the synthesis of carbon nanotubes are seeking a simple and economical method. It has been reported in the literature the use of carbon sources for the synthesis of carbon nanotubes as acetylene [1–4], toluene [5], n-hexane, benzene and cyclopentadiene [6], and so forth, and ultimately alcohols have been used [7–13]. Montoro et al. [13] reported in a large comparative study that acetone is the best candidate to synthesize multiwall carbon nanotubes, based on this type of analysis.

Respect to the catalysts used for the synthesis of the NTC (carbon nanotubes) are used some ceramics such as alumina, impregnated in some salts such as nitrates of cobalt-nickel, cobalt and manganese acetates, and various transition pure metals such as iron, cobalt, zinc, ruthenium/palladium, and Co Na-Y. Inoue and Kikuchi [9] explained that cobalt, compared with iron, exhibits a poor interaction with the coal, but better to decompose ethanol. This also shows a preference between catalysts for the nucleation and growth according to Huh et al. [14, 15], as well as the size of the particle. Furthermore Liu et al. [16] prepared composites based on polyamide and functionalized carbon nanotubes.

In the present study, acetone was selected as a carbon source and nitrate of cobalt as a catalyst, which will assess their performance in the synthesis of carbon nanotubes along the tube to spray synthesis, and it will also consider the ketone and nitrate groups from the breakdown of these compounds on the functionalization of multiwall carbon nanotubes.

2. Experimental Details

A solution was prepared at 0.42M with acetone at 99.6% purity supplied by JT Baker and cobalt nitrate hexahydrate ($Co(NO_3)_2 \cdot 6H_2O$) at 99.0% purity supplied by Acros Organic that has been used as catalyst. A argon flow at 99.999% purity was used as carrier gas.

Figure 1 [5] shows the pyrolytic system employee for the synthesis of the carbon nanotubes multilayer (MWCNT)

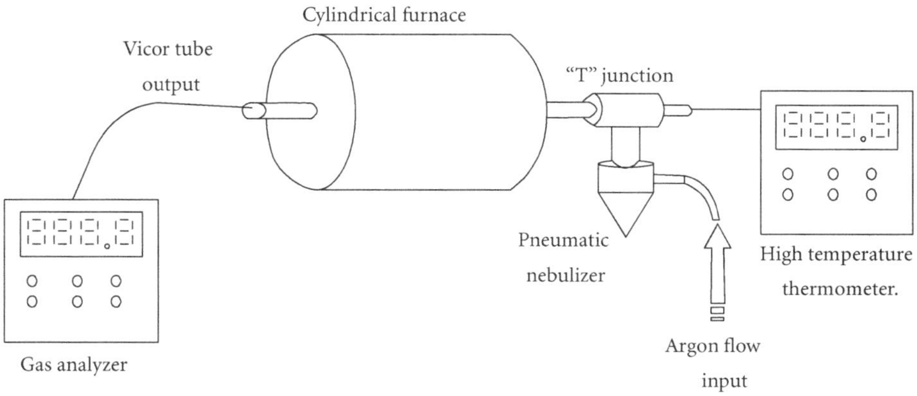

FIGURE 1: Spray pyrolysis system with a gas measurements equipment employed.

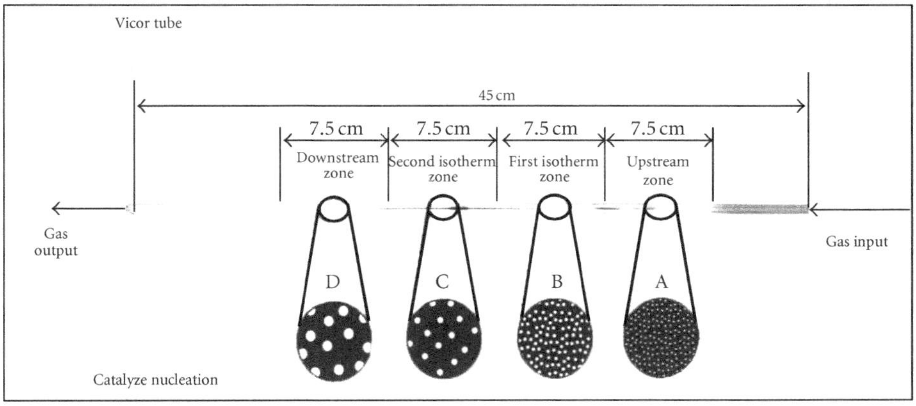

FIGURE 2: Vicor tube image, with measurements and identification of zones.

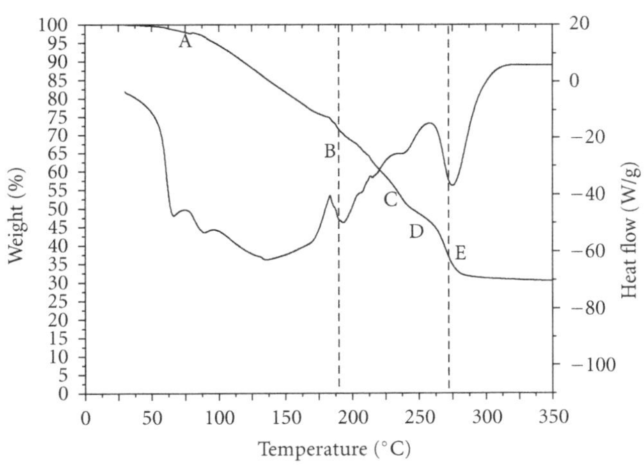

FIGURE 3: Cobalt nitrate decomposition DSC analysis.

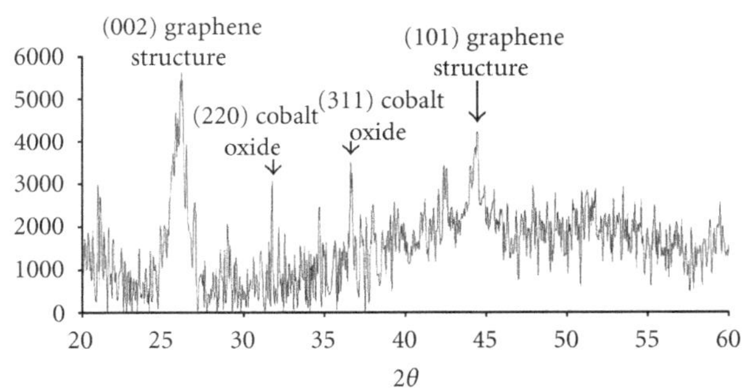

FIGURE 4: RX diffraction of carbon nanotubes.

of acetone-cobalt; in turn it shown the connection way of equipment used to measure inlet temperature, as well as the composition of the flue gases. For this synthesis was used Thermolyne cylindrical furnace 40 cm in length. A vicor tube of 45 cm in length and 9 mm internal diameter was used as the substrate placed inside the oven. Figure 2 shows the division of the zones of the tube considered in the analysis after the reaction has been carried out; the upstream zone is the part at the entry of the tube; in this zone it is where the acetone and the catalyst, which they have previously been nebulized, enter to vicor tube and this is where the decomposition reaction begins that continues along the tube. First isotherm zone, second isotherm zone are the two middle parts of the tube and the downstream zone is the part of the exhaust gas resulting from the decomposition and synthesis of NTC.

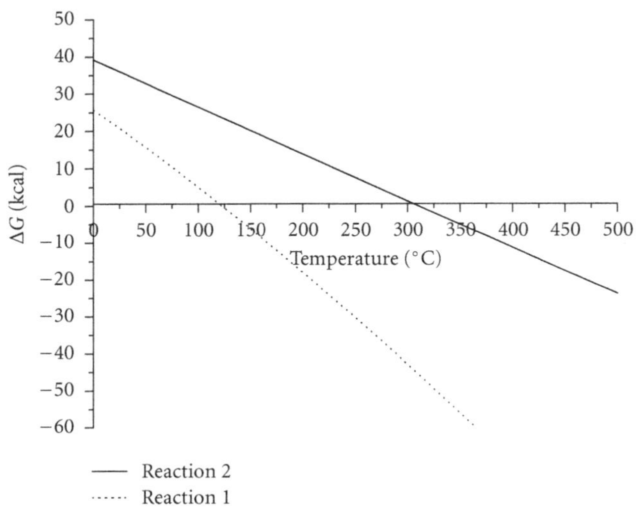

Figure 5: Energy formation ΔG analysis of cobalt nitrate deshidratation and decomposition.

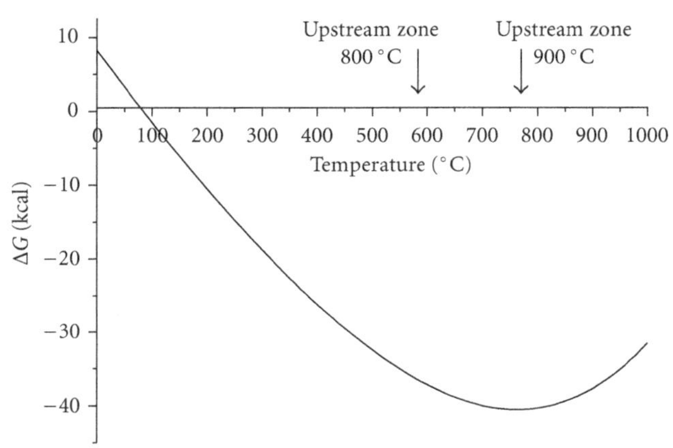

Figure 6: Energy formation ΔG analysis of acetone decomposition.

Table 1: Decomposition of cobalt nitrate (from 25 to 400°C).

Formula	Weight loss (%) Percent weight	Accumulative %
$Co(NO_3)_2 * 6H_2O$	100	—
H_2O	37.140	62.86
NO_2	15.808	47.052
O_2	10.995	—
NO	10.310	25.747
CoO	25.747	

We used a set of gas analyzers CGT-7000 (infrared gas analyzer) and the NOA-7000 (NO_x, O_2 gas analyzer) in order to monitor the exit of the product gases of the reaction to the exit of the vicor tube during the synthesis. In addition a thermocouple was attached in the area to monitor the upstream zone initial temperature at the entrance of the tube and this is where our experience shows the best carbon nanotubes synthesized in this technique (Figure 1).

Table 2: Cobalt nitrate-acetone decomposition. Weight loss.

Cobalt nitrate decomposition during reaction			
Oven temp.	Spraying temperature	NO_x ppm	NO ppm
800°C	30°C	151.7	173.3
900°C	30°C	116.4	123.8

For the synthesis of nanotubes were selected four temperatures 700°C, 800°C, 900°C, and 1000°C and kept constant during the synthesis phase. During the preheating phase, we applied a small argon flow of 0.8 $L\text{-}min^{-1}$. Once the furnace temperature had been reached, the solution was applied through a spray mist whose speed is 1 mL/min of acetone under an argon flow of 2.5 $L\text{-}min^{-1}$. Was nebulized a total 20 mL of acetone catalyst solution into the vicor tube with a flow of 3.5 $L\text{-}min^{-1}$. Once the solution was finished, the heating system was turn off. The assembly was allowed to cool to room temperature keeping during this process a flow of argon 0.8 L/min. The vicor tube was cut into four sections to be named, these sections to the analysis phase have been called according to the vicor tube was cut into four sections each one of 7.5 cm in length and was named in according to its position like upstream zone, first and second downstreamzone Isothermal zones.

In order to observe the growth of carbon nanotubes a small segment of each of these parts has been extracted and analyzed using a scanning microscope JEOL JSM-7401F. Subsequently each of these segments was separately immersed in an acidic solution (1 : 3) nitric/sulfuric [17] and extracted carbon nanotubes from the substrate by vicor ultrasonic treatment. To remove acid residues were washed with distilled water and tri-several rinses in acetone. Finally in the cleanup phase, the specimens of each sections were suspended in acetone in order to be characterized.

A small drop of this solution was placed on a formvar copper TEM grid in order to observed the carbon nanotubes in a transmission microscope FS2200-STEM. The analysis of the presence of nanotubes and waste took place in the PANalytical X-ray spectrometer, model X'Pert PRO MPD X'Celerator detector. For the analysis of the functionalization of carbon nanotubes we got the liquid solution in a range close to 70°C by a day and half to evaporate the solvent for suspension. After several washings with methanol was placed a small drop of the suspension in a Thermoscientific Nicolet FT IR 6700, to observe the IR spectrum. These analyses were performed only on the upstream zone area synthesized at 900° given that the amount of sample collected from all others was not enough for further analysis.

3. Results and Discussions

3.1. Decomposition of Cobalt Nitrate Hexahydrate. In Figure 3 are identified cobalt oxide residues at the end of the test TGA/DSC, and the percentages of weight loss in water, nitrogen oxides, to be corroborated from the calculations in HSC 5.1 [18] of the percentages, are shown in Table 1. We observe a phase denitrification about 25% near 180°C; after this phase, the system no longer holds water in its structure

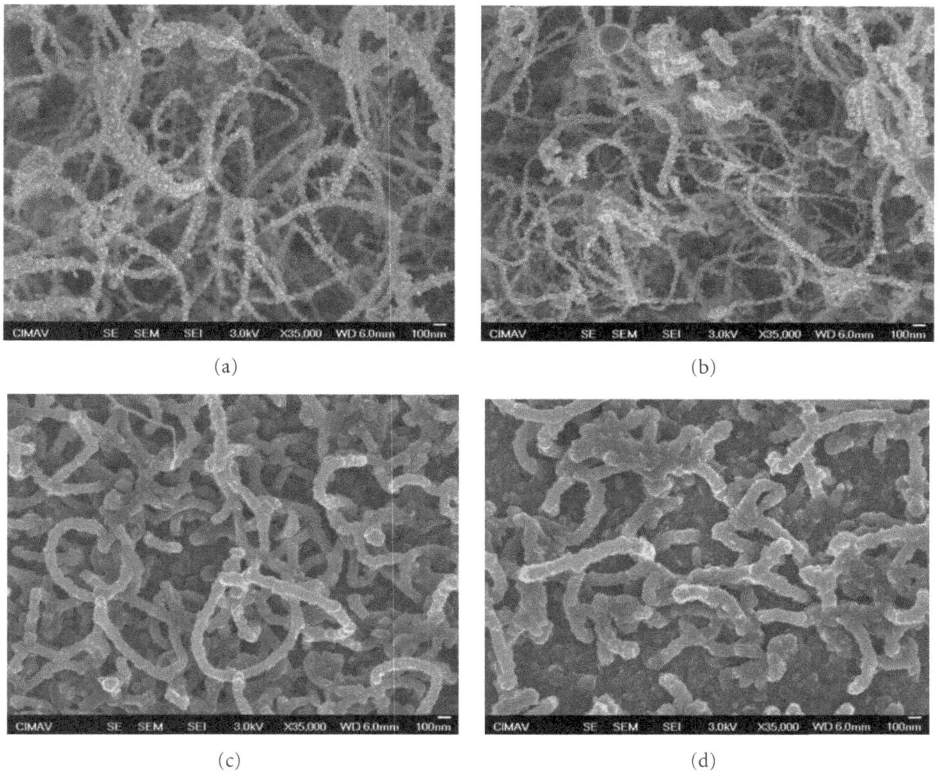

FIGURE 7: SEM images carbon nanotubes synthesized to 800°C of the four zones analyzed.

TABLE 3: Experimental percentage weight of propanone decomposition.

		Acetone decomposition		
Oven temp.	Spraying temperature	Upstream zone temperature	CO ppm	CO_2 vol%
800°C	30°C	560°C	1881	0.02
900°C	30°C	774°C	5754	0.05

and then begins de-nitrification stage; this leads to the loss in weight almost 35% between 182°C to 275°C; however, we see finally that a final residue of 30.37% corresponding to cobalt oxide was detected by XRD analysis, Figure 4. We verified through a thermodynamic model for HSC, which shows the simulated results for the ΔG training. The first reaction predicts the dehydration process:

$$Co(NO_3)_2 6H_2O \longrightarrow Co(NO_3)_2 + 6H_2O \quad (1)$$

The second (Reaction 2) predicts the loss of cobalt nitrate:

$$Co(NO_3)_2 \longrightarrow CoO + NO + NO_2 + O_2 \quad (2)$$

We can say that the production of NO and NO_2 occurs, given that it was possible to detect its presence (Table 2), these results were measured experimentally during the synthesis of NTC. Note also that Reaction 2 carried out effectively begins after the 300°C as predicted in Figure 5.

3.2. Generation of CO, CO_2, and H_2O to Form Carbon Nanotubes. Similarly, Figure 6 illustrates the results of the simulation graphs showing the formation ΔG predicting the possible decomposition of the acetone. The reaction proposal was made based on experimental data contained in Table 3 taken to the exit of the vicor tube during a small complementary experiment about the decomposition of pure acetone. This table shows the presence of carbon monoxide and whose concentration increases with temperature:

$$C_3H_6O \longrightarrow CO + 2C + 3H_2 \uparrow \quad (3)$$

Inside the vicor tube have a temperature gradient caused by the thermocouple was placed in to middle of the tubular furnance. Now we can say that the decomposition reaction of acetone starts to occur around 100°C almost at the entrance of the tube. Complementarily note in Figure 3 that while the acetone is beginning to decompose to CO and hydrogen, cobalt nitrate is just beginning to lose water. This water vapor and CO and CO_2 and this is formed as Nasibulin et al. extra hydrogen [19]. This water vapor, CO and CO_2 formed an extra hydrogen amount Nasibulin et al. [19]. That the unreacted CO (Table 3) reduced cobalt oxides [20] and begins to generate a unstable cobalt carbides. This is

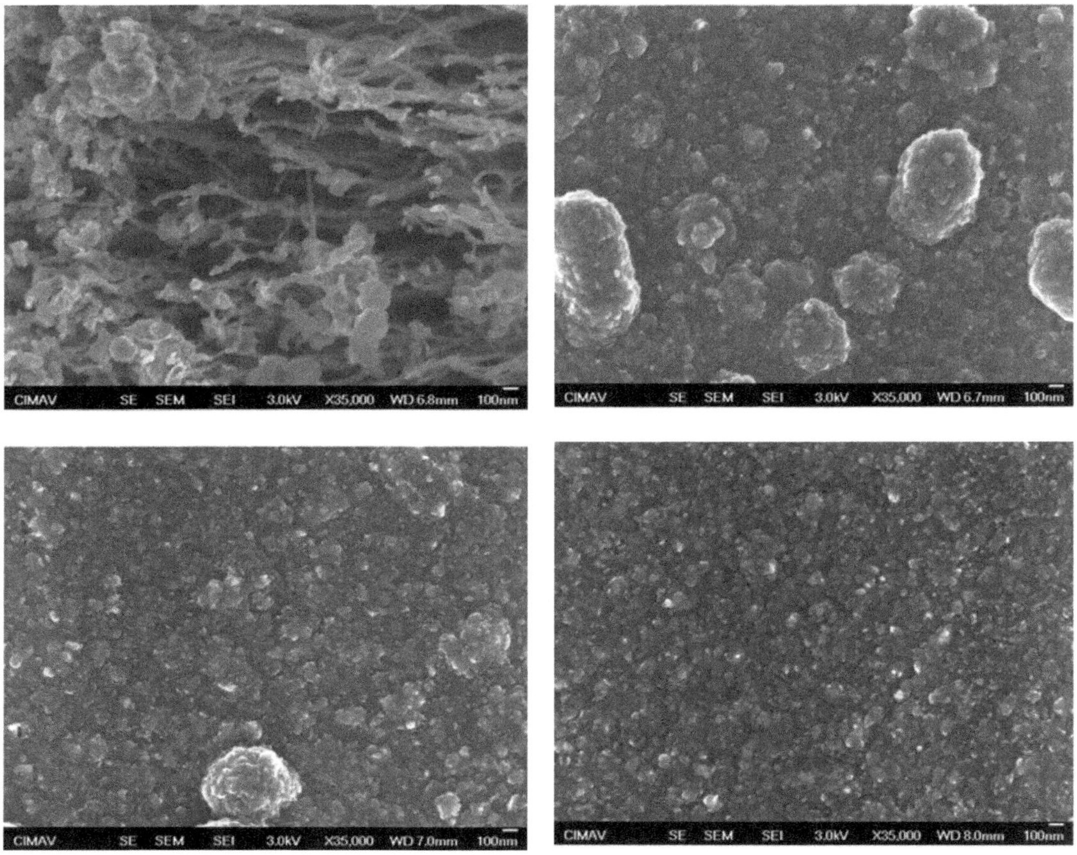

(a) (b)

FIGURE 8: SEM images carbon nanotubes synthesized to 900°C of the four zones analyzed.

FIGURE 9: Well-aligned forest of carbon nanotubes synthesized to 900°C for the upstream zone.

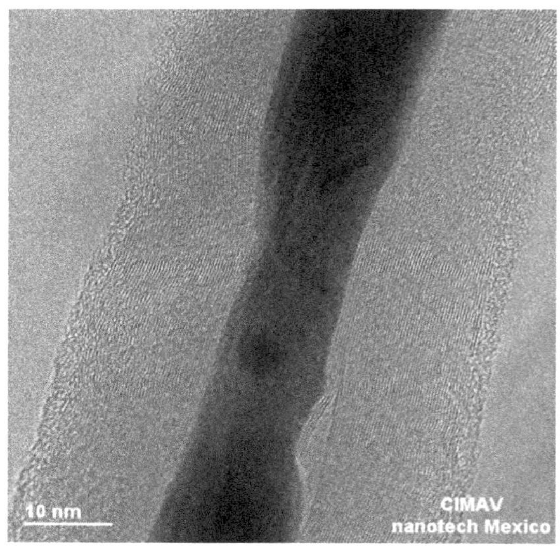

FIGURE 10: TEM images of carbon nanotubes of first isothermal zone synthesized to 900°C.

the initial steps for the synthesis of carbon nanotubes by disproportionalization CO [19, 20] which is the source of carbon for the synthesis of carbon nanotubes. Nasibulin explained that the CO_2 and water "clean" the precipitated amorphous carbon on the surface of the nanotubes, as well as preventing the inhibition of the catalytic particle and maintaining the conditions of the formation of nanotubes by disproportionalization and hydrogenation.

As mentioned, the decomposition reaction begins to occur once the nebulized solution enters the tube synthesis.

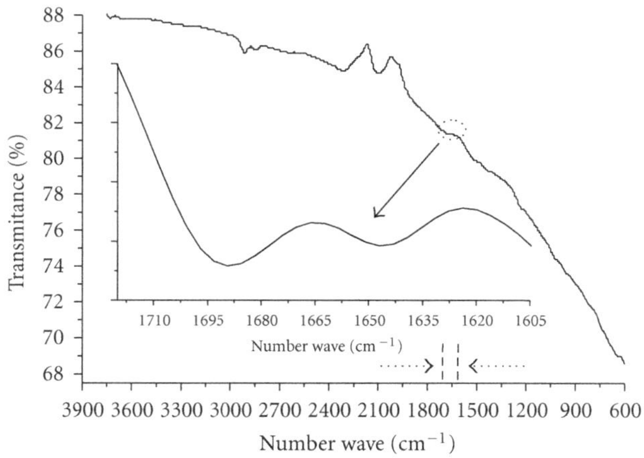

Figure 11: FT IR image of carbon nanotubes of first isothermal zone.

The catalyst particles begin to grow from the inlet to the outlet vicor tube (Figure 2), with a particle size of smaller cobalt at the entrance of the tube and as it moves over the tube, these particle sizes are increasingly large; this is due to that particles have the enough nucleation time. In accordance with what stated by Shin et al. [15], the systematic morphology of the catalyst, such as the grain size and density of the transition metals, controls the diameter and length of the nanotubes, which explains the differences in size and morphology of the nanotubes synthesized at 800°C.

Figure 8 shows SEM images of the four zones synthesized at 800°C, according to Figure 2. Also shown the nanotube formation in the four areas of vicor tube. Figure 7(a) in the image area corresponding to the upstream zone were observed thinner and longer nanotubes with small protrusions attached to hemispherical surface. These bumps are most likely composed of amorphous carbon. In the Figure 7(a) show a first isothermal zone image were observed nanotubes smaller and a bit thicker than in the previous area but fewer. In Figure 7(c) show the second isothermal zone, it is observed that the nanotubes are slightly shorter than in the first isothermal zone, but thicker than in the upstream zone. Finally, in the downstream zone, see Figure 7(d), carbon nanotubes are thicker than the last three areas, but in smaller numbers. Some of these nanotubes appear to be linked to form a "Y".

Figure 8 illustrates the synthesis of NTC at 900°C; we observed the formation of carbon nanotubes only in the upstream zone, while in the other three areas (first isotherm zone, second isotherm zone, and downstream zone) of the tube vicor, there were no structures similar to a nanotube. Indeed, we observe in Figure 8(a), an orderly and uniform growth of these nanotubes has occurred only in the upstream zone at 900°C. For the first isotherm zone in Figure 8(b), were observed certain amorphous carbon conglomerations almost 300 nm in diameter and about one micron in length. For the second isotherm zone 8C, were found scarce conglomerations spherical of near 200 nm radio of amorphous carbon. For upstream zone, 8D, there was no structure whatsoever. Figure 9 illustrates the forest of carbon nanotubes obtained in the upstream zone at 900°C, where appreciate the orderly growth. It is important comment, these nanotubes were longer, thiner and more numerous than those synthesized to 800°C. Nasibulin et al. [19] showed that the reactions that keep the formation of carbon nanotubes (disproportionalization and hydrogenation) are extinguished at temperatures above the 908°C due to the dominant role of these reversible reactions. Due to this fact, we observe that the formation of nanotubes along the tube at 800°C was found in all four segments of vicor tube, whereas in the 900°C a satisfactory growth occurred only in the first zone (Figures 9 and 10), given that there were no carbon nanotubes at 1000°.

Figure 10 shows a TEM image of multiwall carbon nanotubes synthesized, which shows the internal structure containing a small portion of the catalyst; it is presumably of cobalt or carbide. One can also observe a thin outer layer of amorphous carbon which covers it. On Figure 11, in reference of the FT IR spectrum, show the possible evidence of functional groups ($-NO_2$) near the wave numbers 1640 and 1620, and also the ketone groups close to the wave numbers 1725 and 1705. Hanming Lui [16] synthesized functionalized nanotubes where the results are similar in the morphology of carbon nanotubes.

4. Conclusions

The synthesis of carbon nanotubes with cobalt nitrate and acetone using the method of spray pyrolisis resulted, in addition to the carbon source and metallic cobalt as a catalyst, water and CO_2 to remain in the good condition formation of carbon nanotubes. An incomplete decomposition reaction of these two compounds resulted in several functional groups adhered to the graphemic structure of the nanotubes. In which, that was possible to detect a little amount in FT IR equipment.

The water produced by the same cobalt nitrate participates in the formation of CO_2 from CO decomposition product of acetone is proposed where additional work is controlled hydration of cobalt nitrate to observe the effect of growth and quality of carbon nanotubes. In addition to change the way of collection of catalyst particles to increase the production of multiwall carbon nanotubes, we most control the use of the reagents.

References

[1] S. Seelan, D. W. Hwang, L. P. Hwang, and A. K. Sinha, "Synthesis of multiwalled carbon nanotubes by high-temperature vacuum annealing of amorphous carbon," *Vacuum*, vol. 75, no. 2, pp. 105–109, 2004.

[2] Z. X. Xu, J. D. Lin, V. A. L. Roy, Y. Ou, and D. W. Liao, "Catalytic synthesis of carbon nanotubes and carbon spheres using Kaolin supported catalyst," *Materials Science and Engineering B*, vol. 123, no. 2, pp. 102–106, 2005.

[3] E. Terrado, M. Redrado, E. Muñoz, W. K. Maser, A. M. Benito, and M. T. Martínez, "Carbon nanotube growth on cobalt-sprayed substrates by thermal CVD," *Materials Science and Engineering C*, vol. 26, no. 5–7, pp. 1185–1188, 2006.

[4] S. Lim, A. Shimizu, S. H. Yoon, Y. Korai, and I. Mochida, "High yield preparation of tubular carbon nanofibers over

supported Co-Mo catalysts," *Carbon*, vol. 42, pp. 1279–1283, 2004.

[5] G. Alonso-Nuñez, A. M. Valenzuela-Muñiz, F. Paraguay-Delgado, A. Aguilar, and Y. Verde, "New organometallic precursor catalysts applied to MWCNT synthesis by spray-pirolysis," *Optical Materials*, vol. 29, no. 1, pp. 134–139, 2006.

[6] K. Otsuka, Y. Abe, N. Kanai, Y. Kobayashi, S. Takenaka, and E. Tanabe, "Synthesis of carbon nanotubes on Ni/carbon-fiber catalysts under mild conditions," *Carbon*, vol. 42, no. 4, pp. 727–736, 2004.

[7] A. Grüneis, M. H. Rümmeli, C. Kramberger et al., "High quality double wall carbon nanotubes with a defined diameter distribution by chemical vapor deposition from alcohol," *Carbon*, vol. 44, no. 15, pp. 3177–3182, 2006.

[8] S. Inoue, T. Nakajima, and Y. Kikuchi, "Synthesis of single-wall carbon nanotubes from alcohol using Fe/Co, Mo/Co, Rh/Pd catalysts," *Chemical Physics Letters*, vol. 406, no. 1–3, pp. 184–187, 2005.

[9] S. Inoue and Y. Kikuchi, "Diameter control and growth mechanism of single-walled carbon nanotubes," *Chemical Physics Letters*, vol. 410, no. 4–6, pp. 209–212, 2005.

[10] L. Zheng, X. Liao, and Y. T. Zhu, "Parametric study of carbon nanotube growth via cobalt-catalyzed ethanol decomposition," *Materials Letters*, vol. 60, no. 16, pp. 1968–1972, 2006.

[11] D. Nishide, H. Kataura, S. Suzuki, S. Okubo, and Y. Achiba, "Growth of single-wall carbon nanotubes from ethanol vapor on cobalt particles produced by pulsed laser vaporization," *Chemical Physics Letters*, vol. 392, no. 4–6, pp. 309–313, 2004.

[12] G. Ortega-Cervantez, G. Rueda-Morales, and J. Ortiz-López, "Catalytic CVD production of carbon nanotubes using ethanol," *Microelectronics Journal*, vol. 36, no. 3–6, pp. 495–498, 2005.

[13] L. A. Montoro, P. Corio, and J. M. Rosolen, "A comparative study of alcohols and ketones as carbon precursor for multi-walled carbon nanotube growth," *Carbon*, vol. 45, no. 6, pp. 1234–1241, 2007.

[14] Y. Huh, M. L. H. Green, Y. H. Kim, J. Y. Lee, and C. J. Lee, "Control of carbon nanotube growth using cobalt nanoparticles as catalyst," *Applied Surface Science*, vol. 249, no. 1–4, pp. 145–150, 2005.

[15] Y. M. Shin, S. Y. Jeong, H. J. Jeong et al., "Influence of morphology of catalyst thin film on vertically aligned carbon nanotube growth," *Journal of Crystal Growth*, vol. 271, no. 1-2, pp. 81–89, 2004.

[16] H. Liu, X. Wang, P. Fang et al., "Functionalization of multi-walled carbon nanotubes grafted with self-generated functional groups and their polyamide 6 composites," *Carbon*, vol. 48, no. 3, pp. 721–729, 2010.

[17] S. Porro, S. Musso, M. Vinante et al., "Purification of carbon nanotubes grown by thermal CVD," *Physica E*, vol. 37, no. 1-2, pp. 58–61, 2007.

[18] HSC 5.1. Chemistry, http://www.hsc-chemistry.net/.

[19] A. G. Nasibulin, D. P. Brown, P. Queipo, D. Gonzalez, H. Jiang, and E. I. Kauppinen, "An essential role of CO2 and H2O during single-walled CNT synthesis from carbon monoxide," *Chemical Physics Letters*, vol. 417, no. 1–3, pp. 179–184, 2006.

[20] W. Zhang, D. Ma, J. Liu, L. Kong, W. Yu, and Y. Qian, "Solvothermal synthesis of carbon nanotubes by metal oxide and ethanol at mild temperature," *Carbon*, vol. 42, pp. 2329–2366, 2004.

Production of C_{3+} Olefins and Propylene from Ethanol by Zr-Modified H-ZSM-5 Zeolite Catalysts

Megumu Inaba, Kazuhisa Murata, Isao Takahara, and Ken-ichiro Inoue

BTL Catalyst Group, Energy Technology Research Institute, National Institute of Advanced Industrial Science and Technology (AIST), AIST Tsukuba Central 5, 1-1-1 Higashi, Ibaraki, Tsukuba 305-8565, Japan

Correspondence should be addressed to Megumu Inaba, mg.inaba@aist.go.jp

Academic Editor: Wen-Hua Sun

Ethanol conversion to C_{3+} olefins, especially propylene, using Zr-modified H-ZSM-5 catalysts was investigated. Zr-modification to H-ZSM-5 zeolite could improve the initial yield of C_{3+} olefins and propylene and could reduce the initial yield of ethylene. In general, catalysts exhibiting the higher initial yield of propylene showed the steeper decrease in propylene yield as the reaction proceeded. However, Zr-modification to H-ZSM-5 could depress the decrease in propylene yield for aqueous ethanol. As cause of catalytic deactivation, carbon deposition on catalyst and framework collapse of zeolite support can be considered. The addition of water to Zr-modified H-ZSM-5 catalyst could depress carbon deposition in some degree, and, as a result, the decrease in propylene yield could be depressed.

1. Introduction

Major compounds used by the petrochemical industry, for example, ethylene, propylene, benzene, and toluene, are at present synthesized by cracking the naphtha fraction. However, petroleum resources are limited, both quantity-wise and geographically. Moreover, combustion of petroleum produces CO_2, a greenhouse gas. Recently, much attention has been paid to biomass as alternative resources for petroleum, since biomass is renewable and carbon-neutral resource that is readily available.

Ethanol is one of the main biomass products, which can be obtained by fermentation. Catalytic production of hydrocarbons from ethanol has been reported by many researchers [1–26]. In most cases, H-ZSM-5 zeolites have been used as catalysts [1–5] with aromatics being the main products. Saha et al. reported ethanol conversion using Zn- and Ga-doped ZSM-5 zeolite catalysts, which enhanced the formation of aromatics [6].

Recently, production of hydrocarbons from ethanol using modified ZSM-5 has also been reported. For example, Calsavara et al. have reported the effect of conditions for preparation of Fe-modified H-ZSM-5 catalysts on ethanol conversion [7, 8]. Murata et al. have reported that W- and La-modification of H-ZSM-5 zeolite to be effective for enhancing the formation of lower olefins from ethanol [9]. Inoue et al. have reported the conversion of ethanol to propylene using La-modified H-ZSM-5 zeolite with Si/Al_2 ratio of 280, an especially high ratio for production of C_{3+} olefins from ethanol [10]. Fujitani et al. have reported that P- [11] or Zr-modified [12] H-ZSM-5 catalysts show improved catalytic activity and stability for conversion of ethanol to propylene in comparison with H-ZSM-5 without modification. Sano et al. have reported conversion of ethanol to propylene over HZSM-5 type zeolites containing alkaline earth metals [13]. Gayubo et al. had reported ethanol conversion to hydrocarbons over H-ZSM-5 without metals [3–5] and recently reported Ni- [14, 15] or alkali-modified [16, 17] H-ZSM-5 catalysts for ethanol conversion.

On the other hand, production of olefins from ethanol using catalysts other than H-ZSM-5 has also been reported. For example, Berrier et al. have reported the transformation of ethanol into higher hydrocarbons on Co/Al_2O_3 catalysts [18]. More recently, Baba et al. reported the highly selective conversion of ethylene to propylene over SAPO-34; they also found SAPO-34 to be effective for ethanol conversion to

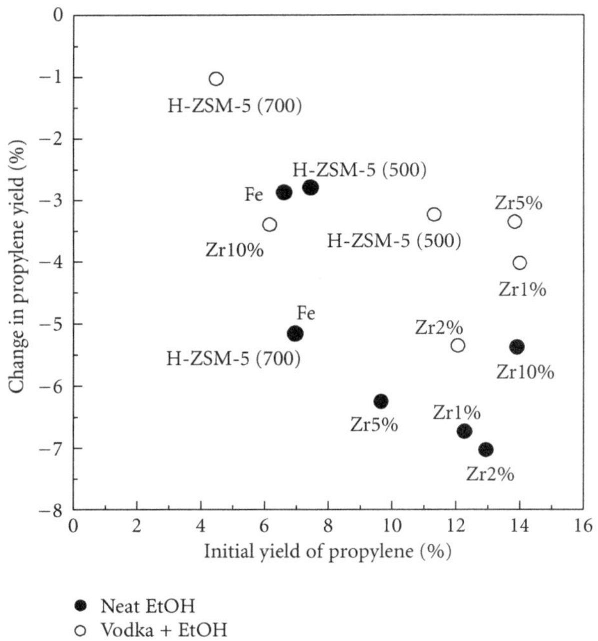

Figure 1: Plots of the initial yield of propylene versus the change in propylene yield over H-ZSM-5 and (Fe or Zr)/H-ZSM-5 catalysts. Numbers in parenthesis denote calcination temperature (°C).

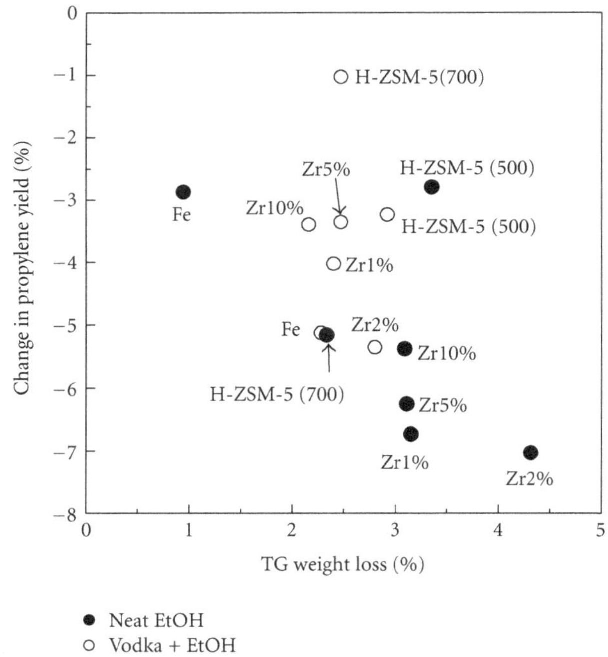

Figure 2: Plots of TG weight loss versus the change in propylene yield over H-ZSM-5 and (Fe or Zr)/H-ZSM-5 catalysts.

propylene [19]. Iwamoto et al. reported that Ni ion-loaded mesoporous silica MCM-41 was effective for producing lower olefins from ethanol [20].

We investigated the production of hydrocarbons from ethanol using modified H-ZSM-5 catalysts and found that Fe-addition to H-ZSM-5 zeolite could improve the selectivity of C_{3+} olefins [21, 22]. Fe is nontoxic and cheap metal. Moreover, olefins, especially propylene, are useful as not only fuels but also chemicals. Polymerized olefins can be used as plastics and can fix CO_2 from the atmosphere when in use. We examined the effect of Fe-loading and reaction temperature to improve selectivity for C_{3+} olefins and propylene and found that lower loading of Fe (1 wt%) and higher reaction temperature (450°C) results in higher selectivity of propylene and catalytic stability [23]. Moreover, to improve catalytic performance, P-addition to Fe/H-ZSM-5 was tried, since it has been reported that phosphorous modification of H-ZSM-5 zeolite can lead to better hydrothermal stability and resistance to coke deposition [11, 24]. We found that P- and Fe-modification to H-ZSM-5 leads to the improved propylene selectivity and stability for water in ethanol in comparison with H-ZSM-5 without modification and Fe-modified H-ZSM-5 catalysts [25].

In this paper, to improve the catalytic activity and stability, the effect of Zr-modification was investigated.

2. Experimental

2.1. Catalyst Preparation. As zeolite support, H-ZSM-5 (Si/Al_2 = 30, Zeolyst) was used. The zeolite was calcined at 500°C in a muffle furnace for 6 h, prior to use as support. H-ZSM-5 was used as not only support but also catalyst: when used as catalyst, additional calcination at 700°C in a muffle furnace for 3 h was carried out after 500°C-calcination. Two kinds of H-ZSM-5 zeolite were used as catalysts: (1) H-ZSM-5 calcined at 500°C (denoted H-ZSM-5 (500)) and (2) H-ZSM-5 calcined at 700°C after 500°C-calcination (H-ZSM-5 (700)). Zr-modified H-ZSM-5 catalysts were prepared by impregnation method. $ZrO(NO_3)_2{\cdot}2H_2O$ (Wako Pure Chemicals) was used as Zr source, and loading of Zr was varied from 1 to 10 wt%. To make a comparison, Fe-modified H-ZSM-5 catalyst was also prepared. $FeCl_3{\cdot}6H_2O$ (Iwai Chemical) was used as Fe source, and loading of Fe was 1 wt%, referring to our previous work [25]. After impregnation, the wet Fe- or Zr-loaded catalysts were dried at 120°C, followed by calcination at 700°C in air-flow for 3 h.

2.2. Catalytic Activity. The catalytic activity was measured in a fixed-bed reactor with atmospheric pressure. The weight of catalyst was 0.15 g. Reaction gas was obtained by mixing N_2 and vaporized EtOH, with the flow rate of N_2 set at 60 $cm^3\,min^{-1}$, resulting in the concentration of N_2 and ethanol being 61.4 and 38.6 vol.%, respectively: WHSV was 28.7 h^{-1}. To distinguish the difference in catalytic activity (i.e., ethanol conversion), WHSV was set higher than our previous work [21–23, 25]: WHSV was 7.45 h^{-1}, and ethanol conversion was always ca. 100%. As ethanol, neat ethanol and bioethanol were used, bioethanol was obtained by mixing of neat ethanol and commercial vodka (alcohol = 50%) in ratio of 1 : 3, giving a concentration of ethanol of approximately 87.5%. Reaction temperature was 450°C, and the change

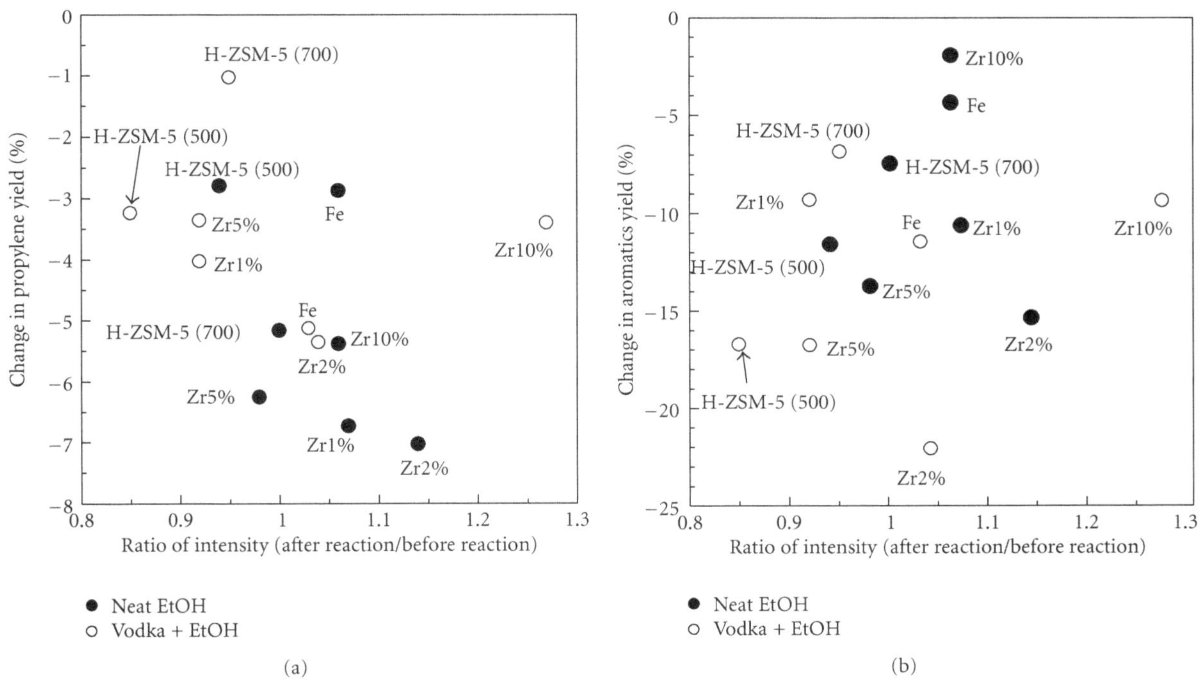

Figure 3: Plots of the ratio of XRD peak intensity at 23.1° (after reaction/before reaction) versus the change in product yield over H-ZSM-5 and (Fe or Zr)/H-ZSM-5 catalysts. (a) Propylene yield and (b) aromatics yield.

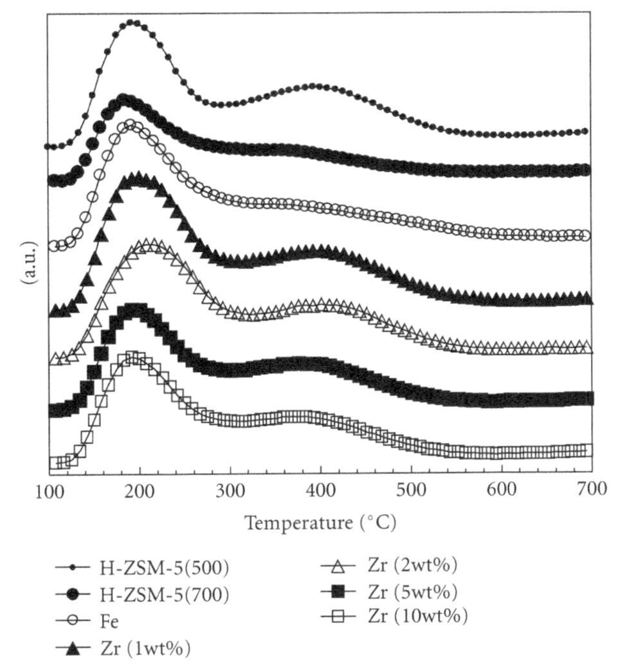

Figure 4: NH_3-TPD profiles of H-ZSM-5 and (Fe or Zr)/H-ZSM-5 catalysts.

in catalytic performance with time-on-stream was observed. Analysis of vodka by gas chromatography (GC) confirmed the chief organic compound in vodka to be ethanol, with only traces of other organic compounds present.

The effluent gas was analyzed by gas chromatography (GC) equipped with Molecular Sieve 5A and Porapak Q columns (Shinwa Chemical Industries). Two detectors were used: thermal conductivity detector (TCD) was used to detect light products (hydrogen, carbon monoxide, and carbon dioxide) and flame ionization detector (FID) to detect hydrocarbons, alcohols, aldehydes, and ethers.

The selectivity for each product was defined as (number of C atoms in the product)/(total number of C atoms in all gaseous products) × 100 (%).

2.3. Characterization. As cause of catalytic deactivation, carbon deposition can be considered, since in the case of sole H-ZSM-5 zeolite used as catalyst, carbon deposition causes catalytic deactivation [4, 5]. Therefore, to estimate carbon deposition on catalysts, thermogravimetric (TG) analyses of catalysts after 7 h reaction were carried out (Mac Science, TG-DTA2000) in flowing air with temperature varied from room temperature to 800°C at 10°C min^{-1}.

Framework collapse of zeolite support, due to dealumination of zeolite framework during reaction, may be considered as another cause of catalytic deactivation [5, 10–17, 22, 23, 25], since solid acidity of zeolite, essential for this reaction, is generated by Al atoms incorporated into zeolite framework. Therefore, X-ray diffraction (XRD) measurements were carried out (Mac Science, M18XHF22-SRA), and intensity of XRD peak of zeolite framework before and after reaction was compared to investigate the framework collapse of zeolite during reaction. Here, intensity of XRD peak at 23.1° was used as index of crystallinity. Fe species were not detected for Fe/H-ZSM-5. Zr species were

Table 1: Catalytic activity of H-ZSM-5 and (Fe or Zr)/H-ZSM-5 zeolite catalysts.

Metal	EtOH	EtOH conv./%	C-selectivity/% Ethylene	C_{3+} olefins	(Propylene)	Paraffins	Aromatics
No (500)	Neat EtOH	86.4	19.1	21.3	8.6	25.3	33.8
	Vodka + EtOH	99.2	17.8	21.8	11.4	14.5	45.6
No (700)	Neat EtOH	84.3	48.9	22.6	8.3	13.7	12.0
	Vodka + EtOH	61.5	45.0	18.9	7.3	11.6	18.6
Fe	Neat EtOH	72.0	48.3	22.9	9.2	13.7	13.1
	Vodka + EtOH	86.3	49.7	21.2	8.2	12.2	15.5
Zr (1 wt%)	Neat EtOH	86.4	26.7	28.8	14.2	19.0	24.6
	Vodka + EtOH	99.2	22.0	29.8	14.2	18.3	29.6
Zr (2 wt%)	Neat EtOH	93.7	29.0	27.3	13.8	18.2	24.5
	Vodka + EtOH	94.8	24.1	26.5	12.8	17.3	31.4
Zr (5 wt%)	Neat EtOH	77.5	31.0	26.2	12.5	17.4	23.4
	Vodka + EtOH	96.7	22.4	30.7	14.3	18.8	26.2
Zr (10 wt%)	Neat EtOH	91.9	25.3	34.8	15.2	21.8	16.7
	Vodka + EtOH	69.0	45.6	20.2	9.0	13.7	18.5

Numbers of 500 and 700 mean calcination temperature (°C).

not detected for Zr/H-ZSM-5 (Zr = 1–5 wt%) but detected for Zr/H-ZSM-5 (Zr = 10 wt%) at about 30°.

NH_3-temperature-programmed-desorption (TPD) ex-peri-ments of H-ZSM-5 and (Zr or Fe)/H-ZSM-5 catalysts were carried out (Nihon Bell, BEL-CAT) to measure the solid acidity. After pretreatment in flowing He at 500°C for 60 min, followed by adsorption of NH_3 at 100°C for 90 min and desorption of weak adsorbed NH_3 in He flow at 100°C for 30 min, TPD profiles were obtained under conditions of He flow with the temperature varying from 100 to 800°C at $10°C\,min^{-1}$.

3. Results and Discussion

3.1. Catalytic Performance. Initial conversion of ethanol and initial selectivity of each product among gaseous products are shown in Table 1: main gaseous products were ethylene, C_{3+} olefins (C_3–C_7 olefins were detected), paraffins, and aromatics (benzene, toluene, xylenes, mesitylene, and pseudocumene). Zr-addition to H-ZSM-5 improved ethanol conversion in comparison with H-ZSM-5 (700), but effect of Fe-addition on ethanol conversion was not clear. Presence of water enhanced ethanol conversion for H-ZSM-5 (500), Fe/H-ZSM-5, and Zr/H-ZSM-5 (Zr = 1–5 wt%) but reduced conversion for H-ZSM-5 (700) and Zr/H-ZSM-5 (Zr = 10 wt%), suggesting that adequate amount of Zr enhances ethanol conversion but excess amount of Zr reduces ethanol conversion by presence of water. In comparison with H-ZSM-5 (700) and Fe/H-ZSM-5, Zr-modification improved selectivity of C_{3+} olefins, propylene, paraffins, and aromatics but reduced ethylene selectivity. In general, catalysts exhibiting high ethylene selectivity showed low selectivity of C_{3+} olefins and propylene, and catalysts exhibiting high selectivity of paraffins and aromatics showed high selectivity of C_{3+} olefins and propylene. In most cases, presence of water contained in ethanol enhanced the aromatics selectivity and reduced the selectivity of ethylene, C_{3+} olefins, propylene, and paraffins.

Ethanol conversion, selectivity, and yield of each product changed with time-on-stream and propylene yield decreased. Yield (%) was defined as (ethanol conversion (%)) × (selectivity of each product (%))/100. Figure 1 shows the plot of the initial yield of propylene versus change in propylene yield during 7 h reaction. In general, catalysts showing the higher initial yield of propylene showed the larger decrease in propylene yield. For neat ethanol, Zr-addition led to the higher initial yield of propylene but led to the larger decrease in propylene yield, while, for aqueous ethanol, 1–5 wt% of Zr-addition improved the initial yield of propylene with inhibiting the decrease in propylene yield, but 10 wt% reduced the initial yield of propylene and the decrease in yield. On the other hand, Fe-addition was not so effective for the increase in propylene yield and the suppression of decrease in propylene yield.

3.2. Carbon Deposition. Figure 2 shows plots of weight loss during TG analyses due to combustion of deposited carbon versus change in propylene yield during 7 h reaction. In general, catalysts exhibiting the larger amount of deposited carbon showed the steeper decrease in yield, suggesting that carbon deposition during reaction may be a cause of catalytic deactivation. Moreover, for neat ethanol, Zr-modification to H-ZSM-5 enhanced carbon deposition in comparison with H-ZSM-5 (700), while, for aqueous ethanol, Zr-modification did not affect the amount of deposited carbon so significantly. These results suggest that water contained in ethanol can reduce the amount of deposited carbon, and, as a result, the decrease in propylene yield can be inhibited in some degree. On the other hand, Fe-addition did not lead to increase in carbon deposition, and the effect to change in propylene yield was not clear.

It has been reported that addition of P [11] or alkali earth metal [13] can inhibit carbon deposition. However, in the case of Zr-modification, different results from these cases were obtained.

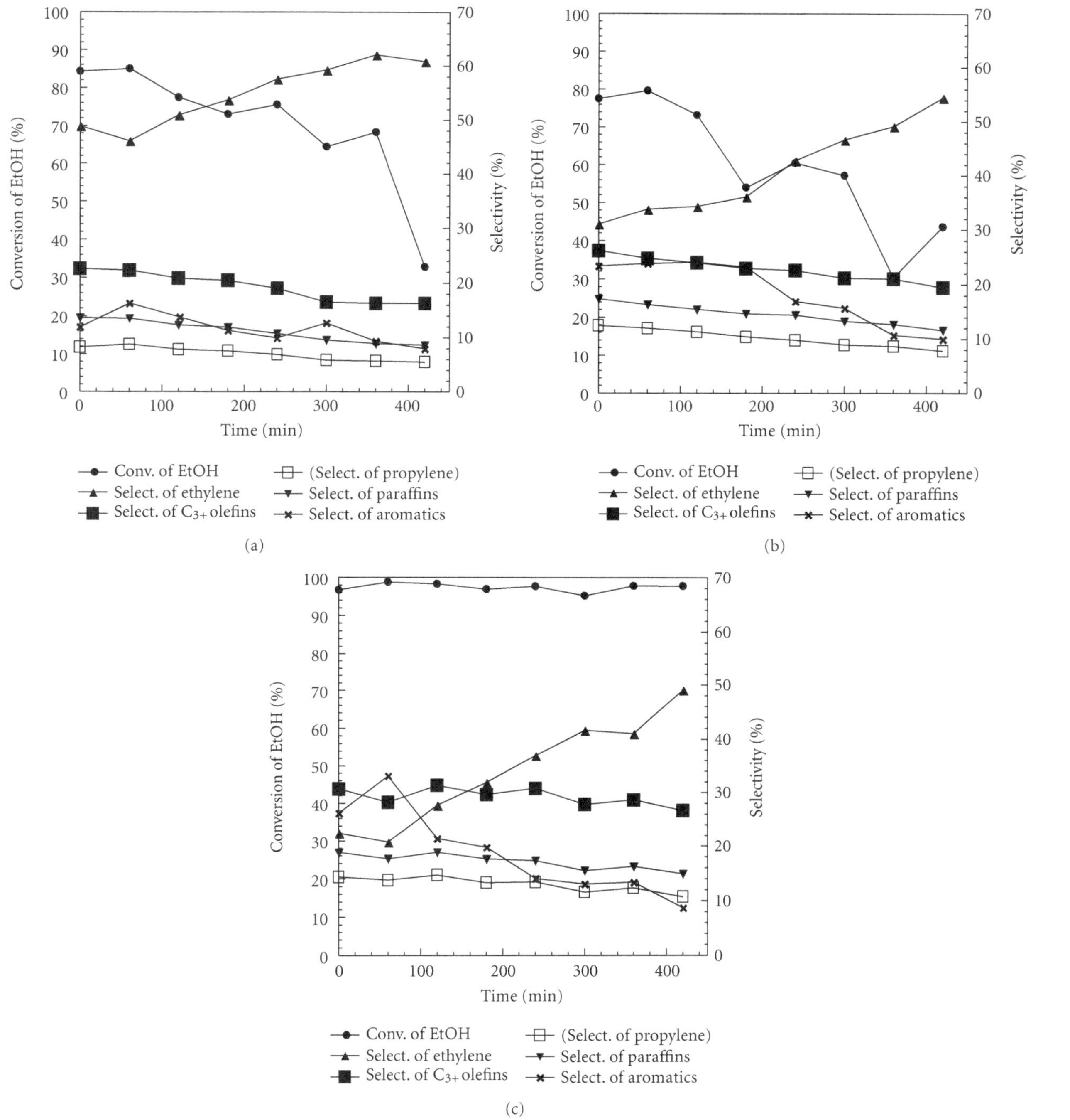

FIGURE 5: Catalytic performance as a function of time. (a) H-ZSM-5 (700) using neat EtOH, (b) Zr/H-ZSM-5 (Zr = 5 wt%) using neat EtOH, and (c) Zr/H-ZSM-5 (Zr = 5 wt%) using mixture of vodka and EtOH.

3.3. Framework Collapse of Zeolite. Figure 3 shows plots of the ratio of XRD peak intensity (after reaction/before reaction) versus change in product yield during 7 h reaction: (a) propylene yield and (b) aromatics yield. As shown in Figure 3(a), except for Zr/H-ZSM-5 (Zr = 10 wt%), water in ethanol enhanced framework collapse. In most cases, Zr-modified catalysts showed higher ratio of peak intensity (after reaction/before reaction) than sole H-ZSM-5, suggesting that Zr-modification can inhibit framework collapse of zeolite support during reaction. These results are consistent with Fujitani's report [12]. There seems to be a weak inverse correlation between the ratio of intensity (after reaction/before reaction) and change in propylene yield: catalysts showing significant framework collapse could inhibit the decrease in propylene yield.

Figure 3(b) shows no clear correlation between the ratio of XRD peak intensity and change in aromatics yield. However, for each catalyst, the decrease in aromatics was larger for aqueous ethanol than for neat ethanol. These results suggest that the decrease in acidic site by framework collapse of zeolite reduces aromatics formation and leads to decrease in propylene yield. Catalysts showing large decrease in aromatics yields could inhibit the decrease in propylene yield relatively.

XRD peak of deposited carbon was not observed. Similar results were obtained by Ouyang's research: bioethanol was converted to ethylene over La-modified H-ZSM-5 catalysts [26]. In the case of addition of P [11] or alkali earth metal [13], also, the framework collapse due to dealumination could be suppressed.

3.4. Results of NH_3-TPD Measurements. Figure 4 shows the NH_3-TPD profiles of H-ZSM-5 and (Fe or Zr)/H-ZSM-5 catalysts. H-ZSM-5 (500) showed NH_3-desorption peak at ca. 400°C. The acidic site showing this desorption peak is essential for the aromatics formation. For 1–5 wt% of Zr-loading, strong desorption peak appeared, while, for 10 wt%, desorption peak was slightly weakened. H-ZSM-5 (700) showed weak desorption peak, while, in the Fe/H-ZSM-5, NH_3-desorption peak was not clear. These results suggest that calcination at 700°C after 500°C-calcination to H-ZSM-5 reduces the amount of acidic site of zeolite by framework collapse due to dealumination, and Fe-addition enhances the framework collapse. However, adequate degree of Zr-modification could inhibit the framework collapse.

3.5. Catalytic Performance as a Function of Time. Catalytic performance as a function of time was investigated. Figure 5 shows the catalytic performance: (a) H-ZSM-5 (700) using neat ethanol, (b) Zr/H-ZSM-5 (Zr = 5 wt%) using neat ethanol, and (c) Zr/H-ZSM-5 (Zr = 5 wt%) using aqueous ethanol.

In the case of H-ZSM-5 (700) using neat ethanol (Figure 5(a)), ethanol conversion decreased from 84.3 to 32.8%. Selectivity of C_{3+} olefins and propylene decreased (22.6 → 16.3% and 8.3 → 5.5%, resp.). Aromatics selectivity decreased from 12.0 to 8.0%, while ethylene selectivity increased from 48.9 to 60.9%.

In the case of Zr/H-ZSM-5 using neat ethanol (Figure 5(b)), ethanol conversion decreased from 77.5 to 43.7%. Selectivity of C_{3+} olefins and propylene decreased (26.2 → 19.5% and 12.5 → 7.8%, resp.). Degree of decrease in ethanol conversion was smaller than H-ZSM-5 (700), while initial conversion of ethanol was slightly lower than H-ZSM-5 (700). Zr-addition enhanced selectivity of C_{3+} olefins and propylene. On the other hand, aromatics selectivity decreased from 23.4 to 10.0%, while ethylene selectivity increased from 31.0 to 54.5%. By Zr-addition, initial selectivity of aromatics increased and that of ethylene decreased. However, change in selectivity with time-on-stream was larger than the case of H-ZSM-5 (700).

In the case of Zr/H-ZSM-5 using aqueous ethanol (Figure 5(c)), ethanol conversion did not change at near 100% (96.7 → 97.8%). Selectivity of C_{3+} olefins and propylene was higher than for neat ethanol, and the selectivity did not decrease so significantly (30.7 → 26.7% and 14.3 → 10.7%, resp.). On the other hand, change in selectivity of aromatics and ethylene (26.2 → 8.7% and 22.4 → 49.1%, resp.) was larger than for neat ethanol.

These results suggest that Zr-addition is effective for ethanol conversion to C_{3+} olefins and propylene, and the effect is more significant for aqueous ethanol than for neat ethanol.

4. Conclusions

In catalytic production of C_{3+} olefins and propylene from ethanol using modified H-ZSM-5 zeolite catalysts, Zr-modification could improve the initial yield of C_{3+} olefins and propylene. In general, catalysts exhibiting the higher initial yield of propylene showed the steeper decrease in propylene yield. However, Zr-addition could depress the decrease in propylene yield for aqueous ethanol. As cause of catalytic deactivation, carbon deposition and framework collapse of zeolite support can be considered. The addition of water to Zr-modified H-ZSM-5 catalyst could depress carbon deposition in some degree, and, as a result, the decrease in propylene yield could be depressed. Though effect of Zr-addition on catalytic activity was reported by Fujitani et al. [12], the effect of presence of water in ethanol was not reported. Therefore, the results in this study show that Zr-modified H-ZSM-5 catalysts are promising candidate for conversion of aqueous ethanol to propylene.

Acknowledgment

This work was partly supported by a New Energy and Industrial Technology Development Organization (NEDO) Grant.

References

[1] S. N. Chaudhuri, C. Halik, and J. A. Lercher, "Reactions of ethanol over HZSM-5," *Journal of Molecular Catalysis*, vol. 62, no. 3, pp. 289–295, 1990.

[2] C. W. Ingram and R. J. Lancashire, "On the formation of C_3 hydrocarbons during the conversion of ethanol using H-ZSM-5 catalyst," *Catalysis Letters*, vol. 31, no. 4, pp. 395–403, 1995.

[3] A. G. Gayubo, A. M. Tarrío, A. T. Aguayo, M. Olazar, and J. Bilbao, "Kinetic modelling of the transformation of aqueous ethanol into hydrocarbons on a HZSM-5 zeolite," *Industrial & Engineering Chemistry Research*, vol. 40, no. 16, pp. 3467–3474, 2001.

[4] A. T. Aguayo, A. G. Gayubo, A. Atutxa, M. Olazar, and J. Bilbao, "Catalyst deactivation by coke in the transformation of aqueous ethanol into hydrocarbons. Kinetic modeling and acidity deterioration of the catalyst," *Industrial & Engineering Chemistry Research*, vol. 41, no. 17, pp. 4216–4224, 2002.

[5] A. T. Aguayo, A. G. Gayubo, A. M. Tarrío, A. Atutxa, and J. Bilbao, "Study of operating variables in the transformation of aqueous ethanol into hydrocarbons on an HZSM-5 zeolite," *Journal of Chemical Technology and Biotechnology*, vol. 77, no. 2, pp. 211–216, 2002.

[6] S. K. Saha and S. Sivasanker, "Influence of Zn- and Ga-doping on the conversion of ethanol to hydrocarbons over ZSM-5," *Catalysis Letters*, vol. 15, no. 4, pp. 413–418, 1992.

[7] N. R. C. Fernandes Machado, V. Calsavara, N. G. C. Astrath, A. M. Neto, and M. L. Baesso, "Hydrocarbons from ethanol using

[Fe,Al]ZSM-5 zeolites obtained by direct synthesis," *Applied Catalysis A: General*, vol. 311, no. 1-2, pp. 193–198, 2006.
[8] V. Calsavara, M. L. Baesso, and N. R. C. Fernandes-Machado, "Transformation of ethanol into hydrocarbons on ZSM-5 zeolites modified with iron in different ways," *Fuel*, vol. 87, no. 8-9, pp. 1628–1636, 2008.
[9] K. Murata, M. Inaba, and I. Takahara, "Effects of surface modification of H-ZSM-5 catalysts on direct transformation of ethanol into lower olefins," *Journal of the Japan Petroleum Institute*, vol. 51, no. 4, pp. 234–239, 2008.
[10] K. Inoue, M. Inaba, I. Takahara, and K. Murata, "Conversion of ethanol to propylene by H-ZSM-5 with Si/Al_2 ratio of 280," *Catalysis Letters*, vol. 136, no. 1-2, pp. 14–19, 2010.
[11] Z. Song, A. Takahashi, I. Nakamura, and T. Fujitani, "Phosphorus-modified ZSM-5 for conversion of ethanol to propylene," *Applied Catalysis A: General*, vol. 384, no. 1-2, pp. 201–205, 2010.
[12] Z. Song, A. Takahashi, N. Mimura, and T. Fujitani, "Production of propylene from ethanol over ZSM-5 zeolites," *Catalysis Letters*, vol. 131, no. 3-4, pp. 364–369, 2009.
[13] D. Goto, Y. Harada, Y. Furumoto et al., "Conversion of ethanol to propylene over HZSM-5 type zeolites containing alkaline earth metals," *Applied Catalysis A: General*, vol. 383, no. 1-2, pp. 89–95, 2010.
[14] A. G. Gayubo, A. Alonso, B. Valle, A. T. Aguayo, M. Olazar, and J. Bilbao, "Hydrothermal stability of HZSM-5 catalysts modified with Ni for the transformation of bioethanol into hydrocarbons," *Fuel*, vol. 89, no. 11, pp. 3365–3372, 2010.
[15] A. G. Gayubo, A. Alonso, B. Valle, A. T. Aguayo, M. Olazar, and J. Bilbao, "Kinetic modelling for the transformation of bioethanol into olefins on a hydrothermally stable Ni-HZSM-5 catalyst considering the deactivation by coke," *Chemical Engineering Journal*, vol. 167, no. 1, pp. 262–277, 2011.
[16] A. G. Gayubo, A. Alonso, B. Valle, A. T. Aguayo, and J. Bilbao, "Kinetic model for the transformation of bioethanol into olefins over a HZSM-5 zeolite treated with alkali," *Industrial & Engineering Chemistry Research*, vol. 49, no. 21, pp. 10836–10844, 2010.
[17] A. G. Gayubo, A. Alonso, B. Valle, A. T. Aguayo, and J. Bilbao, "Selective production of olefins from bioethanol on HZSM-5 zeolite catalysts treated with NaOH," *Applied Catalysis B: Environmental*, vol. 97, no. 1-2, pp. 299–306, 2010.
[18] P. de Werbier d'Antigneul, J. Chami, C. Berrier, M. Blanchard, and P. Canesson, "Selective transformation of ethanol into propene on cobalt catalysts," *Catalysis Letters*, vol. 1, no. 5, pp. 169–175, 1988.
[19] H. Oikawa, Y. Shibata, K. Inazu et al., "Highly selective conversion of ethene to propene over SAPO-34 as a solid acid catalyst," *Applied Catalysis A: General*, vol. 312, no. 1-2, pp. 181–185, 2006.
[20] K. Kasai, T. Haishi, and M. Iwamoto, "Highly selective conversion of ethanol into lower olefins over nickel ion-loaded mesoporous silica MCM-41," in *Europacat 8*, Turku, Finland, August 2007.
[21] M. Inaba, K. Murata, M. Saito, and I. Takahara, "Ethanol conversion to aromatic hydrocarbons over several zeolite catalysts," *Reaction Kinetics and Catalysis Letters*, vol. 88, no. 1, pp. 135–141, 2006.
[22] M. Inaba, K. Murata, M. Saito, and I. Takahara, "Production of olefins from ethanol by Fe-supported zeolite catalysts," *Green Chemistry*, vol. 9, no. 6, pp. 638–646, 2007.
[23] M. Inaba, K. Murata, and I. Takahara, "Effect of Fe-loading and reaction temperature on the production of olefins from ethanol by Fe/H-ZSM-5 zeolite catalysts," *Reaction Kinetics and Catalysis Letters*, vol. 97, no. 1, pp. 19–26, 2009.
[24] K. Ramesh, L. M. Hui, Y. F. Han, and A. Borgna, "Structure and reactivity of phosphorous modified H-ZSM-5 catalysts for ethanol dehydration," *Catalysis Communications*, vol. 10, no. 5, pp. 567–571, 2009.
[25] M. Inaba, K. Murata, I. Takahara, and K. I. Inoue, "Production of olefins from ethanol by Fe and/or P-modified H-ZSM-5 zeolite catalysts," *Journal of Chemical Technology and Biotechnology*, vol. 86, no. 1, pp. 95–104, 2011.
[26] J. Ouyang, F. Kong, G. Su, Y. Hu, and Q. Song, "Catalytic conversion of bio-ethanol to ethylene over La-modified HZSM-5 catalysts in a bioreactor," *Catalysis Letters*, vol. 132, no. 1-2, pp. 64–74, 2009.

Experimental and Theoretical Investigations of the Impact Localization of a Passive Smart Composite Plate Fabricated Using Piezoelectric Materials

M. M. S. Dezfouli,[1] Mohd Roshdi Hassan,[2] Mohd Hafidz Ruslan,[1] Sohif Mat,[1] and B. Bakhtyar[3]

[1] *Solar Energy Research Institute (SERI), National University of Malaysia (UKM), 43600 Bangi, Selangor, Malaysia*
[2] *Department of Mechanical and Manufacturing Engineering, University Putra Malaysia, 43400 Serdang, Selangor, Malaysia*
[3] *SEFB, COB, University Utara Malaysia (UUM), 6010 Sintok, Kedah, Malaysia*

Correspondence should be addressed to M. M. S. Dezfouli; salehi.solar@yahoo.com

Academic Editor: Mohd Sapuan Salit

Two passive smart composite plates are fabricated using one and two PZT patches that are cheaper than the PZT wafer. The composite plate is fabricated in low temperature through the hand lay-up method to avoid PZT patch decoupling and wire spoiling. The locus of the impact point is identified using the output voltage to identify the impact location using one sensor. The output voltages of the sensors are analyzed to identify the impact location using two sensors. The locations of the impacts are determined based on the crossing points of two circles and the origin of an intended Cartesian coordinate system that is concentric with one of the sensors. This study proposes the impact location identification of the passive smart composite using the low-cost PZT patch PIC155 instead of common embedded materials (wafer and element piezoelectric).

1. Introduction

Structural health monitoring systems have recently been developed for various applications. Numerous studies on the different aspects of smart structure systems have been conducted by physicists and engineers in the fields of aerospace, mechanics, and civil engineering [1]. The smart structure system comprises both the host structure, including a laminate composite beam, and the sensing elements (actuations), including piezoceramic patches [2]. The widespread development regarding these systems is caused by the importance and variety of applications of the system in different engineering sciences such as aircraft wing, submarine sterns, and vibration control [3].

One of the most important applications of the systems is the evaluation of the structures of the composite for impact damage and Acoustic Emission (AE), which is applied in damage detection [4, 5]. Numerous researchers have evaluated acoustic wave propagation and damage detection by embedding or bonding PZT (lead zirconate titanate or piezoelectric ceramic material) sensors/actuators in the composite structure. Studies in the field of damage detection on the use of the PZT sensors/actuators can be divided to active and piezocomposite categories [6].

Active systems use a sensor to detect damage and an actuator. The actuator receives an input voltage and actuates the structure of the composite plate. The resulting acoustic wave is propagated in the composite. The PZT sensor shows an output signal after receiving the AE wave. If the composite structure has damage, the output signal of the PZT signal is changed. The changes of the signal are evaluated and analyzed by comparing the signal waves before and after the damage [7, 8]. Passive systems only use the PZT sensor because the AE created by the impact is propagated through the structure of the composite [9]. The damage detection is analyzed after the PZT sensor receives the AE wave [9]. Guan and Yang monitored the impact damages in the composite laminate to evaluate the output acoustic wave [10]. These researchers proved that the AE wave caused by impact includes important information that describes the composite structure based on the availability or unavailability of the damage. Evaluating the impact behavior of the composite laminate to detect the structural properties of the composite is very complicated; thus, different analysis methods were employed by numerous researchers. Sung et al. analyzed [11] the output signal on the time-frequency domains through the wavelet transform (WT) method and by monitoring the impact on a composite laminate that includes a bonded PZT sensor. In this method, the characteristics of the AE waves made by impact were analyzed in the passive sending diagnosis (PSD) system on time-frequency domains. Red et al. evaluated the damage detection by the expressing WT as a signal process technique that includes Fourier transforms. Su et al. [12] evaluated the damage identification system, which includes a bonded PZT on a composite bead, based on the acoustic wave propagated by the impact damage.

Diamanti et al. successfully detected the impact damage on a carbon-fiber-reinforced plastic (CFRP) laminate using a PZT patch attached on the composite surface [13]. These researchers detected impact damage by investigating the propagation of the lamb wave and through finite element analysis. Ghoshal et al. evaluated the damage detection based on stress wave parameters by simulating the acoustic wave propagation on a quasi-isotropic glass-epoxy plate that includes a bonded PZT patch based [14]. The previous studies revealed that the PZT patch bonded on the composite plate and embedded piezoelectric materials [15–20], which were used in the previous studies [20–24], could replace other kinds of piezoelectric materials, such as the wafer, that are more expensive than the PZT patch. Therefore, this study fabricates a piezocomposite plate, including four fiberglass layers with one embedded PZT patch through its two intermediate layers through the cut-out method. This paper also investigates the output voltage from the PZT patch through the impact force on the piezocomposites with one PZT patch and two PZT patches.

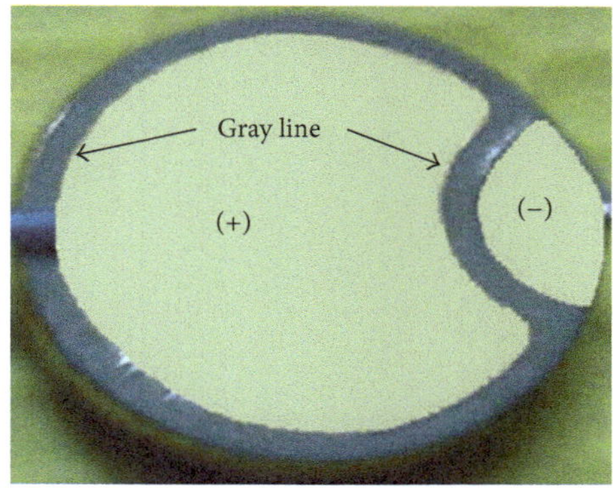

FIGURE 1: PZT patch sensor (PIC155).

TABLE 1: Physical and dielectric properties PIC 155.

Parameter	Abbreviation	PIC 155 (PZT)
Density	ρ (g/cm^3)	7.80
Curie temperature	T_c(°C)	345
Permittivity		
In the polarization direction	$\varepsilon_{33}{}^T$	1450
Perpendicular to the polarity	$\varepsilon_{11}{}^T$	1400
Dielectric loss factor	$\tan\delta$ (10^{-3})	20

TABLE 2: Mechanical properties of epoxy BBT 7892.

Mechanical properties	Amount
Tensile strength (psi)	9,800
Tensile modulus (psi)	420,000
Elongation at break (%)	4.9
Flexural strength (psi)	147,000
Flexural modulus (psi)	510,000
Compress strength (psi)	11,900

2. Material and Methods

Three methods are available for inserting the PZT patch through composite layers:

(1) smart layer;

(2) inserting without cut-cut;

(3) cut-out method.

Previous studies on embedding piezoelectric materials mainly focused on the application of the PZT wafer using the cut-out method and the application of PZT element using the smart layer method. The current study accordingly employs a PZT patch to embed the materials in the composite layers through the cut-out method by presenting a cost effective design. The cut-out method used in this study to fabricate smart plate and composite materials includes PZT patches, fiberglass rowing, and epoxy.

The piezoelectric materials are the PZT patch PIC155—PI Ceramic GmbH (from Germany) is a circular disc with a 1 mm thickness and 10 mm diameter as shown in Figure 1. A gray line shaped like an incomplete circle separates the positive pole (+) of sensor from the negative pole (−).

The interior of the gray circle is the positive pole. The exterior and the sheet behind the sensor are the negative pole. Table 1 shows the physical and dielectric properties of PIC 155.

The fiberglass is an E-glass fiber woven roving (04 XINGDA-DIY Trade-from China) with a 0.25 mm thickness for each of the layers. The epoxy resin is Epoxy BBT 7892 (Berjaya Bintang Timur Sdn, Bhd Malaysia). Epoxy BBT-7892 is a two-component liquid epoxy system specially designed for the wet lay-up laminating process in composite applications. This epoxy resin provides good wetting to most of the composite materials. The details of the properties of BBT-7892 are shown in Table 2.

The process of fabricating the composite plate with the embedded PZT patch has five steps: (i) preparing the mold and epoxy resin, (ii) cut-out fiberglass, (iii) soldering the PZT sensor to the wire, (iv) embedding the PZT sensor into

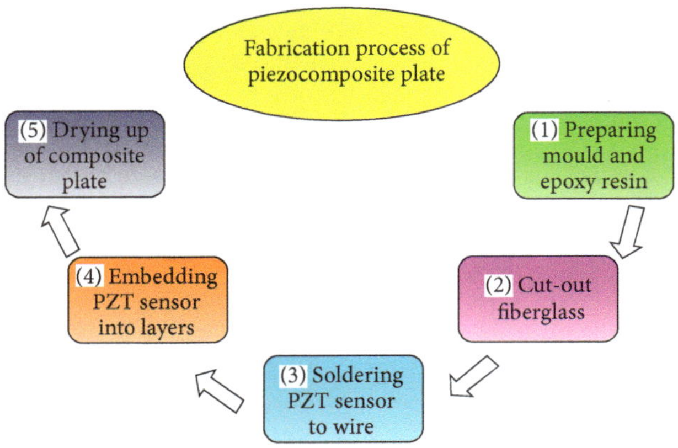

FIGURE 2: Flowchart of fabrication process of piezocomposite plate.

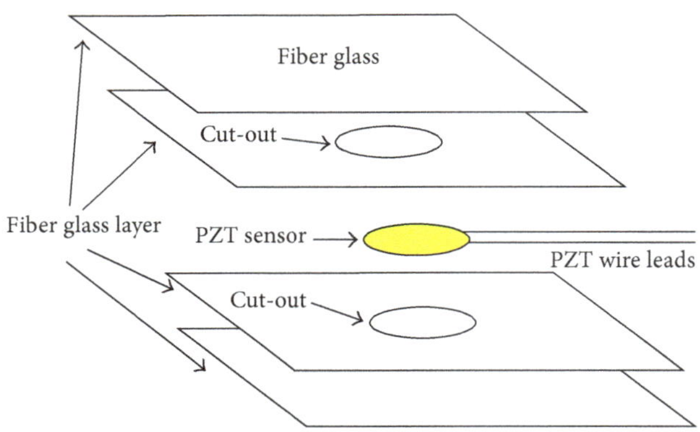

FIGURE 3: Schematic of the embedded piezoelectric into the composite by cut-out method.

layers, and (v) drying up the composite plate. Figure 2 shows the flowchart of the fabrication process of the piezocomposite plate.

Three piezocomposite plates with different specifications were fabricated in this study based on the fabrication process of the composite plate with embedded PZT.

2.1. Fabrication of Piezocomposite Plate with One PZT Patch. The first sample plate included one PZT patch and four layers. The fibers of the first and forth layers had an angle of 45° with the fibers of second and third layers (45°/90°/90°/45°). In this study, this sample will be called "Sample 1."

One symmetric point in the middle layers of the composite plate was purposely cut-out with the same size as the PZT sensors to embed the sensor in the composite plate (Figure 3).

The plastic sheet holder of the first composite layer was placed on the mold. Next, the second and third layers (cut-out layers) were placed on the first layer, where the fibers of second and third layers had an angle of 45° with the fibers of the first layer (Figure 4).

2.2. Fabrication of Piezocomposite Plate with Two PZT Patches. The second sample plate had two PZT patches and four fiberglass layers. The lay-up of layers was similar to that of Sample 1. In this research, this sample will be called "Sample 2". Two symmetric points with a distance of 12.5 cm in the middle layers of the composite plate were purposely cut-out in the same size as the PZT sensors to embed the sensors in the composite plate. The PZT patches were then completely inserted through the empty space between the intermediate layers during lay-up processing. Finally, the last layer was placed on the top. After placing each layer, the empty spaces and the bubbles between the layers were reduced using a pressing roller. The last layering step was the placement of the forth layer on the three layers. The fiber on the forth layer had an angle of 45° with the second and third layers. The lay-up of the composite plate had four layers, and the fibers of the first and the forth layers had an angle of 45° with the fibers of the second and third layers. Finally, the product was dried in environmental temperature (30°C) based on the sensitivity of the PZT sensor to the temperature. After dry-up step, a 14 cm × 19.5 cm composite plate was fabricated; thus, the distance of each sensor to the edge of the composite was 3 cm and the distance between the two sensors was 12.5 cm.

3. Experimental Setup

One of the aims of this study is to investigate the behavior of the output signal because of the impact test on the composite plate with the embedded PZT sensor. The process of the impact test included the production of force using an impact test machine on the composite plate to study the behavior of the composite and the embedded PZT sensor. Figure 5 shows the impact test process, which includes four steps. In the first step, impact was produced using the impact machine on the piezocomposite plate. In the second step, the stress caused by the impact as a mechanical wave was propagated in the plate until reaching the PZT sensor.

In the third step, the crystal structure of the PZT sensor was strained by the effect of the stress because of the impact and made an electrical flow. In the fourth step, the output electrical flow was monitored on the oscilloscope screen in the form of a signal wave (Figure 6).

If an impact happens on point A of the composite with the PZT Patch, stress is produced. This stress is propagated in form of a mechanical wave through the composite plate and reaches point B near the PZT patch. If the pressure on point A and B is equal to P_A and P_B, respectively, P_B is less than P_A ($P_B < P_A$). This relation is proven through an analysis of the deformation of an element in the one-dimensional stress in Section 3.2.

3.1. Impact Load. Figure 7 shows that the impact on the composite plate is produced by releasing rod from a certain height (h) on the composite plate. The approach velocity of the rod can be obtained using (1)

$$V^2 - V_0^2 = 2g(h - h_0), \tag{1}$$

where V is the velocity of the rod striking the composite plate, V_0 is the velocity of rod released at zero (m/s), h is the height

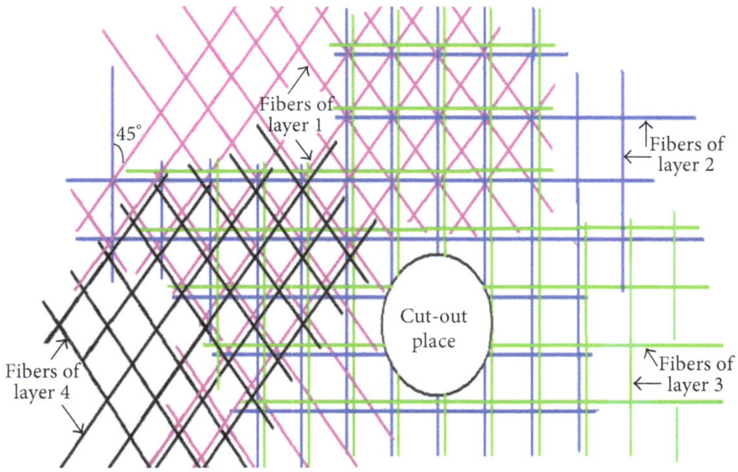

FIGURE 4: Schematic of the Sample 2 structure.

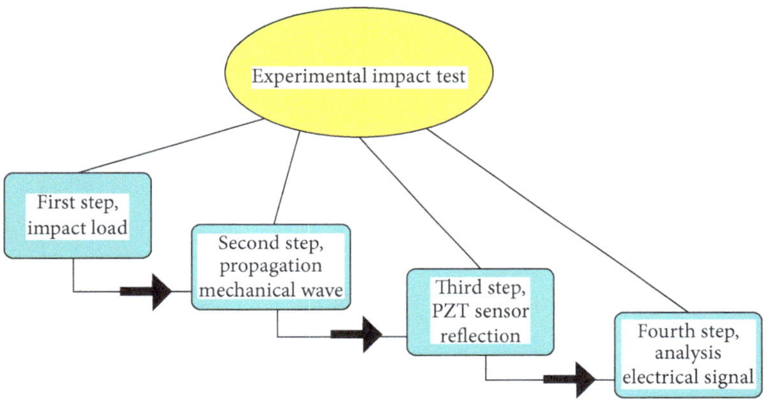

FIGURE 5: Schematic of experimental impact test process.

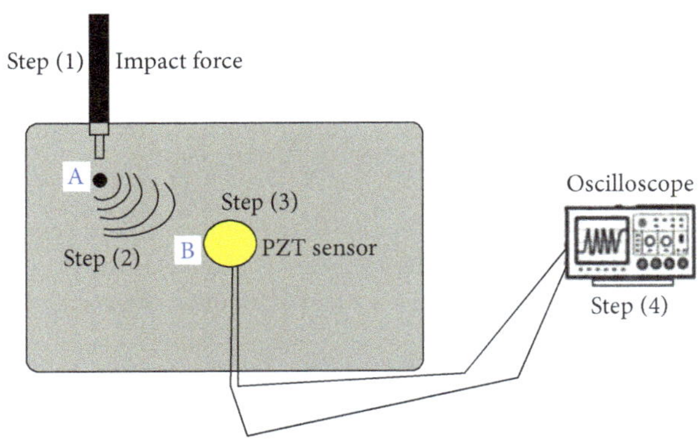

FIGURE 6: Propagation stress wave from impact point (A) to sensor (B).

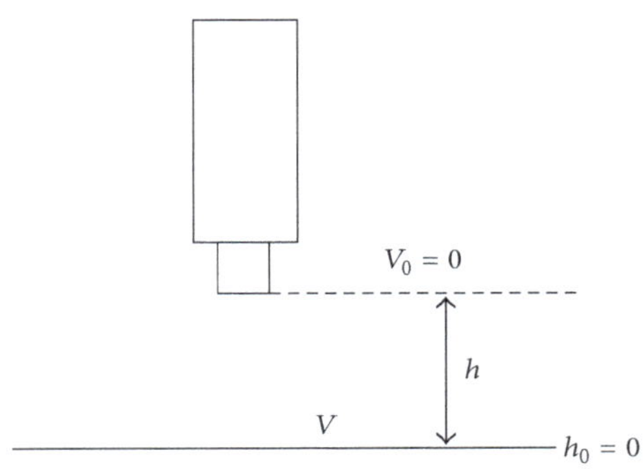

FIGURE 7: Release of the rod on the composite plate.

of rod from the composite plate, and h_0 is the height of the composite plate considered as zero in this study.

$h_0 = 0$ and $V_0 = 0$ are substituted into (1):

$$V = \sqrt{2gh}. \tag{2}$$

Newton's law of motion states that $F = ma$ (force equals mass times acceleration) and $a = V/t$ (acceleration equals velocity divided time). Therefore, Newton's laws of motion can be written as

$$F = m \cdot \frac{V}{t} \quad \text{or} \quad F \cdot t = m \cdot V. \tag{3}$$

Table 3: Force and pressure values at impact point.

h (m)	F (N)	P_0 (Pa)
0.01	0.110	5642.607
0.02	0.157	7979.852
0.03	0.192	9773.282
0.04	0.221	11285.215
0.05	0.248	12617.254
0.06	0.271	13821.509
0.07	0.293	14928.936

In this study, the time value of the impact test for each point is proposed to be 1 s (unit of time).

Therefore, the force equation is obtained by (4) by substituting the mass of rod (in this study mass of rod is $m = 0.25$ kg), time value ($t = 1$ s), and (2) into (3),

$$F = 0.25\sqrt{2gh}. \tag{4}$$

The pressure at the impact point on the composite plate (P_0) can be calculated using (5):

$$P_0 = \frac{F}{A} = \frac{0.25\sqrt{2gh}}{\pi r^2} = 127.3\sqrt{2gh}, \tag{5}$$

where F is force (N), A is the area of the cross-section of the rod (m^2), r is the radius of cross-section of the rod (m), and P_0 is the pressure at the impact point on the composite plate. In this study, the radius of the cross-section of the rod is $r = 0.025$ m.

Table 3 shows the force and pressure values at the impact point by changing the releasing height of the rod from the composite plate based on (4) and (5).

3.2. Propagation Mechanical Wave. The waves in the one-dimensional longitudinal strain are not the only type of wave motion for a purely longitudinal disturbance. The second type is provided by the wave motion in one-dimensional stress, where the longitudinal normal stresse, for example, σ, which is only a function of x and t, is the one nonvanishing stress component. All other stress components vanish [25].

The deformation of an element in the one-dimensional stress is sketched in Figure 8. The figure shows that the cross-section decreases if the element is in tension. $U(x,t)$ is the longitudinal displacement for the case of one-dimensional stress and is related by [25]

$$\sigma = E\varepsilon_x, \tag{6}$$

where E is Young's modulus. Writing the equation of motion for an element leads to

$$\frac{\partial\sigma}{\partial x} = \rho\frac{\partial^2 u}{\partial t^2}. \tag{7}$$

The substitution of (6) into (7) yields

$$\frac{\partial^2 u}{\partial x^2} = \frac{1}{c_b^2}\frac{\partial^2 u}{\partial t^2}, \tag{8}$$

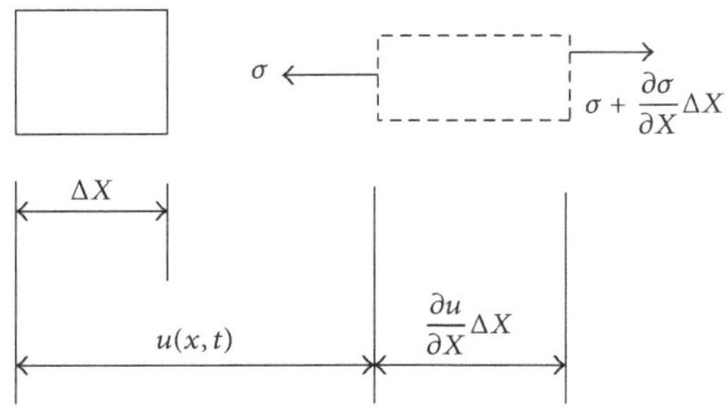

Figure 8: Deformation in one-dimensional stress [26].

where

$$c_b^2 = \frac{E}{\rho}. \tag{9}$$

The velocity c_b is usually called the bar velocity. In this study, Young's modulus of composite plate was achieved through the tensile test, $E = 3.2$ GPa, and the density of composite plate was $\rho = 1230\,\text{kg/m}^3$. Therefore, (8) reveals that the velocity of wave propagation in composite plate is $c_b = 1.62$.

If a semi-infinite thin rod ($x \geq 0$) is subjected at $x = 0$ to a pressure $p(t)$, the resulting stress wave is

$$\sigma = -P_0\left(t - \frac{x}{c_b}\right), \tag{10}$$

where σ is the produced stress using P_0 with a distance of (x) from the impact point, t is the time that the stress reaches the point with a distance of x from an impact point, and c_b is the propagation velocity of the wave in the composite plate. Equation (10) was rewritten as

$$\sigma = -P_0\left(\frac{c_b t - x}{c_b}\right) = -P\left(\frac{\Delta x}{c_b}\right). \tag{11}$$

The substitution of x_A (impact point) instead of x and $c_b t$ instead of x_B yields

$$P_B = \frac{\sigma \cdot c_b}{\Delta x}, \tag{12}$$

where P_B is the pressure at point B, which has a distance of Δx from impact point (A); σ is the stress because of the pressure P_A in the impact point; and Δx is the distance between points A and B.

The Equation of normal stress is [27]:

$$\sigma = \frac{F}{A}, \tag{13}$$

where F is impact force (N) and A is area of cross-section of the rod (m^2). Equation (13) is substituted into (12), which yields (10)

$$P_B = \frac{F \cdot c_b}{\Delta x \cdot A}. \tag{14}$$

In this study $A = 19.625 \times 10^{-6}$. C_b is 1.62. Equation (14) shows that the pressure at point B has an inverse relationship

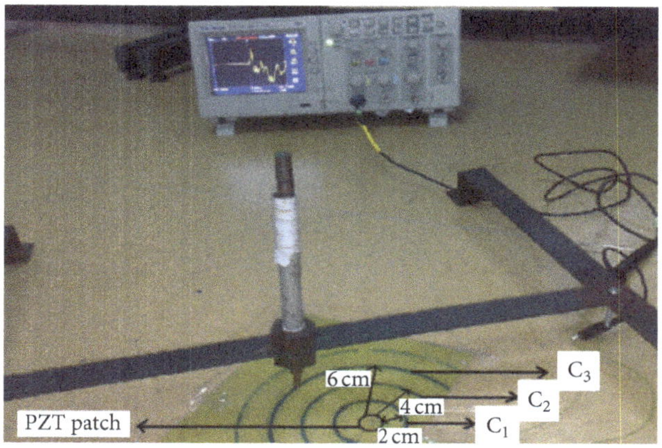

FIGURE 9: Experimental tests with different force values in different points on the composite plate.

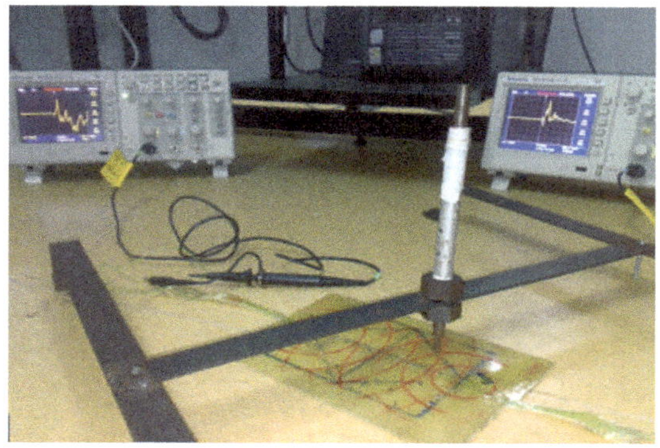

FIGURE 10: Experimental setup on the composite plate with two sensors.

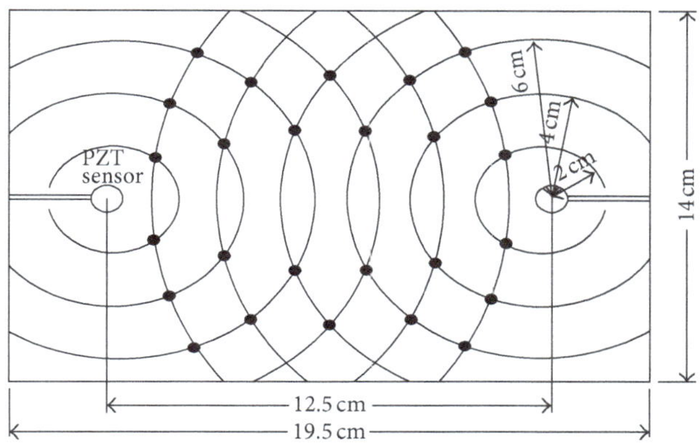

FIGURE 11: Schematic of impact points in the composite plate with two sensors.

with the distance of the impact point from the sensor, which means that the pressure value decreases with the increase of the distance of the impact point (Δx). Therefore, P_B is less than P_A.

3.3. PZT Reflection. The values of the pressure on the sensors caused by the impacts were calculated by substituting the output voltages in [28]:

$$P = \frac{16}{3} \frac{\varepsilon_{33}}{d_{31}} \frac{h}{D^2} \frac{1}{(1 - v)} V, \qquad (15)$$

where ε_{33} is the dielectric constant, d_{31} is the piezoelectric constant, D is the diameter of the PZT sensor, h is the thickness of the PZT sensor, P is the pressure, V is the output voltage, and v is the Poisson ratio. In this study, $\varepsilon_{33} = 2.124 \times 10^{-9}$ F/m, $d_{31} = -1.65 \times 10^{-10}$ (m/V), $h = 0.001$ (m), $D = 0.01$ (m), and $v = 0.3$.

3.4. Electrical Signals from Oscilloscope. The model of the oscilloscope is TEKTRONIX-TDS2012B with a 100 MHz Bandwidth, 1.0 GS/s sample rate, and two channels. The data were collected while the oscilloscope voltage selector was set on 10 mV/div. The sensitivity of TEKTRONIX-TDS2012B is 2 mV/div. The impact tests were done on the floor in fixed conditions.

4. Experimental Procedures

4.1. Composite Plate with One PZT Patch. Figure 9 shows the experimental procedure on the piezocomposite plate. The impact test on the three circles with radii of 2, 4, and 6 cm (C_1 with radii 2 cm, C_2 with radii 4 cm, and C_3 with 6 cm), respectively, with their center as the sensor. The relationship between the force and pressure on the sensor was evaluated. Seven forces were placed on the three points with 2, 4, and 6 cm distance from the sensor by changing the releasing height of the rod from the composite plate ($h_1 = 1, 2, \ldots, 7$ cm).

Each force on each point caused an output voltage from the sensor. Therefore, seven output voltages existed from each point.

4.2. Composite Plate with Two PZT Patches. Figure 10 shows the experimental setup on the composite plate with two sensors. The principal aim of the current study, which is to localize the impacts, was preceded by drawing the circles with radii of 2, 4, 6, 8, 10, and 12 cm (the positions of the centers of the sensors) on the composite plate that holds two embedded sensors.

The circles crossed each other in two points (Figure 11). The impact tests were performed on the crossing points. The sensors calibration process was carried out by detecting the impact voltages at the cross points of the circles. The output voltages of the sensors (V_1, V_2) were recorded using the oscilloscope to manage the possible error in the experiments for sixfold iterations. The acoustic wave is propagated on the surface of the composite as a result of the application of impact on the smart plate. When the stress wave reached the PZT sensor, the crystal structure of the sensor changed and showed an output signal. The PZT sensor had a symmetrical hexagonal crystal structure that changes with the strain of the stress wave (AE). The output is a free electron in form of a voltage. The oscilloscope displays the measured voltage in the

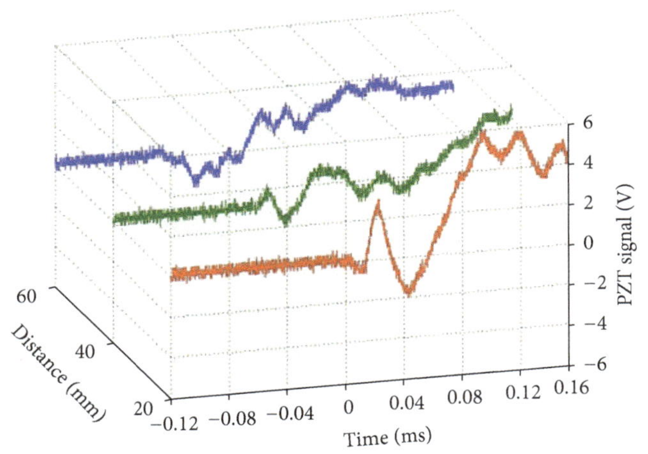

FIGURE 12: First packet of wave output voltage from PZT sensor.

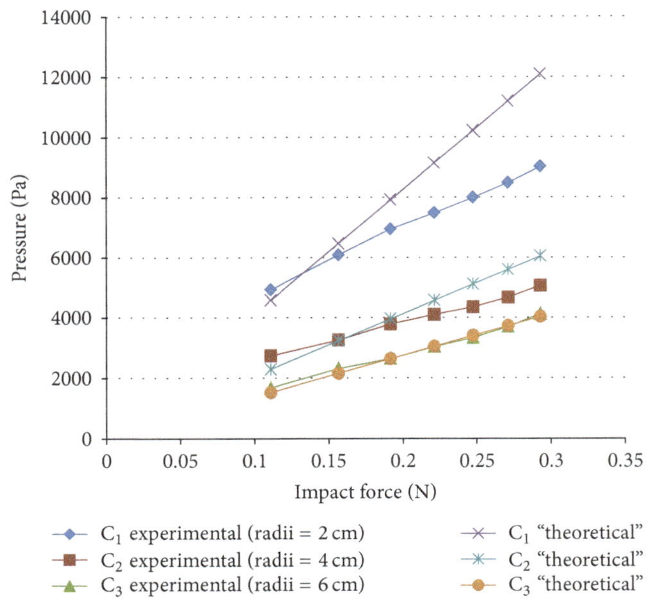

FIGURE 13: Experimental and theoretical results of the pressure on the sensor versus the impact force values.

shape of a sinusoidal waveform. The location of the impact load can be determined through the impact monitoring process. The acoustic wave caused by impact should be evaluated and determined in different parts of smart plate to reach the aim of the study.

5. Results and Discussion

The results of the impact test on the composite plates with embedded PZT patch are analyzed and evaluated. In the first step, the results of the impact on the two fabricated samples (sample 1 and sample 2) are analyzed. The analysis of the two samples aims to study the manner of the output voltage from the embedded PZT patch by the impact test. In the next step, the result of the impact on the composite plate with two embedded PZT patches (sample 3) is analyzed to find the impact location.

Figure 12 shows that the domain of the first packet of wave decreased by performing the impact test and distancing from the PZT sensor. The first packet of wave with a distance of 20 mm from the PZT sensor was the maximum. The first packet of the wave with a distance of 60 mm from the PZT sensor was minimum. Therefore, the domain of the first packet of wave (PK-PK) and the external voltage decreased with the increase of the distance of the impact point from PZT sensor. The cause of this state is demonstrated through the comparison of the theoretical and experimental results.

Figure 13 shows the impact forces versus the pressures on the PZT sensor (experimental results and theoretical results) in which the horizontal axis is the force value for impact and vertical axis is the pressure on the PZT sensor. C_1, C_2, and C_3 are the names of the three circles with radii of 2, 4, and 6 cm from the PZT sensor. On each line, 7 points (force values) were made to impact on each circle (2, 4, and 6 cm from the PZT sensor). The results of the experimental and theoretical impact on the 7 points on each circle (C) are shown as two lines (experimental and theoretical results). Three lines each for the experimental and theoretical results indicate the relationship between the pressure on the sensor and the forces on the points with 2, 4, and 6 cm distances from the sensor.

The theoretical results were attained through the mentioned equations and the force values in Table 2. These lines have an increasing trend, where the pressure on the sensor increases with the increase of the impact force value on different points of the composite plate.

Therefore, the pressure on the sensor has a direct relationship with the force caused by impact. The difference between lines C_1 experimental and C_2 experimental is larger than that between lines C_2 experimental and C_3 experimental. The impact point recedes the sensor (supposing that force is constant). The pressure on the sensor decreases, and the ratio of the reduction of the pressure decreases. Therefore, the pressure on the sensor has an indirect relationship with the distance of impact point from the sensor. A few points can be gleaned by comparing experimental and theoretical results, which are explained later.

The difference between lines C_1 theoretical and C_1 experimental increases with the increase of the impact force value. Increasing the impact force value with a distance of 2 cm from PZT sensor increases the difference of the experimental and theoretical results of pressure on the sensor. Increasing the impact force value also increases the difference between lines C_2 theoretical and C_2 experimental, which means that the difference of experimental and theoretical results of pressure on the sensor also increases by increasing the impact force value with a distance of 4 cm from PZT sensor. The lines of C_3 theoretical and C_3 experimental are almost close to one another, which means that the experimental and theoretical results of pressure on the sensor are almost the same by increasing the impact force value with a distance of 6 cm from PZT sensor. Therefore, the difference of the experimental and

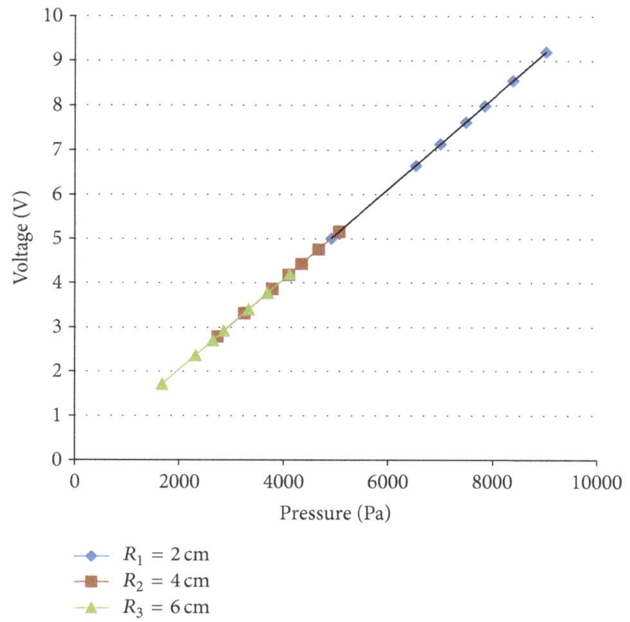

FIGURE 14: Output voltages from the PZT sensor versus pressure on sensor.

theoretical results decreases with the increase of the distance of the impact point from PZT sensor because the sensitivity of the PZT sensor will decrease.

The output voltages from the sensors versus the pressure on the sensor were evaluated using (4) because the pressures on the sensor caused output voltages from the sensor. In Figure 14, the horizontal axis is the pressure on the sensor and the vertical axis is the output voltage. The lines have an increasing trend, which means that the output voltage increases by increasing the pressure on the sensor. The calibration constant is 980.78, which means that $P = 980.78 \times V$.

Therefore, pressure has a direct relationship with voltage. Line R_2 is below line R_1, and line R_3 is below line R_2, which means that the pressure on the sensor and the output voltage decreases when the impact points recede the sensor. Therefore, the output voltage has an indirect relationship with the distance of the impact point from the sensor.

Figure 15 shows that the locus of the impact point can be identified through the output voltage if the impact occurs on random points of the composite to determine location of impact on the composite plate with a PZT Patch. The locus of the impact was a circle with R as the radius, the sensor as the center. R was obtained from (16):

$$R = 0.104V + 10.01, \quad (16)$$

where V is the output voltage from the PZT sensor.

5.1. Composite Plate with Two PZT Patches. The impact test on the composite plate with two sensors showed how the output voltage of the two sensors changed because of the change of the distance of the impact point from two sensors. Figure 16 shows the impact points on the composite plate.

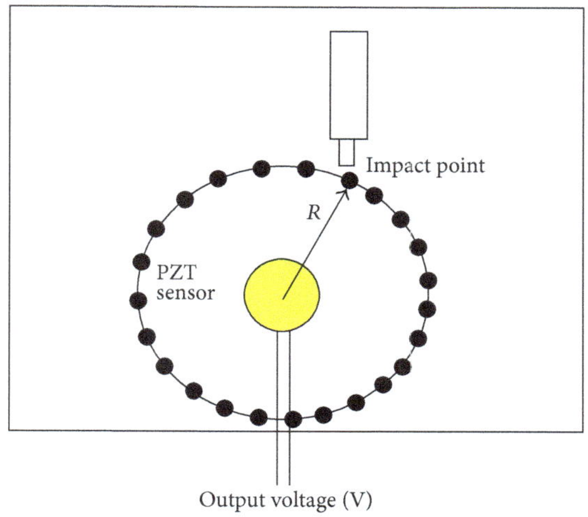

FIGURE 15: Locus of the impact point on the piezocomposite plate with a PZT sensor.

The behaviors of the output voltages of impact points 1 to 6 are similar to those of impact points $1'$ to $6'$. Therefore, the results of impact points 1 to 6 are similar to the results of impact points $1'$ to $6'$. Figures 17(a) and 17(b) show that the output voltage of sensor one (V_1) is the maximum and the output voltage of sensor two (V_2) is the minimum if the impact point is the crossing point of the two circles with radii of 2 cm from sensor one and 12 cm from sensor two. When the impact point distances from sensor one increase and sensor two decrease, V_1 decreases and V_2 increases, respectively. In these figures, the horizontal axis is the radius of the circles and the vertical axis is voltage. When the impact point distance to both sensors is equal, V_1 and V_2 are almost the same.

V_1 and V_2 are the minimum and maximum, respectively, if the impact point is the crossing point of the two circles with radii of 2 cm from sensor one and 12 cm from sensor two. The analysis of the results of the impact test on the crossing points of the circles showed that the output voltage of PZT sensor had an adverse relationship with the distance of the impact point from the sensors. An impact test using two sensors was conducted across the different points of the composite plate with the aim of impact localization through test data analysis. The impact test data include the output voltages of the two sensors in terms of the distances of the impact points from the sensors. In a composite plate with two sensors, the distances of the impact point to sensors 1 (R_1) and 2 (R_2) and the value of the output voltage from the sensors 1 (V_1) and 2 (V_2) are shown in Figure 18. An impact on a point on the page between the two sensors occurred. Two values of voltage from the two sensors were obtained. The results from the impact on different parts of composite plate showed that the voltage values from each sensor could determine the impact point distance to each sensor.

The Output voltage of the first sensor (V_1) indicates the location of the impact point in terms of the radial distance (R_1) from the first sensor. The output voltage of (V_2) indicates the location of the impact point in terms of the radial distance (R_2) from the second sensor. The impact locations

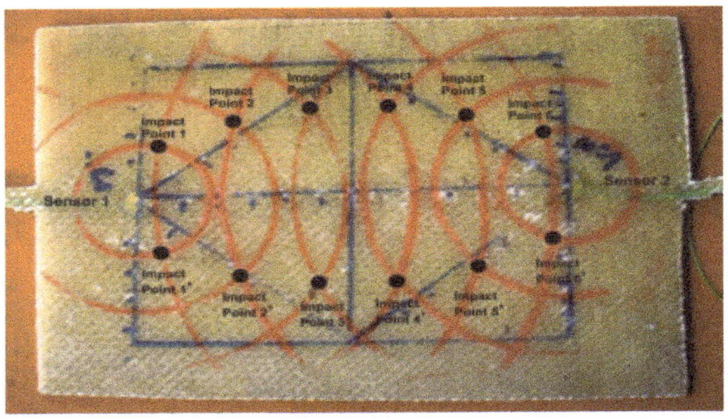

FIGURE 16: Impact points on the composite plate with two PZT sensors.

are obtained as two symmetric points generated because of the intersection of the circles. Therefore, the coordinates of the impact locus point had to be calculated to obtain the position of the impact locus point using two sensors. The origin of the coordinates was intended on the sensor 1; thus, the X-axis was matched on the center line of the two sensors and the Y-axis was perpendicular to the sensor. The distance between sensors 1 and 2 was 12.5 cm. The impact locus points were assumed to be $A(x, -y)$ and $A(x, y)$ to determine these two points.

The equation of the intersection of the two circles intersection was written as (17), where (x, y) is the coordinate of point A, R_1 is the radius to the center of sensor 1, and R_2 is the radius to the center of sensor 2,

$$\begin{aligned} x^2 + y^2 &= R_1^2, \\ (x - 12.5)^2 + y^2 &= R_2^2. \end{aligned} \tag{17}$$

Equation (17) is written as

$$\begin{aligned} y^2 &= R_1^2 - x^2, \\ x^2 + y^2 - 25x + 156.25 &= R_2^2. \end{aligned} \tag{18}$$

The calculation of (18) results to (19) and (20):

$$x = \frac{R_2^2 - R_1^2 - 156.25}{25}, \tag{19}$$

$$y = \pm\sqrt{R_1^2 - x^2}. \tag{20}$$

Therefore, (19) and (20) show that the values of radius R_1 and R_2 had to be determined to determine the values of x and y. The values of the output voltage from the two sensors (V_1 and V_2) were examined and analyzed through two methods.

Method 1. The distance values of the impact point from each sensor were considered using the voltage from the same sensor to determine R_1 and R_2. For instance, the relation between R_1 and V_1 was obtained by analyzing the distance values of the impact point to sensor 1 through the voltage resulting from sensor 1 (V_1),

$$R_1 = -3.807 \operatorname{Ln}(V_1) + 7.3467. \tag{21}$$

Equation (21) shows that the distance of the impact point to sensor 1 was obtained by acquiring the output voltage from sensor 1. The relation of R_2 with V_2 was also determined by analyzing the distance values of the impact point to sensor 2 through the voltage from sensor 2,

$$R_2 = -2.647 \operatorname{Ln}(V_2) + 5.9979. \tag{22}$$

Equation (22) shows that the distance of impact point to sensor 2 was obtained by acquiring the output voltage from sensor 2. Therefore, the distance value of the impact point can be determined from each sensor using the output voltage from each sensor through (21) and (22). R_1 and R_2 according to (21) and (22) are determined from (21) and (22) by obtaining V_1 and V_2 from the impact. The coordinates (x, y) of the impact point are determined by inserting the value of R_1 and R_2 in (19) and (20).

Method 2. The relationship between the distance of the impact point from each sensor with two variables of V_1 and V_2 was determined to obtain an impact location with a higher accuracy than the method 1. For example, the distance values of the impact point to sensor 1 (R_1) obtained by achieving the values from the two sensors (V_1, V_2) were analyzed. Figure 19 shows that the vertical axis is the distance of the impact point to sensor 1 (R_1), the horizontal axis is the output voltage from sensor 1 (V_1), and the perpendicular axis to the page is the output voltage from sensor 2 (V_2). Therefore, the inverse relationship of the voltage change with a change in the distance of the impact point from the sensor in this page holds true. For instance, a decrease in the distance of the impact point from sensor 1 (R_1) from point A to point B caused by the short distance of the impact point to sensor 1 leads to an increase in the output voltage from sensor 1 (V_1). The output voltage from sensor 2 also decreases because the impact point is farther from sensor 2 (R_2). Therefore, the relation of R_1 with the two variables of V_1 and V_1 was analyzed through MATLAB and was written as

$$R_1 = 7.404 - 0.88V_1 + 0.83V_2. \tag{23}$$

Therefore, (23) shows that R_1 can be calculated if V_1 and V_2 are obtained from the impact on each point of composite page. Equation (23) detected the distance of the impact point from sensor 1 (R_1); thus, V_1 and V_2 were the output voltages from sensors 1 and 2, respectively. In the next step, the distance values of the impact point to sensor 2 (R_2), acquired by obtaining the values from the two sensors (V_1, V_2), were analyzed. Figure 20 shows that the vertical axis is the distance of the impact point from sensor 2 (R_2), the horizontal axis is the output voltage from sensor 2 (V_2), and the perpendicular axis on a page is output voltage from sensor 2 (V_1).

The page was drawn using three variables (R_2, V_1, and V_2); thus, the indirect relation of the voltage change with distance change of the impact point from the sensor holds true. For example, the voltage of sensor 2 decreases if the distance of the impact point to sensor 2 from point A to point B increases in this page because the impact point moves farther from sensor 2. The voltage of sensor 1 (V_1) also increases because the impact point moves nearer to sensor 1. Therefore,

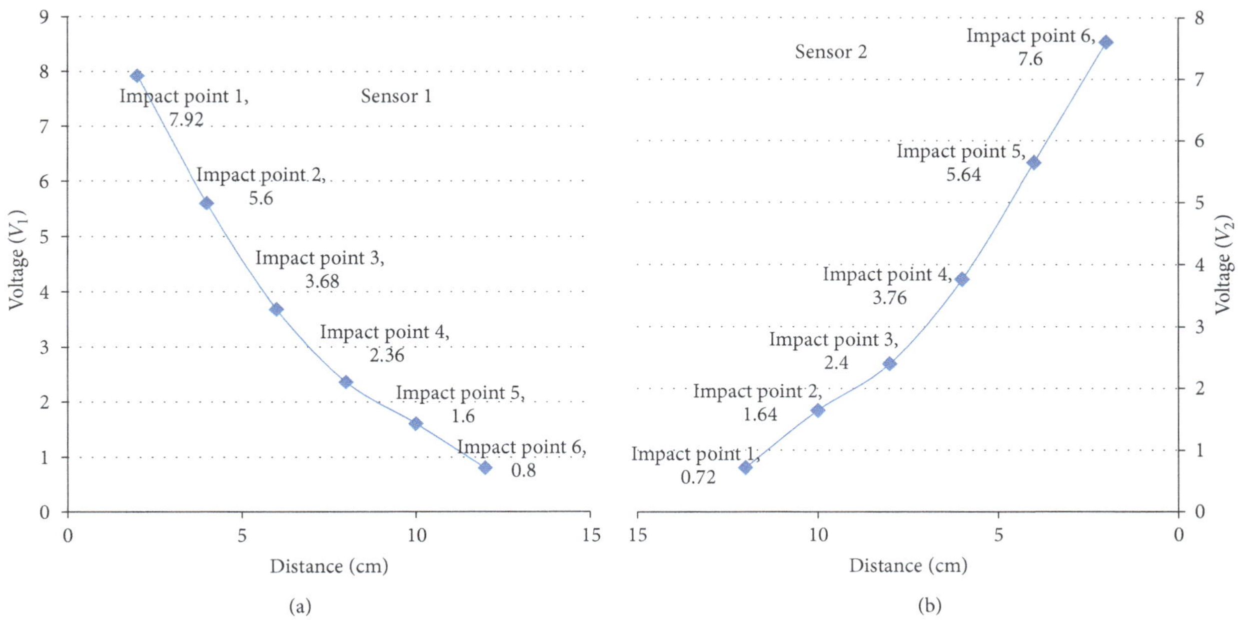

FIGURE 17: Behavior of output voltage from the two sensors.

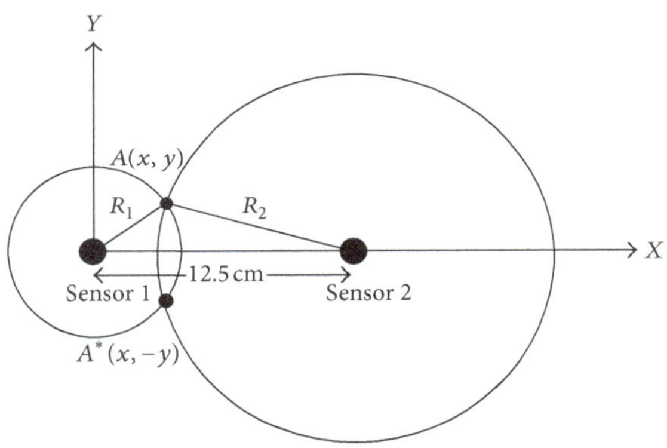

FIGURE 18: Intersection of two circles to detect impact location.

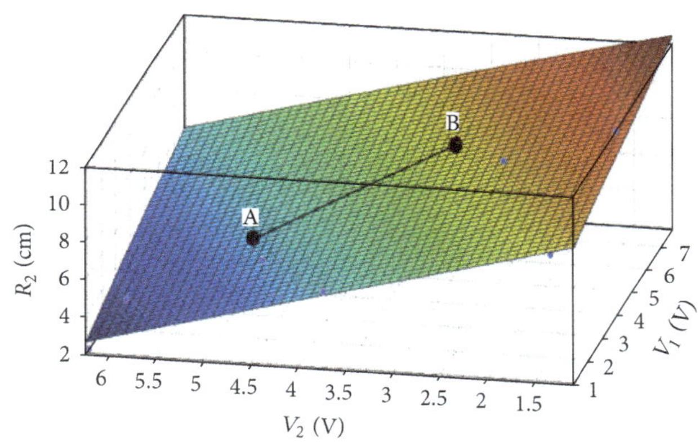

FIGURE 20: Distance of the impact point from sensor 2 (R_2) versus the two output voltages from the two sensors (V_1 and V_2).

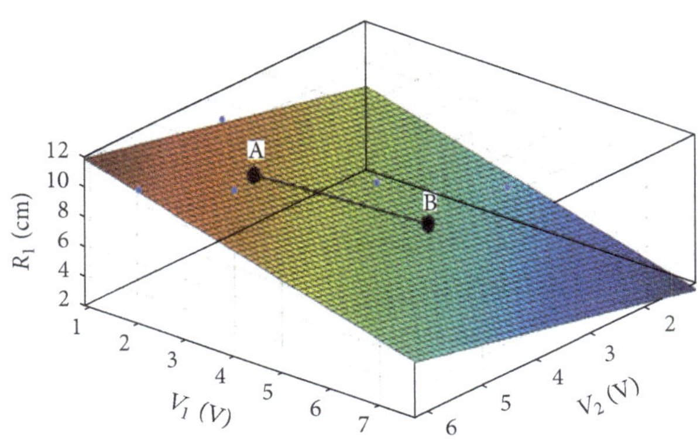

FIGURE 19: Distance of the impact point from sensor 1 (R_1) versus the two output voltages from the two sensors (V_1 and V_2).

the relationship of R_2 with the two variables of V_1 and V_2 was analyzed through MATLAB, which achieved (17):

$$R_2 = 10.2 + 0.476V_1 - 1.286V_2. \quad (24)$$

Therefore, the distance of the impact point from each sensor (R_1, R_2) is obtained by achieving the output voltage from the two sensors (V_1, V_2) and inserted (17) or (24) if an impact is inserted between the two sensors on a composite plate. The coordinates (x, y) or $(x, -y)$ of the impact point can also be achieved by inserting the values of R_1 and R_2 in (19) and (20).

Embedding PZT into the composite technique as a passive smart structure is suitable in damage detection. Passive approaches do not affect any actuators, and receivers are used to "sense and/or hear" any perturbations caused by possible hidden damage [6]. Numerous researchers have studied the passive smart structure using the bonded PZT patch on the composite for damage detection [1–5]. Embedding PZT in composite has some advantages, such as the protection of PZT from environmental effects, prevention of wire spoiling, and the increase of the PZT lifetime, compared with the bonding method. The result in this study shows that embedding the PZT patch method for passive smart structures is applicable to damage detection in a composite plate.

6. Conclusions

This paper presents experimental and theoretical investigations on impact localization on two piezocomposite plates, including one and two embedded PZT patches, through the cut-out method.

The signal from the embedded PZT patch was detected through the impact on the fabricated composite plate with an embedded one PZT patch. The value of external voltage was found to decrease by increasing the distance of the impact point from the PZT sensor. Therefore, the locus of the impact point can be identified using the output voltage if an impact occurs on the random points of the composite. The locus of the impact was a circle with R as the radius and the sensor as the center.

The current study attempts to determine the location of the impact points through the cut-out method and the two embedded PZT patches inside the four-layered composite plate. The fabricating process was conducted at a low temperature using the hand lay-up method to avoid wire spoiling and the decoupling of the PZT patch. The utilized fabrication process is a highly cost-effective manufacturing prototype model considering the cost advantages of using the PZT patch compared with the PZT wafer. The location of the impact points was determined by analyzing the output voltages of the PZT sensors, which were embedded among the composite layers. The applied impact to the page between the sensors resulted in the generation of voltage through the sensors. The output voltage of the sensors determined the distance of each of the imposed impacts from the sensors. The output voltage of the first sensor (V_1) indicated the location of the impact based on the radial distance (R_1) from the first sensor. The output voltage of (V_2) indicated the location of the impact point based on the radial distance (R_2) from the second sensor. The location of the impacts is obtained as two symmetric points and is generated as an intersection of the two circles. Therefore, determining the coordinates of the impact locations is necessary to obtain the locations of impacts. The coordinates of the two symmetrical impact points are defined with the origin of the intended coordinate system as the location of sensor 1.

Abbreviations

PZT: Lead zirconate titanate (Pb[Zr(x)Ti(1-x)]O3)
AE: Acoustic emission
WT: Wavelet transform
PSD: Passive sending diagnosis
CFRP: Carbon-fiber-reinforce plastic
P: Pressure (Pa)
F: Force (N).

Acknowledgment

The authors would like to thank the Mechanical Department of University Putra Malaysia for providing the laboratory facilities and technical support.

References

[1] J. Wang, "Modeling dynamic of composite plate with PZT patches embedded by using FEMLAB3," in *Proceedings of the COMSOL Multiphysics User's Conference 2005 Stockholm*, 2005.

[2] A. Zabihollah, R. Sedagahti, and R. Ganesan, "Active vibration suppression of smart laminated beams using layerwise theory and an optimal control strategy," *Smart Materials and Structures*, vol. 16, no. 6, pp. 2190–2201, 2007.

[3] S. M. Yang, "Design and fabrication of a smart layer module in composite laminated structures," *Smart Materials and Structures*, vol. 14, no. 2, pp. 315–320, 2005.

[4] S. G. Pierce, B. Culshaw, G. Manson, K. Worden, and W. J. Staszewski, "Application of ultrasonic Lamb wave techniques to the evaluation of advanced composite structures," in *Proceedings of the 7th Annual International Symposium on Smart Structures and Materials. International Society for Optics and Photonics (SPIE's '00)*, 2000.

[5] M. M. S. Dezfouli, Z. Hashim, M. H. Ruslan et al., "Experimental investigation of solar hybrid desiccant cooling system in hot and humid weather of malaysia," in *Proceedings of the 10th International Conference on Environment, Ecosystems and Development (WSEAS '12)*, 2012.

[6] M. Yocum, H. Abramovich, A. Grunwald, and S. Mall, "Fully reversed electromechanical fatigue behavior of composite laminate with embedded piezoelectric actuator/sensor," *Smart Materials and Structures*, vol. 12, no. 4, pp. 556–564, 2003.

[7] F. Mustapha, G. Manson, K. Worden, and S. G. Pierce, "Damage location in an isotropic plate using a vector of novelty indices," *Mechanical Systems and Signal Processing*, vol. 21, no. 4, pp. 1885–1906, 2007.

[8] M. M. S. Dezfouli, Z. Azran, M. H. Ruslan, K. Sopian, and B. Bakhtyar, "Experimental investigate of draying chilli by solar assisted heat pump dryer with multifunctional solar thermal collector," in *International Conference on Environment, Ecosystems and Development*, 2012.

[9] W. J. Staszewski, S. Mahzan, and R. Traynor, "Health monitoring of aerospace composite structures—active and passive approach," *Composites Science and Technology*, vol. 69, no. 11-12, pp. 1678–1685, 2009.

[10] Z. Guan and C. Yang, "Low-velocity impact and damage process of composite laminates," *Journal of Composite Materials*, vol. 36, no. 7, pp. 851–871, 2002.

[11] D. Sung, C. Kim, and C. Hong, "Monitoring of impact damages in composite laminates using wavelet transform," *Composites Part B*, vol. 33, no. 1, pp. 35–43, 2002.

[12] Z. Su, L. Ye, X. Bu, X. Wang, and Y. W. Mai, "Quantitative assessment of damage in a structural beam based on wave propagation by impact excitation," *Journal of Civil Structural*

Health Monitoring, vol. 2, no. 1, Article ID 145792, pp. 27–40, 2003.

[13] K. Diamanti, J. M. Hodgkinson, and C. Soutis, "Detection of low-velocity impact damage in composite plates using lamb waves," *Structural Health Monitoring*, vol. 3, no. 1, pp. 33–41, 2004.

[14] A. Ghoshal, W. N. Martin, M. J. Schulz, A. Chattopadhyay, W. H. Prosser, and H. S. Kim, "Health monitoring of composite plates using acoustic wave propagation, continuous sensors and wavelet analysis," *Journal of Reinforced Plastics and Composites*, vol. 26, no. 1, pp. 95–112, 2007.

[15] I. Kim, H. Lee, and J. Kim, "Impact damage detection in composite laminates using PVDF and PZT sensor signals," *Journal of Intelligent Material Systems and Structures*, vol. 16, no. 11-12, pp. 1007–1013, 2005.

[16] J. Han, C. Kim, and J. Kim, "The propagation of Lamb waves in a laminated composite plate with a variable stepped thickness," *Composite Structures*, vol. 76, no. 4, pp. 388–396, 2006.

[17] D. Lee, *Fundamentals of Shock Wave Propagation in Solids*, Springer, Berlin, Germany, 2008.

[18] C. Yoon, S. Lee, S. Lee, H. Kim, and K. Lee, "Transverse 1-3 piezoelectric ceramic/polymer composite with multi-layered PZT ceramic blocks," *Sensors and Actuators*, vol. 134, no. 2, pp. 480–485, 2007.

[19] J. Park, J. Kong, D. Kim, and D. Yoon, "Nondestructive damage detection and interfacial evaluation of single-fibre/epoxy composites using PZT, PVDF and P(VDF-TrFE) Copolymer sensors," *Composites Science and Technology*, vol. 65, no. 2, pp. 241–256, 2005.

[20] J. Barton, "Smart structures and materials Brian Culshaw," in *Optics and Lasers in Engineering*, vol. 27, pp. 337–338, Artech House, Boston, MA, USA, 1997.

[21] M. Arous, H. Hammami, M. Lagache, and A. Kallel, "Interfacial polarization in piezoelectric fibre-polymer composites," *Journal of Non-Crystalline Solids*, vol. 353, no. 47-51, pp. 4428–4431, 2007.

[22] D. Sung, C. Kim, and C. Hong, "Monitoring of impact damages in composite laminates using wavelet transform," *Composites Part B*, vol. 33, no. 1, pp. 35–43, 2002.

[23] S. W. Doebling, C. Farrar, and P. Cawley, "Damage identification and health monitoring of structural and mechanical systems from changes in their vibration characteristics: a literature review," Los Alamos National Laboratory Report LA-13070-M, 1996.

[24] Z. Su, L. Ye, X. Bu, X. Wang, and Y. Mai, "Quantitative assessment of damage in a structural beam based on wave propagation by impact excitation," *Structural Health Monitoring*, vol. 2, no. 1, pp. 27–40, 2003.

[25] J. D. Achenbach, *Wave Propagation in Elastic Solids*, vol. 16, North-Holland, Amsterdam, The Netherlands, 1973.

[26] F.-H. Chou and J. D. Achenbach, "Three-dimensional Vibrations of Orthotropic cylinders," *Journal of the Engineering Mechanics Divisione*, vol. 98, pp. 813–822, 1972.

[27] S. J. Kirk and A. J. Dell'Isola, *Life Cycle Costing For Design Professionals*, McGraw-Hill, New York, NY, USA, 1995.

[28] J. G. Webster, *The Measurement, Instrumentation, and Sensors Handbook*, Springer, 1999.

11

Interpenetrating Polymer Network Hydrogels Based on Gelatin and PVA by Biocompatible Approaches: Synthesis and Characterization

Eltjani-Eltahir Hago[1] and Xinsong Li[2]

[1] *Department of Chemistry and Chemical Engineering, Southeast University, Nanjing 210096, China*
[2] *School of Chemistry and Chemical Engineering, Southeast University, Nanjing 211189, China*

Correspondence should be addressed to Eltjani-Eltahir Hago; eltjanihago@gmail.com

Academic Editor: Delia Brauer

In this work, a new approach was introduced to prepare interpenetrating polymer network PVA/GE hydrogels by cross-linking of various concentration gelatin in the presence of transglutaminase enzyme by using the freezing-thawing cycles technique. The effects of freezing-thawing cycles on the properties of morphological characterization, gel fraction, swelling, mechanical, and MTT assay were investigated. The IPN PVA/GE hydrogels showed excellent physical and mechanical Properties. MTT assay data and the fibroblasts culture also showed excellent biocompatibility and good proliferation. This indicates that the IPN hydrogels are stable enough for various biomedical applications.

1. Introduction

Polyvinyl alcohol (PVA) hydrogels are a gel polymer known for the quality of biocompatibility that have been used in many biomedical applications, for example, as implants [1], artificial devices [2], contact lenses [3], drug delivery devices [4], and also wound dressings [5–7]. The following methods were used on a large scale: chemical cross-linking [8]; glutaraldehyde as the cross-linking agent [9]; cross-linking through γ radiation [10, 11], ultraviolet radiation [12], and by using cycles of freezing-thawing of successive [13, 14]. The last one is very convenient and biocompatible, while the molecular bonds (hydrogen bonds), which form through the freezing-thawing process of aqueous solutions of PVA, act as an efficient crosslinks [15].

The extent of poly(vinyl alcohol) (PVA) hydrogels for various biomedical applications has been realized because of their excellent biocompatibility and chemical stability. The freezing-thawing techniques used in the preparation of the PVA hydrogels showed good mechanical strength and have been examined for use as intervertebrate disk intended [16], artificial meniscus [17], and a contact lens [18, 19]. However, manufactured polymers, due to the lack of bioactive moieties, have no biological activity compared with natural polymers. The concept of integrating artificial materials with cell locations approved sites of naturally derived materials is very attractive. To achieve this result, efforts were made to incorporate cell adhesion of synthetic biomaterials [20].

Gelatin (GE) is a protein produced by partial denaturalization of collagen extracted from the bones, connective tissues, organs, and some intestines of animals such as domesticated cattle, porcine, and horses. Gelatin presents biological activities because of the natural origin, which makes it suitable for use of an ingredient of wound dressing, scaffolds for tissue engineering, and drug delivery carriers. Gelatin hydrogels are common for the applications in medical areas [21]. Hydrogels are insoluble hydrophilic polymers having a high-water content and tissue like mechanical properties that make them highly attractive scaffolds for implantation in empty tubular nerve prosthesis or for direct injection at the lesion site to enhance cell attachment and growth. Gelatin hydrogels have often crossed-linked by chemical approach using crosslinks, such as glutaraldehyde, to improve elasticity, consistency, and stability. However, the chemical cross-linker may lead to toxic effect on physiological environment, due to the presence of residual crosslinks. Therefore, it is highly demanded to develop biocompatible mild cross-linking methods to prepare hydrogels [20].

Transglutaminases are widely distributed in various organisms, including vertebrates, invertebrates, plants, and microorganism, and are reportedly responsible for certain biological events such as epidermal keratinization, blood coagulation, and regulation of erythrocyte membranes. A microbial transglutaminase isolated from the culture medium of *Streptoverticillium mobaraense* has become commercially available [22]. Unlike TGases from many sources, the mTG possesses many features, including Ca^{2+} independence, a broader substrate specificity for acyl donors, a smaller molecule size, and a higher reaction rate, which makes them suitable for industrial applications. Currently, this mTG has been successfully applied in the food industry for improving the physical properties and texture of protein-related foods [23]. More recently, the use of mTG to modify gelatin has also been reported [24, 25]. The enzymatic cross-linking is a biocompatible approach to enhance the mechanical properties of gelatin hydrogels. An interpenetrating polymer network (IPN) is a composite of polymers, exhibiting varied characteristics, which is obtained when one polymer network is synthesized or cross-linked independently in the instantaneous presence on the other [26]. The IPN is a combination of at least two polymer chains each in network form, which is synthesized and/or cross-linked in the immediate presence on the other without any covalent bonds between them [27] or the two or more networks can be envisioned to be entangled in such a way that they are concatenated and cannot be pulled apart but not bonded to each other by any chemical bond [28]. Many researchers have been focusing on the development of polymer blends with IPN structure. Recently, hydrogels with IPN structure have attracted much attention because of their potential application in the biomedical area. Kurokawa et al. invented double network hydrogels with superior toughness [29].

In this paper, a new approach was introduced to prepare interpenetrating polymer network PVA/GE hydrogels by a combination of enzymatic and physical methods, used freezing-thawing process, and in situ with synthesis of gelatin/mTG in PVA solution. The influence on the contents and dispersed condition of gelatin in PVA matrixes on the hydrogels mechanical strength was investigated in order to obtain applicable hydrogels. The morphology and crystalline structures of interpenetrating polymer network PVA/GE were also observed by some experimental analysis techniques, such as scanning electronic microscope (SEM). Moreover, in order to understand the initial behavior of fibroblasts cells, proliferation was assessed in vitro using fibroblast like L 929 cell culture.

2. Experimental

2.1. Materials. PVA with a degree of polymerization of 1750 ± 50 and hydrolysis degree values of greater than 99% was purchased from Beijing Organic Chemical Plant, China. Gelatin (GE) (type A, 300 bloom from porcine) was obtained (Sigma-Aldrich, St. Louis, MO, USA). The source of transglutaminase was the commercial product obtained from Yiming Biological Products Co., Ltd. (Jiangsu, China). As determined by a colorimetric hydroxamate method [30], the enzyme activity of mTG was 102 (U/g) of powder. All other regents used in the paper were of analytical grade.

TABLE 1: Composition of IPN PVA/GE hydrogels.

Hydrogels	IPN-1	IPN-2	IPN-3	IPN-4
Gelatin %w/v	0	2	5	7
PVA %w/v	15	15	15	15

The concentration of mTG at a range of 5–20 U/mL.

2.2. Moisture Determination of Gelatin. In order to know the accurate concentration of the gelatin solution prepared subsequently, it is necessary to know the moisture content of the gelatin. The moisture of gelatin was determined according to Chinese standards GB/T 5009.3-2003 [31].

2.3. Preparation of Gelatin and mTG Blends. A concentrated gelatin solution was prepared by adding 50 gm of gelatin to 100 mL of warm deionized water and mixing at 50°C until the protein was dissolved. Aliquots from this concentrated gelatin solution were then mixed with deionized water to prepare solutions with differing gelatin concentrations (0%, 2%, 5%, and 7% w/v). The gelatin solutions were stored at 4°C. A concentrated enzyme solution was prepared for each experiment by mixing 0.5 gm of mTG per 5 mL of deionized water at room temperature. This enzyme solution was stored at 4°C until it was ready for use.

2.4. Preparation of IPN PVA/Gelatin Hydrogels. IPN PVA/GE hydrogels were prepared by cyclic freezing-thawing method. For this purpose, aqueous solutions containing 15% by weight PVA with different amounts of gelatin, (i.e., 0%, 2%, 5%, and 7%) by weight, were used. The (15% w/v) of PVA solution was prepared at 95°C. A varying amount of solution's gelatin and mTG was mixed slowly to the PVA solution at room temperature to induce cross-linking. The mixture was cast between glass slides with 3 mm thick spacers then incubated at 45°C for 4 h to achieve complete cross-linking. Then, they were physically cross-linked by three freeze-thaw cycles, which consisted of freezing at −20°C for 24 h and thawing at room temperature 21°C for 24 h, respectively, to produce IPN hydrogels. The IPN hydrogels films had a thickness of 3 mm. Table 1 shows the composition of IPN PVA/GE hydrogels.

2.5. Morphological Characterization. For morphological characterization, hydrogels after swelling (equilibrium) in water were freeze-dried using a freeze drier (Christ, Germany, Alpha 1-2) at −52°C for 12 h. Transverse sections were cut from freeze-dried film samples using a cold knife. Samples were then examined by a scanning electron microscope (SEM JSM-6360LV, a voltage of 20 KV, China). The working face of the samples was sprayed with gold in advance. The observed morphologies for each SEM fractograph were analyzed using

Sigma Scan Pro software. Quantitative analysis of the pore size was obtained from structural indices measured from scaffold samples. The density of the solid material (film) was calculated according to the mass fractions (M_x where x refers to the constituent material) and densities of the constituent materials (ρ_x), by assuming that the total mass and volume remain the same before and after reaction [32], using (1)

$$\rho_s = \left[\frac{(M_{\text{PVA}} + M_{\text{GE}} + M_{\text{mTG}})}{[(M_{\text{PVA}}/\rho_{\text{PVA}}) + (M_{\text{GE}}/\rho_{\text{GE}}) + (M_{\text{mTG}}/\rho_{\text{mTG}})]} \right]. \quad (1)$$

2.6. Gel Fraction. The weight ratio of the dried hydrangeas in rinsed and unrinsed conditions or gel fraction can be assumed as an index of the degree of cross-linking. Therefore, the gel fraction of samples can be calculated as follows (2):

$$\text{Gel fraction (\%)} = \left(\frac{W_f}{W_i} \right) \times 100, \quad (2)$$

where W_f and W_i are the weights of the dried hydrogel after and before rinsing and extraction. To perform gel fraction measurement, preweighed slice of each sample was dried under vacuum at room temperature until observing no change in its mass. Nearly identical weight of another slice of the same sample was immersed into excess of distilled water for 4 days to rinse away amorphous or soluble part. Subsequently, the immersed sample was removed from distilled water and dried at room temperature under vacuum until the dried mass showed constant weight.

2.7. Equilibrium Degree of Swelling and Equilibrium Water Content. For swelling experiments, the vacuum-dried films were immersed in excess distilled water at room temperature (~30°C) until reaching to an equilibrium state. Then, the swelled samples were withdrawn from distilled water and weighed after gentle surface wiping using absorbent paper. The equilibrium degree of swelling (EDS) and equilibrium water contents (EWC) were calculated, respectively, as follows [33]:

$$\begin{aligned} \text{EDS (\%)} &= \left[\frac{(W_s - W_d)}{W_d} \right] \times 100, \\ \text{EWC (\%)} &= \left[\frac{(W_s - W_d)}{W_s} \right] \times 100, \end{aligned} \quad (3)$$

where W_s is the swollen weight of the sample of the equilibrium state and W_d is the final dry weight of the extracted sample.

2.8. Mechanical Properties. The samples were cut into a slice shape with a thickness of 3 mm and their mechanical properties, including tensile strength at break, were determined. The mechanical properties were measured at a temperature of 17°C and humidity of 60%, at a crosshead speed of 50.000 (mm/min) using Instron tester—4466, type: 42/43/4400. The elastic moduli of the IPN PVA/GE hydrogels were determined by performing constant strain-rate compression measurements on an Instron 4466 mechanical tester at room temperature. The hydrogel sample, 15 mm in diameter and 10 mm in height, was tested at a rate of 5.000 (pts/secs) and humidity of 40%, at a crosshead speed of 1.5000 mm/min. The results are expressed as mean value ± standard deviation with the confidence level of 95%.

2.9. Cell Viability Assay. Cellular survivability evaluation against cross-linked PVA, GE, and IPN PVA/GE blend has been evaluated by MTT (3-[4,[5-dimethylthiazol-2-yl]-2,5-diphenyl tetrazolium bromide) assay [34]. This standard test method is based on exposing L929 cells to the fluid extract of the test materials and control materials. The extract solution of the PVA, GE, or cross-linked IPN PVA/GE blend was prepared according to the protocol as outlined in ISO 10993 [35]. In brief, to determine the cytotoxicity of osteoblast-like L929 cells on the IPN hydrogels samples, samples were incubated with DMEM media at 37°C for 72 h in an incubator. The extract solutions were diluted serially with the media (0%, 12.5%, 25%, 50%, and 100%). For the control, 100% media were used to compare among the dilution, and then, 200 mL of extracting and diluted solution was added in a 96-well plate. The 96-well plate was previously (24 h before adding the extract solutions) incubated and coated with fibroblasts cells (1×10^4 cells/well). The plate was then incubated in a CO_2 incubator at 37°C for 1 day, 3 days, and 5 days. In MTT, assay cell survivability is measured through the production of purple colors by the reaction of dehydrogenase enzymes of living cells and MTT. Metabolically active and viable cells produce mitochondrial dehydrogenase enzymes during incubation in media, which can be read out through the change of color intensity by a spectrophotometer. The optical density (OD) corresponds to the viable cell numbers. Therefore, cell survival and proliferation at the different dilutions and time points were quantified by adding 100 μL of the MTT solution (20 μg/100 mL) to each of the wells. After 4 h of incubation, the OD values of the solution were measured using an BIORAD reader (Model 680, Microplate reader) at a wavelength of 595 nm.

2.10. Optical Microscope Study. In order to observe directly whether the viable cell numbers increases after treating the osteoblast like L929 cells with sample extracts after three freezing-thawing cycles and day 5, an inverted light microscope (Olympus, 1 × 71) attached with LCD monitor was used. After 70–80% confluent growth in the subculture flask, cells were trypsinized (0.25% Trypsin-EDTA), detached, and counted. Approximately, 1×10^3 cells/mL media was pipette into the wells of a Microtiter plate containing 24 wells. The plate was then incubated in a CO_2 incubator (5% CO_2, 37°C) for 24 h. after seeding cells into the wells of Microtiter plates, media were removed carefully and replaced with the extract solution of the samples, and the Microtiter plate was kept back into the CO_2 incubator either for 5 days. After finishing each of the incubation periods, cell growth in the Microtiter plate was observed using an inverted light microscope.

2.11. Statistical Analysis. All experiments were done in triplicate, and the results were expressed as mean ± SD, and the means were analyzed by one-way ANOVA at a P value of 0.05.

3. Results and Discussion

3.1. Moisture of Gelatin. The result of moisture in gelatin used in the study is 11.51%. As the gelatin solutions were prepared, the moisture content of the gelatin was taken into account.

3.2. Morphological Characterization. In applied freezing-thawing process, the structure of the PVA and PVA/GE IPN hydrogels was determined by SEM. The densities of the cross-linked IPN hydrogels containing gelatin contents from 0 wt.% to 7 wt.% were investigated by density measurement, SEM. Tables 2 and 3 show the densities of IPN PVA/gelatin-mTG films (ρ_s) calculated by (1). The density of the IPN PVA/gelatin-mTG hydrogels was found to increase monotonously with increasing the gelatin content, namely, from 1.45 Mg m^{-3} to 1.49 Mg m^{-3}. Figure 1 shows the scanning electron micrographs of transversal sections of the films after 1 (Figures 1(a1), 1(a2), 1(a3), and 1(a4))) and 3 (Figures 1(b1), 1(b2), 1(b3), and 1(b4)) freezing-thawing cycles. These micrographs show that increasing the concentration of gelatin blends and number of freezing-thawing cycles results in higher network arrangement, which leads to porous structure on the surface of IPN hydrogels. The internal porosity of samples shows significant correlation between number of freezing-thawing cycles, the chemical structure of additive and the pore size, and arrangement. Before adding the gelatin/mTG blends as modifying agent to PVA blend, the PVA hydrogel produced smooth structured films and exhibits a few irregular pores, and the morphology is not significantly changed as compared to the IPN PVA hydrogel Figures 1(a1) and 1(b1). As illustrated in Figures 1(b2), 1(b3), and 1(b4), IPN hydrogels after 3 freezing-thawing cycles shows highly irregular porous structure with the large pores size in the range of 50–100 μm. Figures 1(a1) and 1(b1) after 1 and 3 cycles showed a fewer and smaller pore size in the range of 5–10 μm, as a consequence of the preparation technique. In the freeze-thaw process, pore size is in fact controlled by the size of the ice crystals, which could be adjusted by varying the rate of freezing. In Figures 1(a2), 1(a3), and 1(a4), IPN hydrogels after 1 cycle showed lower and smaller irregular porous size structure in the range of 20–40 μm than those obtained with Figures 1(b2), 1(b3), and 1(b4). More significant change in sample morphology is observed after introducing the gelatin/mTG and increased the cycles as cross-linking agent. Therefore, it might be an optimum candidate for biomedical applications. In the films produced with PVA/GE mixtures, the presence of cracks and empty spaces increased with an increasing proportion of gelatin in the mixture.

3.3. Gel Fraction. Enzymatic cross-linking of gelatin and cyclic freezing-thawing of pure PVA leads to the formation of insoluble and entangled polymeric network in the IPN PVA/GE hydrogels. A typical dependency of the gel fractions to the quantity of gelatin incorporated into hydrogels is given in Tables 2 and 3 using (2). As seen, the gel fraction of samples is increased by increasing the number of freezing-thawing cycles and amounts of gelatin up to 7 (%) by weight after (1 and 3) cycles. For example, the gel fraction of IPN hydrogel increases to 74 (%) by adding 7% weight of gelatin after three cycles compared with the other IPN hydrogels after one cycle. On the other hand, Tables 2 and 3 exhibit a semilinear relationship between gel fraction, number of freezing-thawing cycles, and gelatin weight fraction in mixture hydrogels. The increase of the gel fraction may be attributed to the enzymatic cross-linked gelatin and the additional interactions between PVA and gelatin; besides, the bonds existed between PVA long neighboring chains induced crystallization, which caused an increase in gel fraction.

3.4. Degrees of Swelling and Water Content. In this work, the equilibrium degree of swelling (EDS) and equilibrium water content (EWC), as important swelling characteristics of hydrogels, were calculated using (3). These characteristics indicated the ability of absorption of fluids and exudates. Tables 2 and 3 demonstrate EDS and EWC of IPN PVA/GE hydrogels as a function of the amount of gelatin and the number of freezing-thawing cycles. Both parameters showed nearly similar decreasing trends by increasing the quantity of the gelatin content and the number of freezing-thawing cycles. The samples treated for three freeze-thaw cycles showed a much more swollen structure than others treated with one freeze-thaw cycle. The highly swollen hydrogel represents a high tight structure and a higher degree of cross-linking when compared with the hydrogels with lower swelling ratio. The comparison of the EDS and EWC data for the gel fraction values indicates that there is a logic relationship between these swelling characteristics and gel fraction, that is, more gel fractions lead to less EDS or EWC. Although the swelling characteristics of IPN PVA/GE hydrogels decrease due to presence of gelatin in comparison with pure PVA hydrogel, it seems that there have been high enough swelling capacity to be used as a suitable dressing even for healing exudative wounds.

3.5. Mechanical Properties. As mentioned before, the main interest in the production of IPN PVA/gelatin hydrogels is to achieve the materials with better mechanical properties. The investigated mechanical properties, that is, tensile strength, as well as strain at break were found to depend on the amount of gelatin and number of freezing-thawing cycles. As expected, cross-linking led to an improvement in the tensile strength. Based on these analyses, the composition of IPN hydrogels and three freezing-thawing cycles was considered the optimum conditions for the preparation of IPN hydrogels. The irregular arrangement within PVA hydrogel modified with gelatin/mTG after 3 freezing-thawing process (Table 3) has strong influence on mechanical properties. Independent of the enzymatic modification, the values for tensile strength of the films produced from IPN PVA/GE mixtures was significantly different varied as a function of the gelatin concentrations and number of freezing-thawing cycles in the mixture Tables 2 and 3. Table 3 shows that the highest values of compressive stress and tensile strength increases with increasing the concentration of gelatin and number of freezing-thawing cycles. It is observed that the tensile strength for pure PVA is (0.41 $\pm$ 0.07 MPa) and the elastic modulus is (0.45 $\pm$ 0.60 MPa). By addition of 2% w/v gelatin into PVA blend, the tensile strength increases to (0.66 $\pm$ 0.12 MPa)

TABLE 2: The properties of IPN PVA/GE hydrogels after one freezing-thawing cycle.

Samples	IPN-1	IPN-2	IPN-3	IPN-4
Gel fraction (%)	25 ± 1.20	36 ± 1.35	49 ± 1.50	58 ± 1.65
EDS (%)	135 ± 4.50	132 ± 3.75	130 ± 3.45	130 ± 3.50
EWC (%)	45 ± 2.00	44.5 ± 2.25	42.5 ± 2.50	42 ± 2.75
Stress (MPa)	0.11 ± 0.14	0.15 ± 0.18	0.25 ± 0.24	0.28 ± 0.22
Strain (%)	18.23 ± 4.60	25.67 ± 7.80	34.56 ± 12.90	47.78 ± 14.65
Elastic modulus (MPa)	0.35 ± 0.45	0.41 ± 0.49	0.56 ± 0.62	0.65 ± 0.64
Tensile strength (MPa)	0.30 ± 0.05	0.42 ± 0.09	0.52 ± 0.15	0.58 ± 0.19
Solid density (ρ_s) (Mg m^{-3})	1.13	1.20	1.29	1.34

TABLE 3: The properties of IPN PVA/GE hydrogels after three freezing-thawing cycles.

Samples	IPN-1	IPN-2	IPN-3	IPN-4
Gel fraction (%)	44 ± 1.30	58 ± 1.60	67 ± 1.70	74 ± 1.90
EDS (%)	165 ± 3.50	158 ± 3.10	149 ± 2.80	142 ± 2.40
EWC (%)	62 ± 1.98	61 ± 1.92	60 ± 1.87	59 ± 1.82
Stress (MPa)	0.25 ± 0.12	0.43 ± 0.20	0.64 ± 0.34	0.67 ± 0.38
Strain (%)	25.50 ± 6.40	31.85 ± 10.50	48.64 ± 14.00	68.70 ± 17.50
Elastic modulus (MPa)	0.45 ± 0.60	0.45 ± 0.60	0.87 ± 1.75	0.98 ± 2.45
Tensile strength (MPa)	0.41 ± 0.07	0.66 ± 0.12	0.74 ± 0.16	0.75 ± 0.18
Solid density (ρ_s) (Mg m^{-3})	1.27	1.45	1.47	1.49

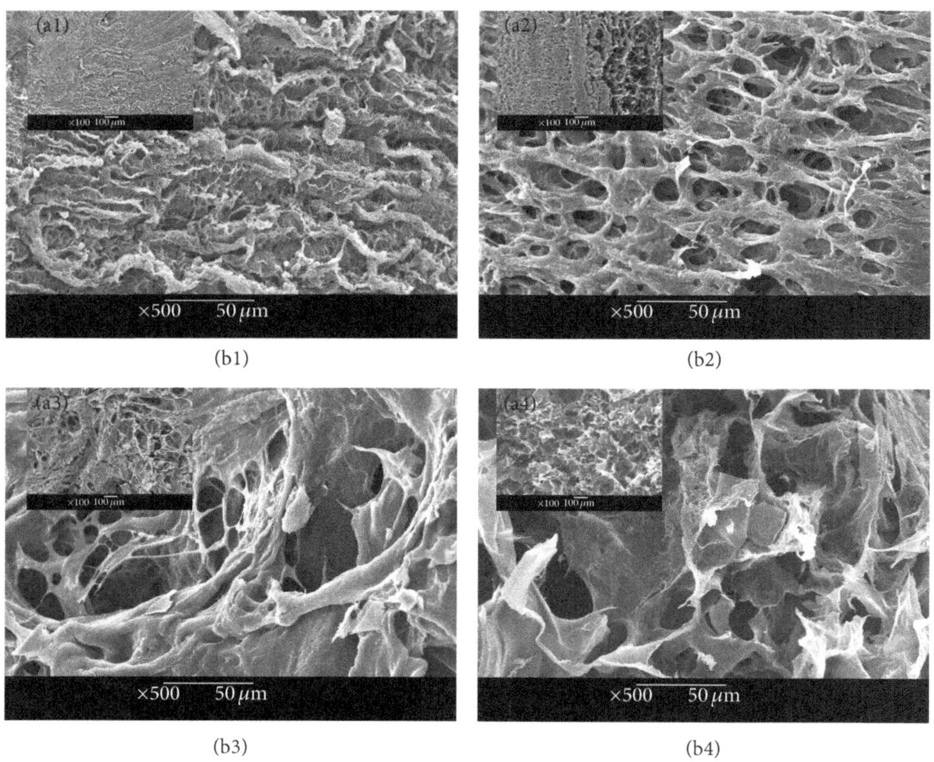

FIGURE 1: SEM micrographs of the surface sections of PVA hydrogel and PVA with additives: ((a1), (b1)) IPN-1 with 15% w/v PVA; ((a2), (b2)) IPN-2 with 2% w/v gelatin; ((a3), (b3)) IPN-3 with 5% w/v gelatin; ((a4), (b4)) IPN-4 with 7% w/v gelatin; after 1 and 3 of freezing-thawing cycles, respectively. Scale bar (a1)–(a4): 100 (μm); (b1)–(b4): 50 (μm).

with an elastic modulus of (0.80 ± 1.50 MPa). The IPN hydrogels containing 7% w/v gelatin display the highest tensile strength compared to the other samples. Table 2 shows a tensile strength of (0.75 ± 0.18 MPa) and the elastic modulus is (0.98 ± 2.45 MPa). The increment of elastic modulus and tensile strength is attributed to an increase in rigidity, because the IPN hydrogels a hard inorganic component. The extraordinary asset value of strength at break for IPN hydrogels is undoubtedly due to the presence of gelatin, which shows higher gel fraction and more entangled structure in comparison with pure gel. To examine the correlation between mechanical property changes and IPN hydrogels, constant strain-rate compression tests were performed on the IPN hydrogel in order to determine their elastic moduli. Representative stress-strain curves of the IPN hydrogels are presented in Tables 2 and 3. These results displayed the ability of gelatin concentration and number of cycles to individually affect the compressive modulus of hydrogels. Based on the mechanical property results for our biocomposite IPN hydrogels which are able to attain high tensile strength, it is concluded that there is an opportunity for applications as potential candidate for scaffolds tissue engineering.

3.6. Cytotoxicity of Photocrosslinked IPN PVA/GE Hydrogels. The MTT assay, which is a rapid, standardized, sensitive, and inexpensive method to determine cell viability and proliferation or whether a material contains significant quantities of biologically harmful extracts, is the first step used to screen the biocompatibility of a biomaterials. The effect of PVA, GE, uncross PVA/GE, and cross-linked PVA/GE IPN hydrogels on viability (cytotoxicity) and proliferation of L929 cells was examined using the MTT assay, which is a colorimetric assay that measures the metabolic activity of viable cells. The cytotoxicities of L929 cells on the PVA, GE, uncross PVA/GE, and cross-linked PVA/GE are shown in Figure 2. Based on cytotoxicity results, a fairly acceptable viability of the cells was observed on PVA (>88% at 100% extract solution), uncross PVA/GE (>94% at 100% extract solution), and cross-linked PVA/GE (>98% at 100% extract solution), while GE membrane showed a slightly higher viability (109% at 100% extract solution). Although the biocompatibility of GE was high, the general survival rate of the cells for photocrosslinked PVA/GE scaffold was satisfactory; therefore, the IPN photocrosslinked PVA/GE scaffolds were not toxic to the L929 cells and could be used for tissue engineering scaffolds.

3.7. Proliferation of Fibroblast Cells L 929 on IPN PVA/GE Blend. The proliferation of fibroblasts on PVA, GE, uncross PVA/GE, and cross-linked IPN PVA/GE blend was evaluated using MTT assay after 1, 3, and 5 days of incubation as shown in Figure 3. A significantly higher cell proliferation (day 5) in GE and PVA/GE blend compared to PVA was exhibited in Figure 3. After 1 day, the number of cells had exceeded the number of cells initially seeded (10^4 cells/well). As clearly shown in Figure 3, the number of fibroblasts that grew was greater than the number of cells initially seeded even after only 3 days in culture. From 1 day to 5 days in culture, the L 929 cells continually proliferated on including the cross-linked IPN PVA/GE.

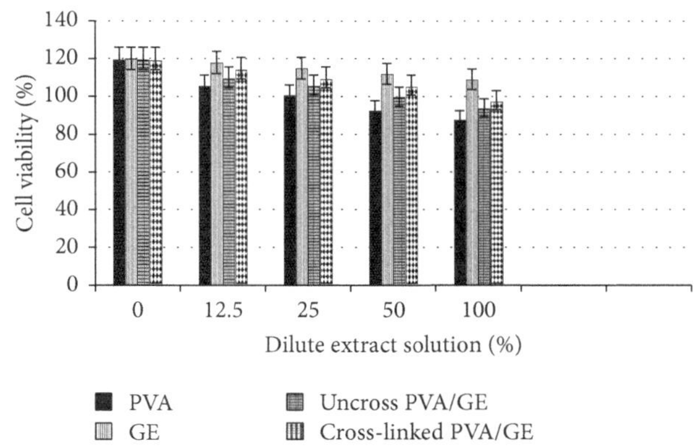

FIGURE 2: Cell viability of PVA, GE, uncross PVA/GE, and cross-linked PVA/GE IPN blend. The MTT assay was used to measure the viability of L929 cells at various dilute extract solution concentrations. Media at 100% were used as a control. Standard errors were expressed as bar diagram.

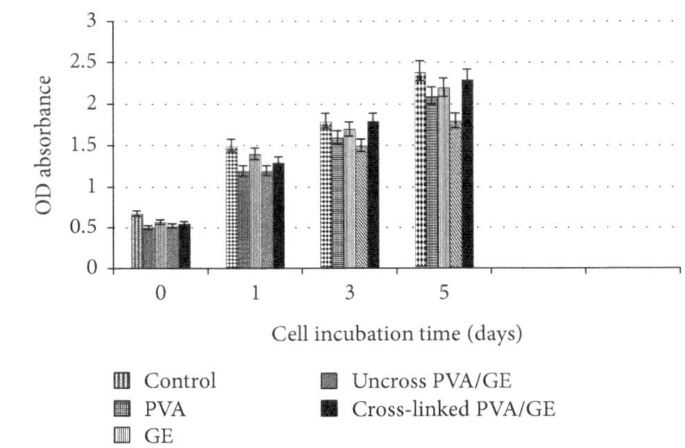

FIGURE 3: Fibroblast cellular proliferation of PVA, GE, uncross PVA/GE, and cross-linked IPN PVA/GE blend after 3 cycles and 1, 3, and 5 days using the MTT assay. Standard errors were expressed in the bar diagram. Media at 100% were used as a control and optical density (OD) corresponded to the cell survivability after 1, 3, and 5 days.

Cellular viability assay by MTT was used to find out the percentage of viable cell numbers per extract dilutions after certain periods. However, to observe the increased cell numbers and their real time morphologies, light microscopic studies were conducted. Observation of the morphological changes of cells is a very important means for predicting post-cellular responses: spreading, proliferation, and survival. An ideal scaffold must provide the contact guidance for controlling cell adhesion and directing cell migration that influences cell proliferation [36, 37]. Figure 4 shows the light microscopic images of cells treated with extract solutions for day 5. For the morphologies' fibroblasts at day 1, only a little amount of cells was observed. However, the amount of cell numbers was found to increase gradually at day 5 for all samples. Especially, at day 5, cellular growth was almost confluent. However, in case of cross-linked PVA/GE composed of cells, growth pattern was better than that of the

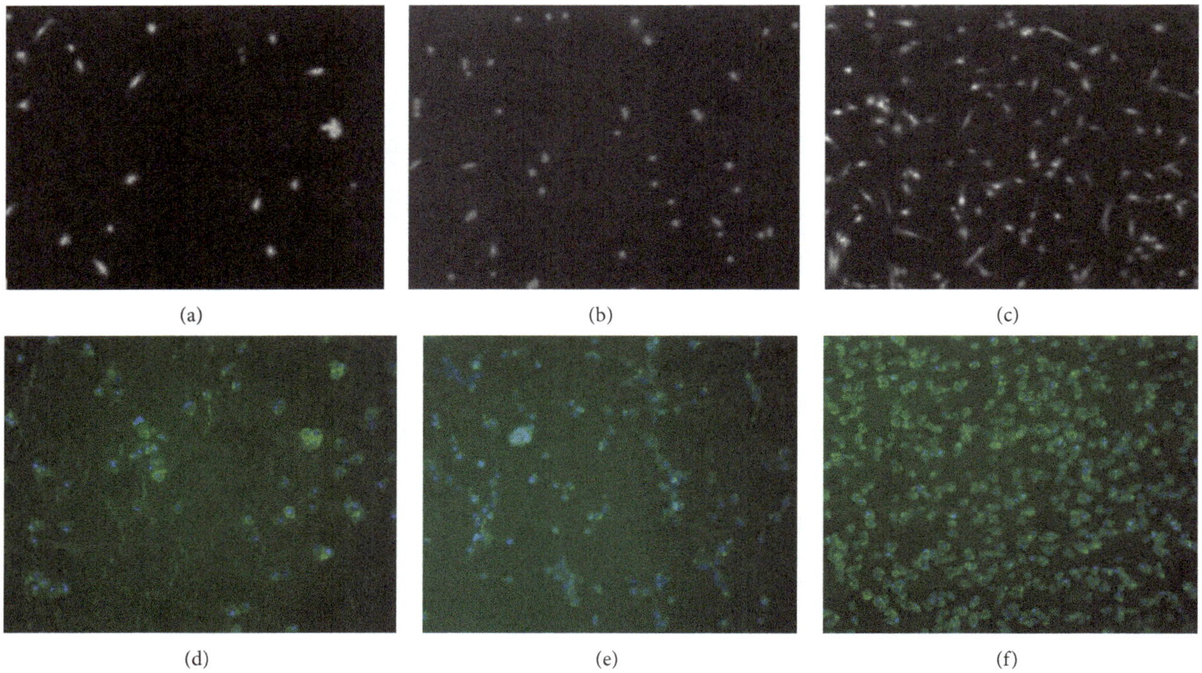

FIGURE 4: Fibroblast cellular proliferation of uncross PVA/GE and cross-linked IPN PVA/GE blend after 1 ((a), (d)), 3 ((b), (e)), and 5 ((c), (f)) days using the optical microscope observation.

uncrossing PVA/GE. In fact, this result corresponds with the cell viability and proliferation results very well.

4. Conclusion

PVA/GE hydrogels with IPN structure were prepared successfully by enzymatic and cyclic freezing-thawing method. The size and arrangement of these pores are the result of number of freezing-thawing cycles and the presence of cross-linking agents. The IPN PVA/GE hydrogels have the highest values of mechanical properties as evaluated from compression tests, also the cross-linking density showed the highest values. The SEM micrographs of these hydrogels show that the majority of the pores is opened and irregular. These interactions become stronger with increasing number of freezing-thawing cycles and results in more regular internal structure. The quantity of gelatin was the key factor to obtain IPN PVA/GE hydrogels with desirable properties. The IPN PVA/GE hydrogels showed excellent physical and mechanical properties, which met the essential requirements for ideal medical applications. Based on swelling measurements, they exhibited high capability in absorbing fluid, so recommended for exudative wounds. In addition, fibroblasts that grew over the cells treated with extract solutions exhibited the appropriate morphology and displayed good proliferation; this indicates that the cross-linked network structure of obtained gels is stable enough, suggesting that developer scaffolds might be used for tissue engineering applications.

Acknowledgment

This paper is supported by the Project of the National Natural Science Foundation of China (51073036).

References

[1] J.-H. Juang, S. Bonner-Weir, Y. Ogawa, J. P. Vacanti, and G. C. Weir, "Outcome of subcutaneous islet transplantation improved by polymer device," *Transplantation*, vol. 61, no. 11, pp. 1557–1561, 1996.

[2] D.-H. Chen, J.-C. Leu, and T.-C. Huang, "Transport and hydrolysis of urea in a reactor-separator combining an anion-exchange membrane and immobilized urease," *Journal of Chemical Technology and Biotechnology*, vol. 61, no. 4, pp. 351–357, 1994.

[3] S. H. Hyon, W. I. Cha, Y. Ikada, M. Kita, Y. Ogura, and Y. Honda, "Poly(vinyl alcohol) hydrogels as soft contact lens material," *Journal of Biomaterials Science*, vol. 5, no. 5, pp. 397–406, 1994.

[4] J. K. Li, N. Wang, and X. S. Wu, "Poly(vinyl alcohol) nanoparticles prepared by freezing-thawing process for protein/peptide drug delivery," *Journal of Controlled Release*, vol. 56, no. 1–3, pp. 117–126, 1998.

[5] M. T. Razzak, D. Darwis, Z. Zainuddin, and S. Sukirno, "Irradiation of polyvinyl alcohol and polyvinyl pyrrolidone blended hydrogel for wound dressing," *Radiation Physics and Chemistry*, vol. 62, no. 1, pp. 107–113, 2001.

[6] F. Yoshii, Y. Zhanshan, K. Isobe, K. Shinozaki, and K. Makuuchi, "Electron beam crosslinked PEO and PEO/PVA hydrogels for wound dressing," *Radiation Physics and Chemistry*, vol. 55, no. 2, pp. 133–138, 1999.

[7] F. Yoshii, K. Makuuchi, D. Darwis, T. Iriawan, M. T. Razzak, and J. M. Rosiak, "Heat resistance poly(vinyl alcohol) hydrogel," *Radiation Physics and Chemistry*, vol. 46, no. 2, pp. 169–174, 1995.

[8] D. A. Ossipov and J. Hilborn, "Poly(vinyl alcohol)-based hydrogels formed by 'click chemistry'," *Macromolecules*, vol. 39, no. 5, pp. 1709–1718, 2006.

[9] H. K. Purss, G. G. Qiao, and D. H. Solomon, "Effect of "glutaraldehyde" functionality on network formation in poly(vinyl

alcohol) membranes," *Journal of Applied Polymer Science*, vol. 96, no. 3, pp. 780–792, 2005.

[10] Z. Ajji, "Preparation of poly(vinyl alcohol) hydrogels containing citric or succinic acid using gamma radiation," *Radiation Physics and Chemistry*, vol. 74, no. 1, pp. 36–41, 2005.

[11] S. Benamer, M. Mahlous, A. Boukrif, B. Mansouri, and S. L. Youcef, "Synthesis and characterisation of hydrogels based on poly(vinyl pyrrolidone)," *Nuclear Instruments and Methods in Physics Research B*, vol. 248, no. 2, pp. 284–290, 2006.

[12] P. Martens and K. S. Anseth, "Characterization of hydrogels formed from acrylate modified poly(vinyl alcohol) macromers," *Polymer*, vol. 41, no. 21, pp. 7715–7722, 2000.

[13] N. A. Peppas and N. K. Mongia, "Ultrapure poly(vinyl alcohol) hydrogels with mucoadhesive drug delivery characteristics," *European Journal of Pharmaceutics and Biopharmaceutics*, vol. 43, no. 1, pp. 51–58, 1997.

[14] T. Hatakeyema, J. Uno, C. Yamada, A. Kishi, and H. Hatakeyama, "Gel-sol transition of poly(vinyl alcohol) hydrogels formed by freezing and thawing," *Thermochimica Acta*, vol. 431, no. 1-2, pp. 144–148, 2005.

[15] R. Ricciardi, C. Gaillet, G. Ducouret, F. Lafuma, and F. Lauprêtre, "Investigation of the relationships between the chain organization and rheological properties of atactic poly(vinyl alcohol) hydrogels," *Polymer*, vol. 44, no. 11, pp. 3375–3380, 2003.

[16] T. Koyano, N. Koshizaki, H. Umehara, M. Nagura, and N. Minoura, "Surface states of PVA/chitosan blended hydrogels," *Polymer*, vol. 41, no. 12, pp. 4461–4465, 2000.

[17] W.-Y. Chuang, T.-H. Young, C.-H. Yao, and W.-Y. Chiu, "Properties of the poly(vinyl alcohol)/chitosan blend and its effect on the culture of fibroblast in vitro," *Biomaterials*, vol. 20, no. 16, pp. 1479–1487, 1999.

[18] T. Chandy and C. P. Sharma, "Prostaglandin E1-immobilized poly(vinyl alcohol)-blended chitosan membranes. Blood compatibility and permeability properties," *Journal of Applied Polymer Science*, vol. 44, no. 12, pp. 2145–2156, 1992.

[19] J. M. H. Yang, M. J. Huang, and T. S. Yeh, "Preparation of poly (acrylic acid) modified polyurethane membrane for biomaterial by UV radiation without degassing," *Journal of Biomedical Materials Research*, vol. 45, pp. 133–139, 1999.

[20] H. S. Koh, T. Yong, C. K. Chan, and S. Ramakrishna, "Enhancement of neurite outgrowth using nano-structured scaffolds coupled with laminin," *Biomaterials*, vol. 29, no. 26, pp. 3574–3582, 2008.

[21] H.G.l, "United States Patent," 4055554, 1977.

[22] U. Gerber, U. Jucknischke, S. Putzien, and H.-L. Fuchsbauer, "A rapid and simple method for the purification of transglutaminase from Streptoverticillium mobaraense," *Biochemical Journal*, vol. 299, no. 3, pp. 825–829, 1994.

[23] K. Yokoyama, N. Nio, and Y. Kikuchi, "Properties and applications of microbial transglutaminase," *Applied Microbiology and Biotechnology*, vol. 64, no. 4, pp. 447–454, 2004.

[24] C. W. Yung, L. Q. Wu, J. A. Tullman, G. F. Payne, W. E. Bentley, and T. A. Barbari, "Transglutaminase crosslinked gelatin as a tissue engineering scaffold," *Journal of Biomedical Materials Research A*, vol. 83, no. 4, pp. 1039–1046, 2007.

[25] M. Carmen Go'mez-Guillén, A. Isabel Sarabia, M. Teresa Solas, and P. Montero, "Effect of microbial transglutaminase on the functional properties of megrim (*Lepidorhombus boscii*) skin gelatin," *Journal of the Science of Food and Agriculture*, vol. 81, no. 7, pp. 665–673, 2001.

[26] A. P. Rokhade, S. A. Agnihotri, S. A. Patil, N. N. Mallikarjuna, P. V. Kulkarni, and T. M. Aminabhavi, "Semi-interpenetrating polymer network microspheres of gelatin and sodium carboxymethyl cellulose for controlled release of ketorolac tromethamine," *Carbohydrate Polymers*, vol. 65, no. 3, pp. 243–252, 2006.

[27] A. P. Rokhade, S. A. Patil, and T. M. Aminabhavi, "Synthesis and characterization of semi-interpenetrating polymer network microspheres of acrylamide grafted dextran and chitosan for controlled release of acyclovir," *Carbohydrate Polymers*, vol. 67, no. 4, pp. 605–613, 2007.

[28] L. H. Sperling, "Interpenetrating polymer networks and related materials," *Journal of Polymer Science Macromolecular Reviews*, vol. 12, pp. 141–180, 1977.

[29] T. Kurokawa, H. Furukawa, W. Wang, Y. Tanaka, and J. P. Gong, "Formation of a strong hydrogel-porous solid interface via the double-network principle," *Acta Biomaterialia*, vol. 6, no. 4, pp. 1353–1359, 2010.

[30] J. E. Folk, "[127] Transglutaminase (guinea pig liver)," *Methods in Enzymology*, vol. 17, pp. 889–894, 1970.

[31] Chinese standards GB/T, 5009. 3-2003, D.o.m.i.f.

[32] M. Frydrych, C. Wan, R. Stengler, K. U. O'Kelly, and B. Chen, "Structure and mechanical properties of gelatin/sepiolite nanocomposite foams," *Journal of Materials Chemistry*, vol. 21, no. 25, pp. 9103–9111, 2011.

[33] P. R. Hari and K. Sreenivasan, "Preparation of polyvinyl alcohol hydrogel through the selective complexation of amorphous phase," *Journal of Applied Polymer Science*, vol. 82, no. 1, pp. 143–149, 2001.

[34] G. Mickisch, S. Fajta, G. Keilhauer, E. Schlick, R. Tschada, and P. A lken, "Chemosensitivity testing of primary human renal cell carcinoma by a tetrazolium based microculture assay (MTT)," *Urological Research*, vol. 18, no. 2, pp. 131–136, 1990.

[35] American National Standard, *Biological Evaluation of Medical Devices-Part 5: Tests for Cytotoxicity*, 1999.

[36] J. Fukuda, A. Khademhosseini, Y. Yeo et al., "Micromolding of photocrosslinkable chitosan hydrogel for spheroid microarray and co-cultures," *Biomaterials*, vol. 27, no. 30, pp. 5259–5267, 2006.

[37] A. J. Horobin, K. M. Shakesheff, and D. I. Pritchard, "Promotion of human dermal fibroblast migration, matrix remodelling and modification of fibroblast morphology within a novel *3D* model by *Lucilia sericata* larval secretions," *Journal of Investigative Dermatology*, vol. 126, pp. 1410–1418, 2006.

12

Uniaxial Bond Stress-Slip Relationship of Reinforcing Bars in Concrete

Sungnam Hong and Sun-Kyu Park

Department of Civil and Environmental Engineering, Sungkyunkwan University, Suwon 440-746, Republic of Korea

Correspondence should be addressed to Sun-Kyu Park, skpark@skku.edu

Academic Editor: Gai-Fei Peng

This paper documents a study carried out on the estimation of the bond stress-slip relationship for reinforced concrete members under axial tension loading. An analytical model is proposed that utilizes the conventional bond stress-slip theories as well as the characteristics of deformed bar and concrete cross-sectional area. An equation for the estimation of the bond stress is formulated as the function of nondimensional factors (e.g., bond stress, slip, etc.). The validity, accuracy, and efficiency of the proposed model are established by comparing the analytical results with the experimental data and the JSCE design codes, as well as the analytical models given by Ikki et al. and Shima. The analytical results presented in this paper indicate that the proposed model can effectively estimate the bond stress-slip relationship of reinforced concrete members under axial tension loading.

1. Introduction

Bond of reinforcing bars to the surrounding concrete influences the behavior of reinforced concrete structures in many ways [1]. It can be a key element for the ultimate load-carrying capacity of reinforced concrete structures since it affects the anchorage of bars and the strength of lap slices. Moreover, the deformation capacity of the members, and hence the redistribution capacity in statically indeterminate structures, is directly influenced by the bond. For these reasons, it can be said that a fundamental issue for reinforced concrete structures is the bond between the reinforcing bars and the concrete [2].

One of the major drawbacks in reinforced concrete bond research is the absence of a generalized method for determining bond strength. This leads comparisons between various researches and test results on bond difficult. Most investigators have used pull-out tests that are commonly adopted in reinforced concrete bond studies [3–11]. In most research, not a great deal of attention has been focused on the embedment length and the stress state in the concrete. The embedment lengths and the stress states have, therefore, not coincided with those in real reinforced concrete members. Because this has not been recognized by many researchers, results often show a significant difference to real behavior.

In general, bond action in reinforced concrete members is represented by the bond stress-slip relationship. Many bond stress-slip relationships have been proposed and some of these have been formulated. These relationships have been used in the finite-element method (FEM) [12–14] and cracking [15, 16] analysis. However, most of the proposed relationships were derived from pull-out tests and differ from each other. The main limitation of the pull-out test is that it does not simulate the actual conditions in a reinforced concrete flexural member. Therefore, utilizing relationships derived from the pull-out test might be considered to contain inevitable problems. Consequently, it is necessary to establish a realistic bond stress-slip relationship that takes into account the actual conditions in a reinforced concrete flexural member, which is usually long embedment length and axial tension force at the cracking section.

This article presents an analytical model to estimate the bond stress-slip behavior of reinforced concrete members based on the axial tension test. The models suggested for the estimation of the bond stress-slip relationship in the JSCE code [17] and by Ikki et al. [18, 19] and Shima [20] are

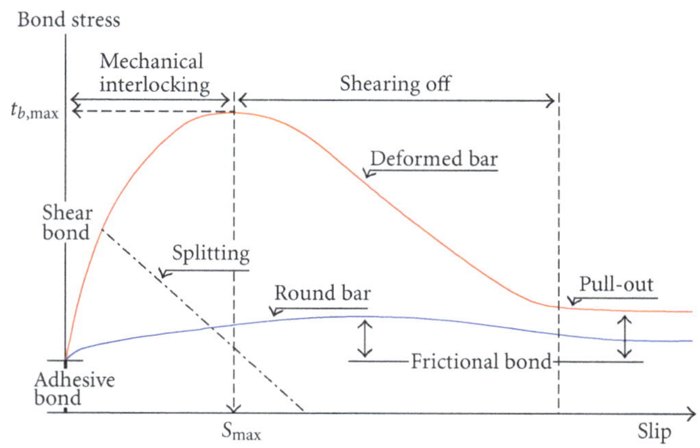

FIGURE 1: Bond stress-slip relationship.

extended in order to analyze the bond stress-slip behavior of reinforced concrete flexural members. An analytical equation is formulated as an exponential function of the relative rib area and the nondimensional slip. The validity, accuracy, and efficiency of the proposed model are established by comparing the results of the present study with the results obtained from the analytical and experimental studies. In addition, a parametric study was performed to evaluate the effects of the bond factors. The results of analysis presented in this paper indicate that the proposed model can be used effectively to estimate the bond stress-slip relationship.

2. Bond Mechanism

2.1. Interaction between Reinforcing Bar and Concrete. With the growth of bond researches, a reinforced concrete structure was known as a composite structure [21], whereby the slip occurs along the reinforcing bar under loading and the bond action is a function of the slip. To put it concretely, when the external force is progressively applied to a reinforced concrete member, interfacial stresses between the reinforcing bar and the concrete are created and the capacity of the interface to transmit stresses begins to weaken at certain load levels. These irreparable damages spread to the surrounding concrete. As a result of this process, the capacity of the interface to transmit stresses gradually deteriorates and a slip of both materials inevitably occurs.

As shown in Figure 1, the stress transfer mechanisms, which refer to the bond action, are usually expressed by the bond stress-slip relationship obtained from pull-out tests. The bond actions are comprised of an adhesive bond, a frictional bond, and a shear bond. In the case of deformed bars, the bond resistance capacity is mainly governed by the mechanical interlocking action.

2.2. Relative Rib Area. In fact, the bar geometry, and more specially the rib geometry, governs to a high degree the general bond behavior and determines the bond resistance [2]. In particular, as shown in Figure 2, the bar diameter (d_s), the rib height (h_d), and the rib spacing (l_d) are found to be the most important parameters [1]. In addition, deformed bars

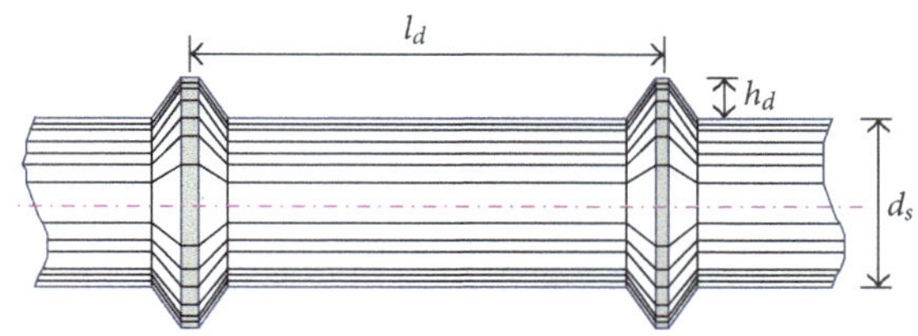

FIGURE 2: Details of deformed bar.

are currently mainly used for reinforcement in reinforced concrete structures. The best bond actions can be obtained by means of an appropriate combination of the 3 above mentioned parameters.

Rehm [22] showed that the relative rib area could be regarded as the ruling criterion for the bond action of deformed bars. The experimental evidence also proves that bond actions are relatively the same, provided that the relative rib areas are the same, and the rib face angle is greater than 30°. The generally accepted values ranging between 0.05–0.15 represent a good compromise for the relative rib area, in terms of ultimate bond strength and splitting ability, and so forth [1].

The relative rib area, the so-called bond index, is defined as

$$f_R = \frac{A_R}{\pi \times d_s \times l_d}, \tag{1}$$

where A_R: area of the projection of a single rib on the cross-section of a deformed bar.

Yamao et al. [23] pointed out that bond stress-slip relationships in long embedment and short embedment lengths differ significantly, as explained in Figure 3. In addition, the maximum bond stress obtained from specimens with short embedment show a steady increase that coincides with the increase of the relative rib area.

Also, when the relative rib area increases, the maximum bond stress obtained from specimens with long embedment will have smaller values than that obtained from specimens with short embedment. However, even in cases where the embedment length is sufficient, an increase is shown in

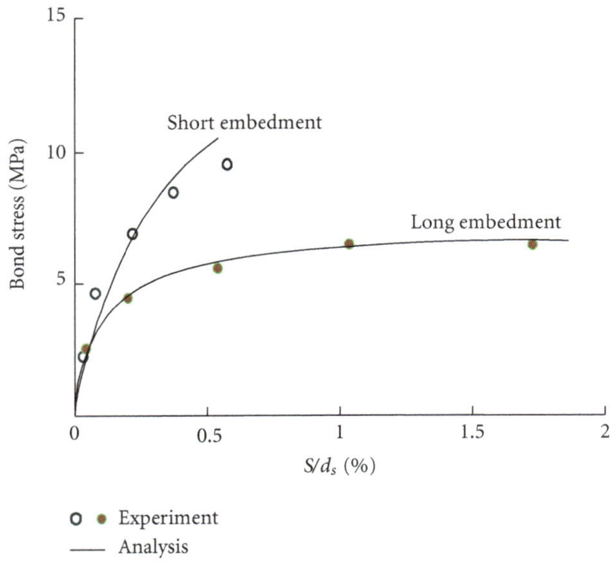

FIGURE 3: Bond stress-slip relationships of short embedment and long embedment.

proportion to the relative rib area, even though the maximum bond stress is smaller than that of cases where the embedment length is short. Hence, the effect of the relative rib area on the bond action in the case of long embedment length needs also be considered.

2.3. Stress State by Boundary Conditions. Generally, test methods used to measure bond stress and slip can be categorized into two methods, namely, pull-out tests and axial tension tests. The formulation of the bond-slip relationship has been developed for the following typical bond problems, as shown in Figure 4: (a) one side pull-out; (b) both sides pull-out. In these tests, the main limitation is that they do not simulate the actual conditions in a structural member. Due to the effect of the compressive force (C), concrete is in compression and the reinforcing bar is in tension. In reality, in the tensile zone of a reinforced concrete member, the concrete and reinforcing bar are both in tension. The presence of a lateral stress modifies the stress distribution shown in Figures 4(a) and 4(b) to a considerable degree.

In the axial tension tests shown in Figure 4(c), a monotonically increasing tensile force (F) is applied to the two protruding ends of the reinforcing bar embedded in the concrete prism. The distribution of tensile stresses induced in the reinforcing bar and concrete will be varied as shown in Figure 4(c). This distribution is very similar to that of the tension face of a reinforced concrete flexural member. It can, therefore, be considered that utilizing the bond-slip relationships obtained from the axial tension tests in the analysis is more realistic than utilizing those obtained from the pull-out tests.

3. Bond Models

3.1. Existing Models. The bond behaviors of reinforced concrete members have been studied extensively in the last century. A number of bond stress-slip relationships and corresponding bond models have been proposed in the last three decades. A number of popular bond models are introduced in the following paragraphs.

In 1957, Rehm [22] carried out pull-out tests with specimens that have a short embedment. He also proposed the following bond model, which is a function of slip, and comparatively analyzed the bond stress distribution with the proposed model as

$$\tau_b = c_1 S^a + c_2 S, \tag{2}$$

where τ_b: bond stress, S: slip, c_1, c_2, and a: experimentally obtained values.

In 1967, based on the pull-out tests and axial tension tests, Mugurama and Morita [24, 25] proposed the bond model as

$$\tau_b = \tau_{b,\max} \times \exp\left[\frac{\ln\{(\exp - 1) S/S_{\max} + 1\}}{(\exp - 1) S/S_{\max} + 1}\right], \tag{3}$$

where $\tau_{b,\max}$: maximum bond stress determined by experiment, $S_{\max}$: slip at the maximum bond stress determined by experiment.

They also estimated the bond behavior under three different boundary conditions using (3). Results from the above analysis concurred with the experimental results. However, in this model, it needs to be assumed that the slip corresponding to the maximum bond stress and the maximum bond stress obtained from the long specimens under axial tension is significantly smaller than that of the short pull-out specimens.

In 1987, Shima [20] reported the following bond models that can predict the tension stiffening effect and that can be used under any boundary condition. Equation (4) were proposed for the bar with a long embedment and for the bar with a short embedment, respectively, while considering the strain effect in accordance with the boundary conditions. This model is beneficial when the concrete is under the compressive stress condition at the reinforcing bar level and can be applied to the analysis of the behavior of a reinforced concrete member without any experiment in order to determine experimental factors. For this reason, this model is the representative bond model that is currently being used in the analysis and design of reinforced concrete members as

$$\tau_b = 0.9 f_c'^{2/3} \times \left[1 - \exp\left\{-40\left(\frac{S}{d_s}\right)^{0.6}\right\}\right],$$

$$\tau_b = 0.73 f_c'^{2/3} \times \left[\ln\left\{1 + 5000\left(\frac{S}{d_s}\right)\right\}\right]^3 \times \frac{1}{1 + 10^5 \times \varepsilon_s}, \tag{4}$$

where f_c': compressive concrete strength, d_s: bar diameter, ε_s: strain of the bar.

However, (4) was obtained from the experimental results of a pull-out test for an anchored bar embedded in the footing of a reinforced concrete pier or column. The applicability of this equation may, therefore, be limited because restrained cracks occurred in the specimens. In addition, the

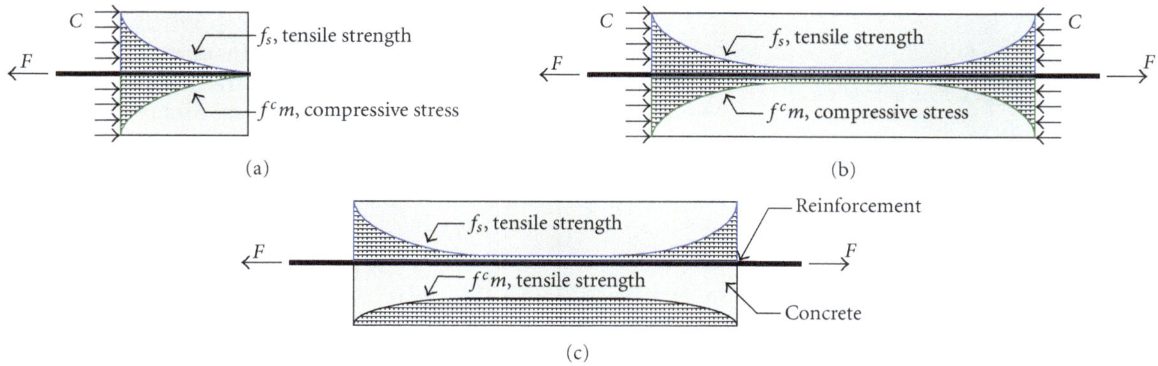

FIGURE 4: Typical stress distributions for (a) one side pull-out, (b) both sides pull-out, and (c) axial tension.

bond behavior must be influenced by many factors such as the concrete cover, the amount of confining reinforcement, or the axial force of the column.

While taking into consideration all the above-mentioned problems, Ikki et al. [18, 19] performed axial tension tests which were evaluated by using the following bond model, modified from (4) given by Shima [20]. Generally, (5) was comparatively accurate at predicting the bond-slip relationship of test specimens. However, the maximum bond stress corresponding to the slip cannot be calculated by the following (5) as

$$\tau_b = k_{sf} \times k_d \times 0.9 f_c'^{2/3} \times \left\{ 1 - \exp\left[-40 \left(\frac{S}{d_s} \right)^{0.6} \right] \right\}, \quad (5)$$

where k_{sf}: coefficient of the concrete stress condition (1.0 when the concrete is compressed, 0.7 when the concrete is tensioned), k_d: coefficient of the bar direction during concrete casting (1.0 for a vertically cast reinforcing bar, 0.9 for a horizontally cast reinforcing bar).

A review of earlier literature indicates that most of the bond models were obtained from pull-out tests and, therefore, might not be suitable for reinforced concrete flexural members. Furthermore, the maximum bond stress corresponding to the slip cannot be calculated. This is very important because the maximum bond stress is a main factor in the determination of crack spacing, anchorage strength, and the strength of lap slices. Therefore, it is not possible to accurately calculate the maximum crack width and the lengths of anchorage and lap slices using the bond models proposed by the previously mentioned researchers.

It should be noted that the models given by Ikki et al. [18, 19] and Shima [20], among the aforementioned models, were compared with the proposed model. This is because some coefficients of the models given by Rehm [22] and Mugurama and Morita [24, 25] must be decided experimentally.

3.2. Newly Proposed Model. In order to formulate a bond model which works successfully under axial boundary conditions as shown in Figure 4(c), the following factors are considered to express the bond stress. In research carried out by Shima [20], it is discovered that the bond strength increases lineally according to the compressive concrete strength ($f_c'^{2/3}$). Therefore, the effect of the compressive concrete strength is considered by using a nondimensional bond stress ($\tau_b/f_c'^{2/3}$). Also, the effect of the bar diameter is considered by using a nondimensional slip (S/d_s). This is because the slip is proportional to the bar diameter [23, 26]. These nondimensional indexes were used to formulate the bond stress. In addition, the relative rib area was introduced in order to consider the effects of the rib geometry of the deformed bars. With some modification of the Ikki et al. [18, 19] model, the bond model is proposed as

$$\tau_b = k \times f_c'^{\,2/3} \times \left\{ 1 - \exp\left[-4500 \left(\frac{S}{d_s} \right)^{1.45} \right] \right\}^{0.5} \times \exp\left[-5 \left(\frac{S}{d_s} \right) + 5.5 f_R^{\,0.9} \right], \quad (6)$$

where k: coefficient that accounts for the effects of the proposed model on bond stress, f_R: relative rib area.

From the position of the reinforcing bar during casting, and depending on whether or not a stirrup is used, k can be classified as

$$k = 0.2 \times \exp\left\{ \left[-4.5 + 55 \left(3.06 f_R - 0.24 \right) \right] \times \frac{100}{A_c} \right\} \quad \text{(vertically cast bar)},$$

$$k = 0.2 \times k_{sh} \times \exp\left\{ \left[-4.5 + 55 \left(3.06 f_R - 0.24 \right) \right] \times \frac{100}{A_c} \right\} \quad \text{(vertically cast bar with stirrups)},$$

$$k = 0.2 \times \exp\left\{ \left[-4.5 + 55 f_R \right] \times \frac{100}{A_c} \right\} \quad \text{(horizontally cast bar)},$$

$$k = 0.2 \times k_{sh} \times \exp\left\{ \left[-4.5 + 55 f_R \right] \times \frac{100}{A_c} \right\} \quad \text{(horizontally cast bar with stirrups)}, \quad (7)$$

where A_c: cross sectional area of the concrete, k_{sh}: coefficient which expresses the effect of the stirrups (1.0 for a vertically

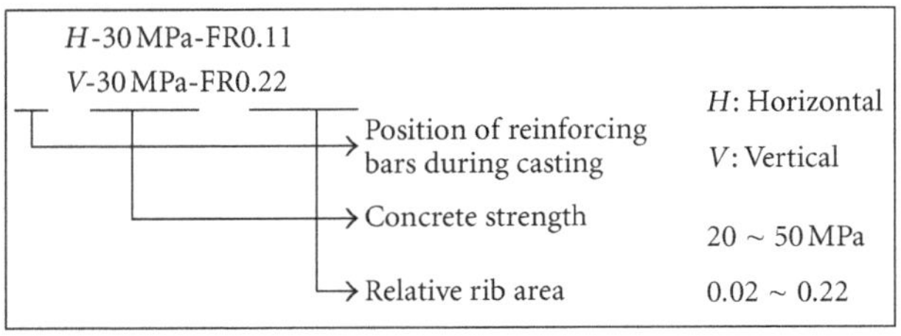

Figure 5: Key to specimen numbering system.

cast stirrup, 0.85 for a horizontally cast stirrup), f_R: relative rib area.

Although the Equations k are divided into a horizontally cast bar and a vertically cast bar, the change in bond capacity due to the position of the bar during casting was insignificant. This is because the depth of the concrete below a horizontal bar of reinforced concrete members was usually smaller than 30 cm. As shown in research carried out by Rehm [22], bars that are horizontally cast 30 cm above the bottom of the specimen show an inferior bond behavior compared to bars that are cast in a horizontal position marginally above the bottom of the specimen.

In addition, the reciprocal value of the cross-sectional area of concrete ($100/A_c$) has the same meaning as the average stress of the corresponding cross sectional area of concrete falling under the unit load. In this study, it is believed that the size effects of the cross section need to be considered in accordance with the cross sectional area of concrete, which is actually loaded.

As reflected in the k_{sh}, the effect of lateral confinement on bond behavior or bars in structural members should be considered, since the degree of confinement of a bar affects both the magnitude of the ultimate bond stress and the bond stress-slip relationship. Ikki et al. [19] reported significant results that showed that the bond stress was slightly higher in specimens that did not have stirrups than it was in specimens that did have stirrups. Even though Ikki et al. [19] may not have determined the fundamental cause for the previously described phenomena, it is assumed that the bond-slip relationship does not rely on the in-plane restraint (which is the perpendicular reinforcing bar axis), until splitting cracks occur. This is because splitting cracks had not occurred within the range for measuring the maximum bond stress.

Furthermore, the bond stress at any location along a reinforcing bar is proportional to the slope of the steel stress distribution curve at that point. The bond stress can be obtained at any point by using the bond model proposed by many researchers. However, the bond model is very sensitive, since the slope of the steel stress distribution curve is a differential term. Hence, the dispersion of the bond stress test data is currently unavoidable.

3.3. Parametric Study. In order to estimate the various factors of the proposed model, a parametric study was performed by transmuting nondimensional slip (S/d_s) to $1 \sim 4\%$, relative rib area (f_R) to $0.02 \sim 0.22$, compressive concrete strength (f_c') to $20 \sim 50$ MPa, and the reciprocal of the cross sectional area of concrete ($100/A_c$) to 0.111%, 0.0625%, and 0.04%, respectively. Each specimen is identified by a numbering system containing a code to help identify the characteristics of that specimen. The specimen numbering system is explained in Figure 5.

Figures 6 and 7 demonstrate the effect of the aforementioned factors on the bond stress-slip relationship for the specimens with a 0.0625% reciprocal of the cross-sectional area of concrete and 30 MPa and 40 MPa f_c' compressive concrete strength, respectively. Due to the scope of this paper, not all figures are shown.

When using deformed bar, the uniform variation of the bond stress noted for a general range of the relative rib area (f_R) was $0.10 \sim 0.15$. However, a very high variation was seen at 0.16 and above. Based on the above results, the proposed model in (6) was varied by a fixed ratio for the nondimensional slip (S/d_s) and the compressive concrete strength (f_c') with small sensitivity for the relative rib area (f_R) at a general range of $0.10 \sim 0.15$. However, sensitive variation was verified when the size of the concrete cross-section is relatively small. Nevertheless, such effects would not be large because the size of the cross-section in a real application is usually sufficient.

Rehm [27] reported that the upper limit of bond resistance capacity is controlled through shear failure in the case of $f_R = h_d l_d > 0.15$. This fact was unexpectedly proved through the parametric study for the general range imposed on the value f_R by the relevant standards or by approval documents. The results of the parametric study verify that the parameters considered in the proposed model appropriately simulate the bond behavior of a reinforced concrete member under axial tension boundary conditions.

4. Experimental Program

4.1. Specimen and Experimental Conditions. For verification of the proposed model, the test results performed by Ikki et al.[18, 19] were used. The experimental conditions and properties of each specimen are shown in Table 1.

The specimens for the axial tension tests are made up of two series with different properties, namely, the T-Series and the D-Series. The concrete cover of all specimens is 40 mm, regardless of the concrete cross-section. In the T-Series, the square section was used, with the exception of specimen TV0-50, which was used in the circular section. The length of each specimen was 764 mm as shown in Figure 8. Concrete was cast in a horizontal direction perpendicular to the embedded reinforcing bars. However, specimens TV0-45 and TV0-45 were cast in a vertical direction parallel to the embedded reinforcing bars. In addition, a stirrup was arranged only in specimens TH-30 and TH-45. The tensile reinforcing bar was embedded at the center of the square and circular sections. Here, D19 bar was used as the tensile reinforcing bar for all specimens, while the Young's modulus of the bar is $E_s = 181.3$ GPa, and the yield stress is $f_y = 357.8$ MPa. The test details of specimen TH1-45 are shown in Figure 8. In the D-Series, two types of circular sections

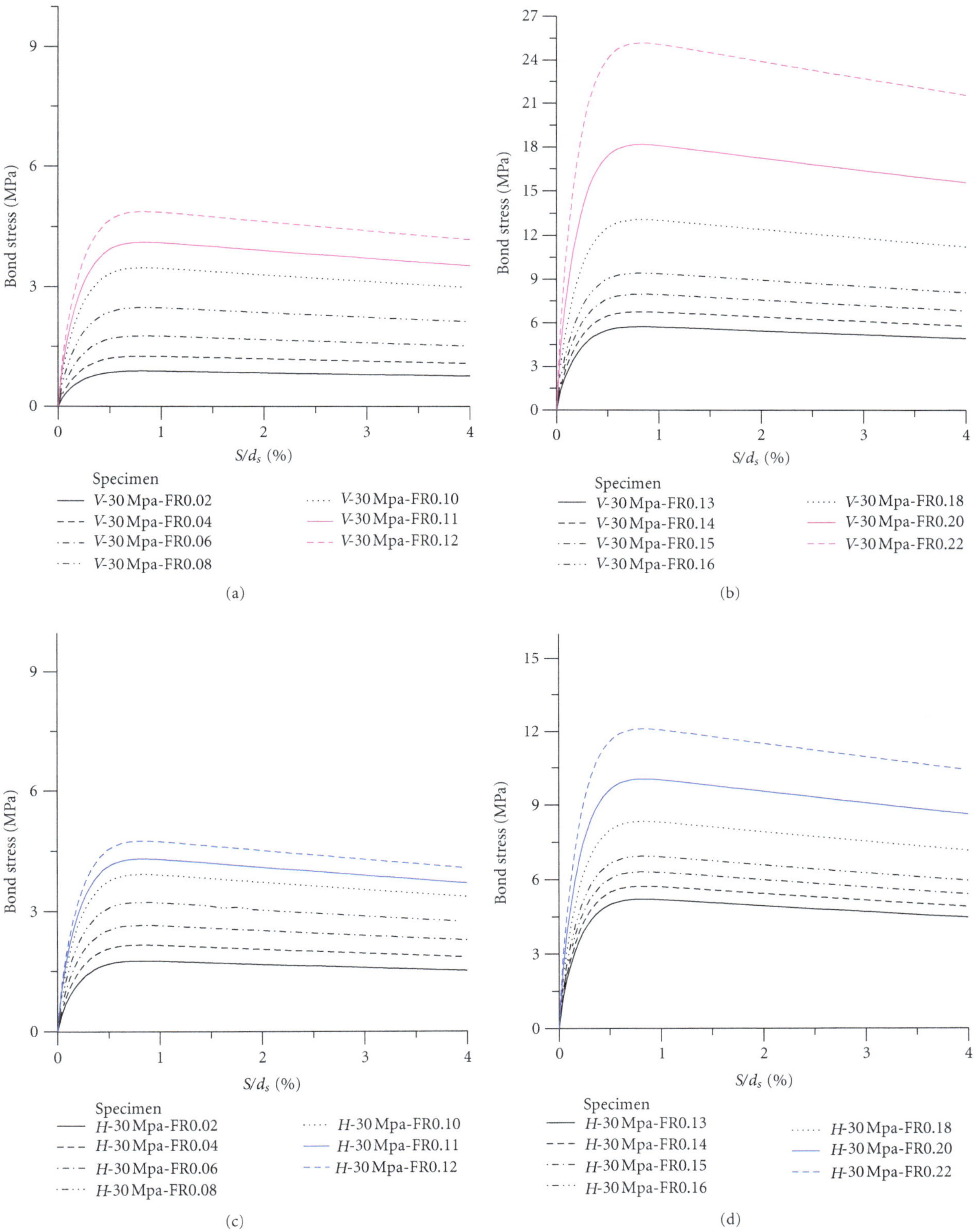

FIGURE 6: Effects of bond factors ($f_c' = 30$ MPa, $100/A_c = 0.0625\%$).

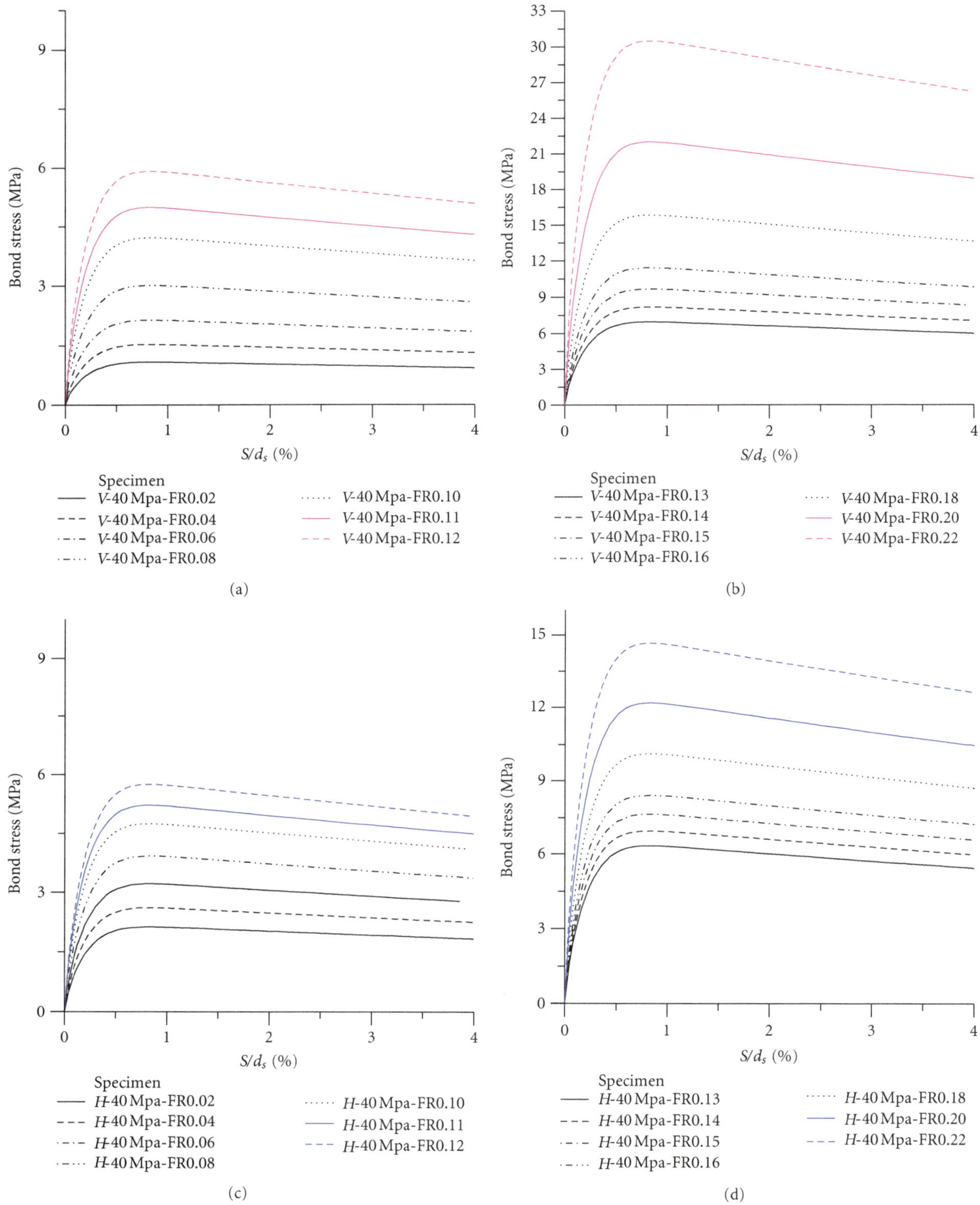

FIGURE 7: Effects of bond factors ($f_c' = 40$ MPa, $100/A_c = 0.0625\%$).

Table 1: Properties of test specimen.

Type (Series)	Specimen	Direction of bar on casting	Existence of stirrup (D6)	Sectional property [mm]
(1)	(2)	(3)	(4)	(5)
T-Series	TH1-30	Horizontal	Yes	300 × 300
	TH1-45	Horizontal	Yes	450 × 450
	TH0-30	Horizontal	No	300 × 300
	TH0-45	Horizontal	No	450 × 450
	TV0-45	Vertical	No	450 × 450
	TV0-50	Vertical	No	ϕ 500 mm
D-Series	D-30	Vertical	No	ϕ 300 mm
	D-45	Vertical	No	ϕ 450 mm

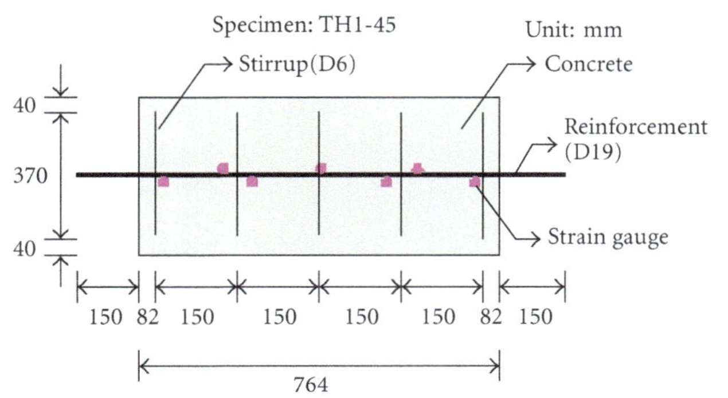

Figure 8: Sectional properties of the specimen TH1-45 and positions of strain gauges.

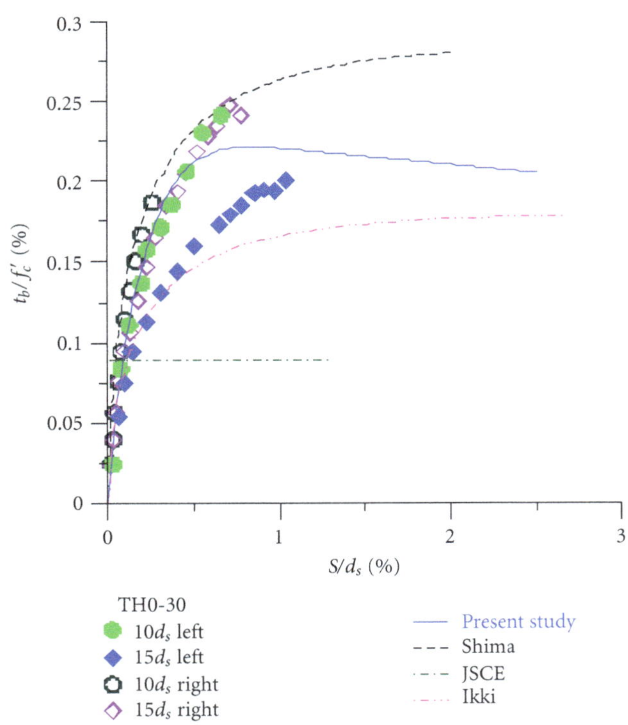

Figure 9: Bond stress-slip relationship of TH0-30 specimen.

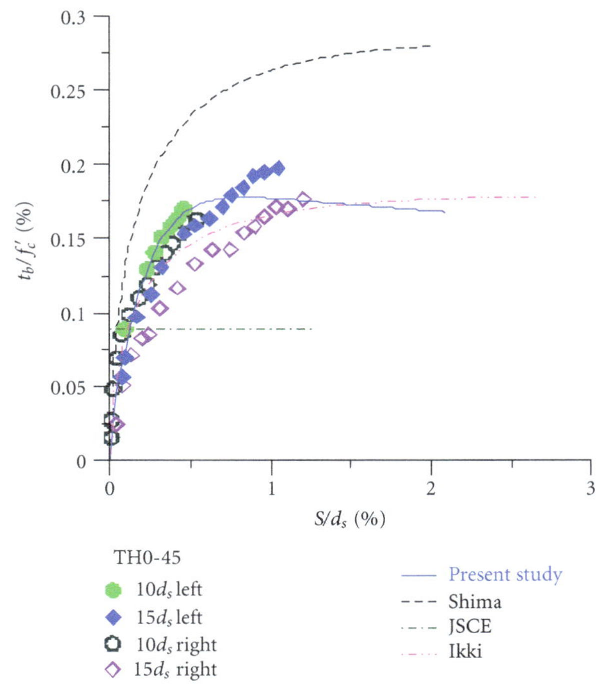

Figure 10: Bond stress-slip relationship of TH0-45 specimen.

were used, with diameters of 30 cm and 45 cm, respectively. The length of each specimen was 964 mm. Concrete was cast in a vertical direction. No stirrup was arranged and the tensile reinforcing bar was also embedded at the center of the circular section. Here, D16 bar was used as the tensile reinforcing bar, with the Young's modulus of the bar E_s = 192 GPa, and the yield stress f_y = 390 MPa.

4.2. Determination of Bond Stress and Slip. To obtain the bond stress and slip from the measured strain, the method proposed by Yamao et al. [23] was utilized. At any point, the bond stress derived from an equilibrium condition can be written as

$$\tau_b = \frac{d_s}{4}\frac{df_s}{dx} = \frac{d_s E_s}{4}\frac{d\varepsilon_s}{dx}, \tag{8}$$

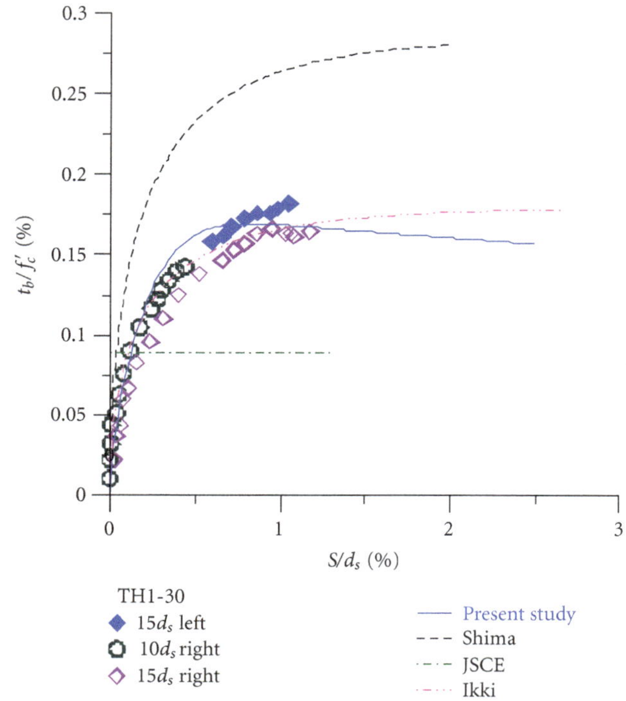

FIGURE 11: Bond stress-slip relationship of TH1-30 specimen.

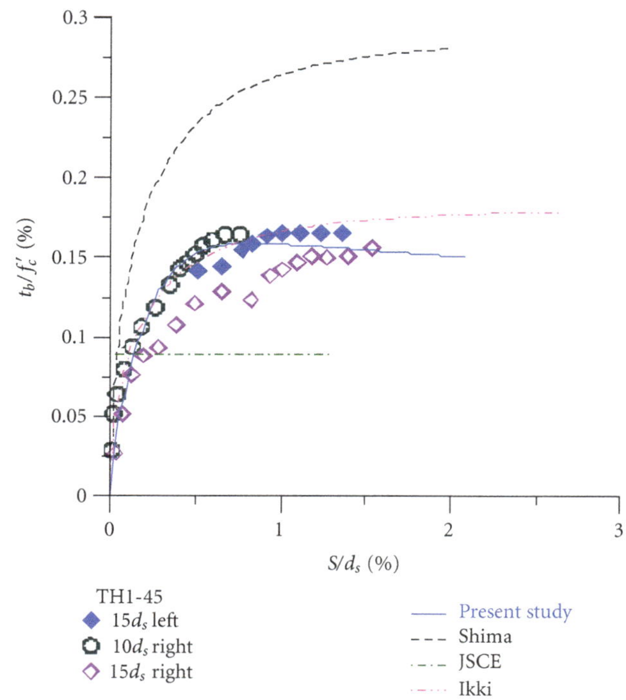

FIGURE 12: Bond stress-slip relationship of TH1-45 specimen.

where E_s: Young's modulus of the bar, $d\varepsilon_s/dx$: slope of the strain distribution curve.

The function of strain $\varepsilon_s = \varepsilon(x)$ is assumed as the strain distribution curve obtained by connecting every three neighboring point with a second degree polynomial function. Here, if the three neighboring points are set as $(x_{i-1}, \varepsilon_{i-1})$, (x_i, ε_i), and $(x_{i+1}, \varepsilon_{i+1})$, the strain curve can be expressed and the obtained strain curve shows the strain distribution from point $(i-1)$ to point (i) as

$$\varepsilon_s = \varepsilon(x) = a_i + b_i x + c_i x^2, \tag{9}$$

where a_i, b_i, c_i: numbers, called coefficients.

By substituting (9) into (8), $\tau_{bi}(x)$ can be written as:

$$\tau_{bi} = \frac{E_s d_s (b_i + 2c_i x)}{4}, \tag{10}$$

where τ_{bi}: bond stress corresponding to point i.

In the axial tension test, the slip at any point could be obtained by the integration of the strain from the certain point $(i-1)$ with zero slip to the point concerned i as

$$S_i(x) = S_{i-1}(x) + \int_{x_{i-1}}^{x_i} (a_i + b_i x + c_i x^2)\,dx. \tag{11}$$

5. Validation of the Proposed Model

All the test specimens are calculated by using the proposed model and the results obtained are compared with the experimental data given by Ikki et al. [18, 19] and analytical solutions provided by eminent researchers [18–20] and the JSCE code [17]. For notational convenience, the analytical solutions have been designated according to the surname of the first authors (Ikki, Shima) and of design codes (JSCE). For example, Shima represents the solution provided by Hiroshi Shima. Also, the solution obtained from the proposed model is denoted by the present study.

The following equation is the concrete design bond strength proposed by the JSCE code [17] as

$$f_{bdj} = 0.28 f_c'^{2/3} \left(f_{bdj} \le 3.2\ \text{MPa} \right). \tag{12}$$

In Figures 9–16, analytical solutions are shown compared to the experimental data and the proposed model at different locations along a reinforcing bar. From these figures, it is verified that the bond stress-slip relationship in a specimen is the same, independent of the locations along a reinforcing bar. However, the bond stress in the data obtained from the right sides of specimens D-30 and D-45 shown in Figures 15 and 16 is underestimated compared to that of the left sides. For this reason, Ikki and Kiyomiya [18, 19] explained that the local hardness of the concrete layer in contact with the ribs of indentation had decreased due to the bleeding of the concrete.

The important basic factor for the analysis of deformational behavior and the calculation of crack width in reinforced concrete members is the constitutive law for reinforcing bar and concrete. The constitutive law of bond action between reinforcing bar and concrete is also important for FEM analysis. Consequently, the JSCE model [17] is always consistently conservative.

The Shima [20] model yields an over-predicted overall bond-slip behavior for all specimens, which is not desired.

TABLE 2: Comparison results for the maximum bond stress and the associated slip.

Method Type	Data of Experiment		Present Study		Ikki	Shima	JSCE
	S/d_s [%]	$\tau_{b,\max}/f_c'$	S/d_s [%]	$\tau_{b,\max}/f_c'$	$\tau_{b,1\%}/f_c'$	$\tau_{b,1\%}/f_c'$	τ_{bdj}/f_c'
(1)	(2)	(3)	(4)	(5)	(6)	(7)	(8)
TH1-30	1.013	0.189	0.852	0.169	0.166	0.264	0.0891
TH1-45	0.813	0.179	0.845	0.159	0.166	0.264	0.0891
TH0-30	0.834	0.230	0.857	0.221	0.166	0.264	0.0891
TH0-45	1.050	0.197	0.850	0.178	0.166	0.264	0.0891
TV0-45	0.847	0.197	0.865	0.201	0.185	0.264	0.0891
TV0-50	0.832	0.196	0.809	0.190	0.185	0.264	0.0891
D-30	0.771	0.199	0.831	0.205	0.207	0.295	0.0999
D-45	0.930	0.180	0.850	0.172	0.207	0.295	0.0999
Mean	0.886	0.196	0.845	0.187	0.181	0.272	0.092

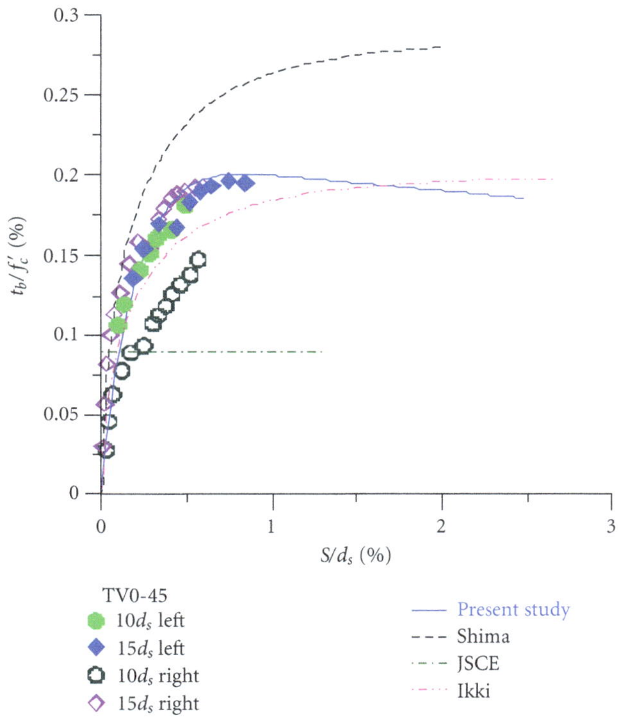

FIGURE 13: Bond stress-slip relationship of TV0-45 specimen.

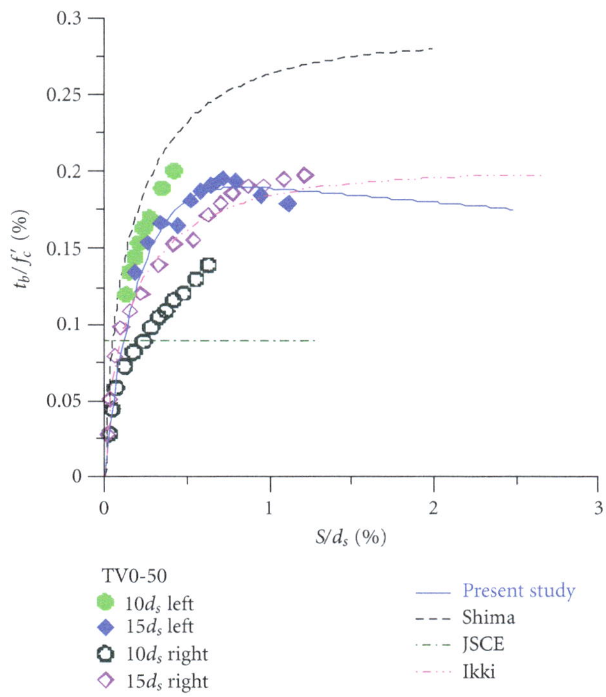

FIGURE 14: Bond stress-slip relationship of TV0-50 specimen.

Moreover, the model predicted the maximum bond stress of $\tau_{b,1\%}/f_c'$ as significantly more than the experimental values for all test specimens. This trend can also be seen in Table 2. This model is quite erroneous when used for predicting the bond stress-slip behavior for the axial tension members. This may be due to the fact that the model was developed based on the pull-out tests.

The Ikki et al. [18, 19] model, in general, predicts comparatively closer to the experimental values for all the test specimens. For the square specimens, this model slightly underestimated the maximum bond stress $\tau_{b,1\%}/f_c'$, yet the predictions were very close to the test values. Conversely, this model slightly overestimated the circular specimens. As for the Shima [20] model discussed above, this model also did not give the maximum bond stress and the associated slip.

The model proposed in this study is quite accurate at predicting the bond stress-slip relationship of test specimens and its accuracy is not affected by changes in test variables. This is important since the test variables change rapidly among different types of reinforcing bars and various concrete strengths, and so forth. Also, the benefit of this model is that the maximum bond stress corresponding to the associated slip can be obtained, which is a core factor in the estimation crack spacing.

It should be noted that the peak value of the nondimensional bond stress curve $\tau_b/f_c'^{2/3}$, compared to the nondimensional slip index curve S/d_s, is of greater significance when identifying the maximum crack-width-related aspects of reinforced concrete members. That is, the maximum bond stress is a main factor in the estimation of crack spacing, so

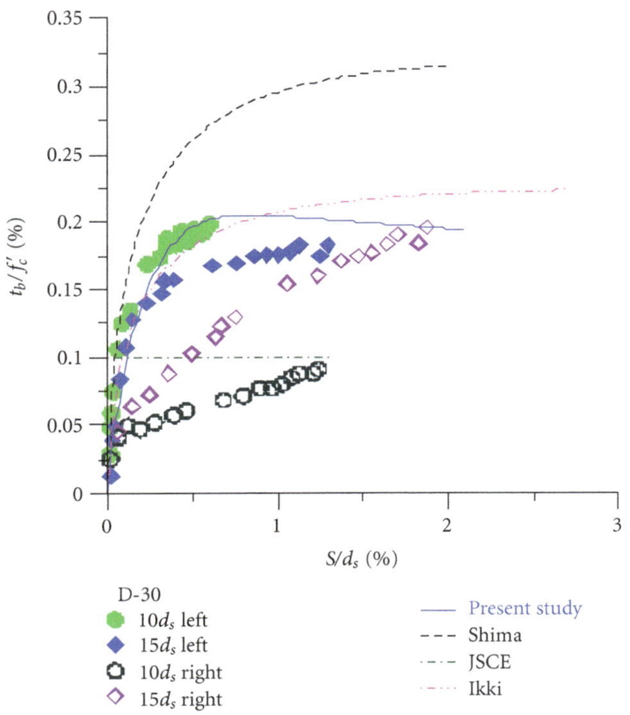

FIGURE 15: Bond stress-slip relationship of D-30 specimen.

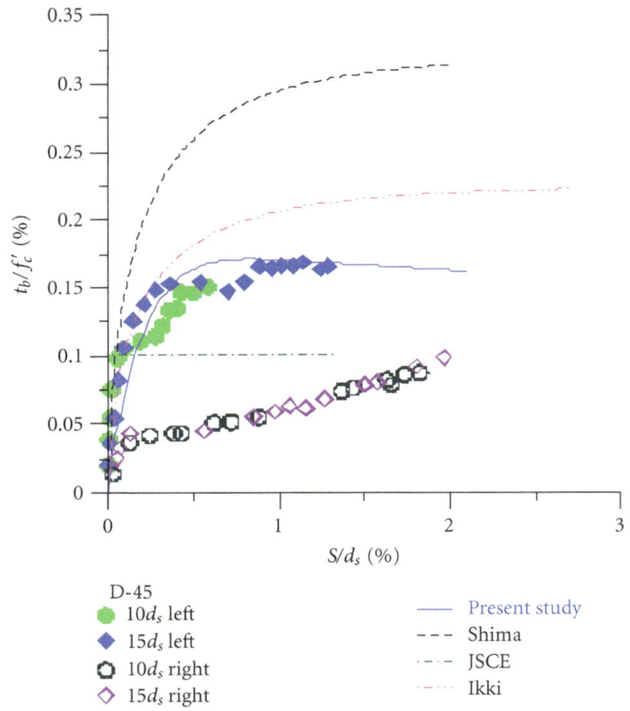

FIGURE 16: Bond stress-slip relationship of D-45 specimen.

direct calculation of the maximum crack width is possible from the estimation of the maximum bond stress.

Furthermore, in order to consider the predictive assessment in respect of the analytical solutions including the proposed model, the calculations were carried out for the nondimensional maximum bond stress and the corresponding nondimensional slip for all specimens. The comparison results can be seen in Table 2.

However, the bond stress $\tau_{b,1\%}$ when $S/d_s = 1\%$ was used, since the estimation of the maximum bond stress by using the formula proposed by Ikki et al. [18, 19] and Shima [20] was difficult. Based on the comparison results, the model proposed by the author showed that the mean values for the nondimensional maximum bond stress and the corresponding nondimensional slip were 0.845 and 0.187, respectively. These values are the nearest values for the experimental values. Consequently, the model proposed in this paper is the most accurate model for estimating the bond behavior of axial tension members.

6. Conclusion

In this paper, an analytical model was proposed to estimate the bond stress-slip relationship for reinforced concrete members under axial tension loadings. By extending the conventional bond stress-slip theory and considering the relative rib area of deformed bars and the concrete cross-sectional area, an equation to compute the bond stress was formulated as a function of the nondimensional factors. To estimate the various factors of the proposed model, a parametric study has been performed on each of the bond factors. Also, based on the results of the parametric study, the factors considered in the proposed model appropriately simulate the bond behavior of reinforced concrete members under axial tension boundary conditions.

The results of the analysis presented in this paper indicate that the proposed model can be used to effectively estimate the bond stress-slip relationships of reinforced concrete flexural members. The results compared well with the experimental data and validation was obtained.

Although the current design codes and the bond models proposed by different researchers provide design criteria for the bond behaviors of reinforced concrete members, these criteria are not well defined and there is no concurrence among the different variables affecting the bond behaviors. Therefore, the proposed model could be of benefit to structural engineers when analyzing and designing reinforced concrete members under bending.

Acknowledgment

This work was supported by the Postdoctoral Research Program of Sungkyunkwan University (2011).

References

[1] FIP, "Bond of reinforcement in concrete: state-of-art report," Fib Bulletin No. 10, CEB-FIP, Lausanne, Switzerland, 2000.
[2] CEB-FIP, *Structural Concrete—Textbook on Behaviour, Design and. Performance, Updated Knowledge of the CEB-FIP Model Code 1990*, vol. 1, Sprint-Druck, Stuttgart, Germany, 1999.
[3] J. Dai, T. Ueda, and Y. Sato, "Unified analytical approaches for determining shear bond characteristics of FRP-concrete

interfaces through pullout tests," *Journal of Advanced Concrete Technology*, vol. 4, no. 1, pp. 133–145, 2006.

[4] R. Eligehausen, E. P. Popov, and V. V. Bertero, "Local Bond stress-slip relationships of deformed bars under generalized excitations," Report UCB/EERC-83/23, University of California, Berkeley, Calif, USA, 1983.

[5] A. P. Fantilli and P. Vallini, "A cohesive interface model for the pullout of inclined steel fibers in cementitious matrixes," *Journal of Advanced Concrete Technology*, vol. 5, no. 2, pp. 247–258, 2007.

[6] P. M. Ferguson and J. N. Thompson, "Development length for large high strength reinforcing bars," *Journal of the American Concrete Institute*, vol. 62, no. 1, pp. 71–93, 1965.

[7] C. A. Menzel, "Some factors influencing results of pull-out bond tests," *Journal of the American Concrete Institute*, vol. 35, no. 6, pp. 517–542, 1939.

[8] S. Morita and T. Kaku, "Local bond stress-slip relationship under repeated loading. Resistance and ultimate deformability of structures acted on by well-defined repeated loads," in *Proceedings of the IABSE Symposium*, pp. 221–227, Lisbon, Portugal, 1973.

[9] B. H. Oh and S. H. Kim, "Realistic models for local bond stress-slip of reinforced concrete under repeated loading," *Journal of Structural Engineering*, vol. 133, no. 2, pp. 216–224, 2007.

[10] H. Sezen, J. P. Moehle et al., "Bond-slip behavior of reinforced concrete members. Concrete structures in seismic regions," in *Proceedings of the Fib Symposium*, Athens, Greece, May 2003.

[11] E. Vos and H. W. Reinhardt, "Influence of loading rate on bond behaviour of reinforcing steel and prestressing strands," *Materials and Structures*, vol. 15, no. 1, pp. 3–10, 1982.

[12] D. Ngo and A. C. Scordelis, "Finite element analysis of reinforced concrete beams," *Journal of the American Concrete Institute*, vol. 64, no. 6, pp. 152–163, 1967.

[13] A. H. Nilson, "Nonlinear analysis of reinforced concrete by the finite element method," *Journal of the American Concrete Institute*, vol. 65, no. 9, pp. 757–766, 1968.

[14] H. Noguchi and N. Inoue, "Analytical techniques of shear in reinforced concrete structures by finite element method," in *Proceedings of JCI Colloquium on Shear Analysis of RC Structures*, pp. 57–96, JCI-C4E.

[15] G. L. Balazs, "Cracking analysis based on slip and bond stresses," *ACI Materials Journal*, vol. 90, no. 4, pp. 340–348, 1993.

[16] S. De Silva, H. Mutsuyoshi, and E. Witchukreangkrai, "Evaluation of shear crack width in l-shaped prestressed reinforced concrete beams," *Journal of Advanced Concrete Technology*, vol. 6, no. 3, pp. 443–458, 2008.

[17] JSCE, *Standard specifications for concrete structures: structural performance*, Japan Society of Civil Engineers, Tokyo, Japan, 2002.

[18] N. Ikki and O. Kiyomiya, "Effect of axial concrete stress on bond strength of deformed bar," *Proceedings of Japan Concrete Institute*, vol. 21, no. 3, pp. 373–378, 1996.

[19] N. Ikki, O. Kiyomiya, and M. Yamada, "Experimental study on the effects of numerous factors on bond-slip relationship," *Journal of Materials, Concrete Structures and Pavements*, vol. 33, no. 550, pp. 73–83, 1996.

[20] H. Shima, *Bond models micro and macro models for bond in reinforced concrete*, Ph.D. thesis, Tokyo University, Tokyo, Japan, 1987.

[21] R. Park and T. Paulay, *Reinforced Concrete Structures*, John Wiley & Sons, New York, NY, USA, 1975.

[22] G. Rehm, "The fundamental law of bond," in *Proceedings of the Symposium on Bond and Crack Formation in Reinforced Concrete*, pp. 491–498, RILEM, Stockholm, Sweden, 1957.

[23] H. Yamao, L. Chou, J. Niwa et al., "Experimental study on bond stress slip relationship," *Proceedings of Japan Society of Civil Engineers*, vol. 343, pp. 219–228, 1984.

[24] H. Muguruma and S. Morita, "Fundamental study on bond between steel and concrete-part 1: basic laws of bond stress distribution I," *Journal of Structural Construction Engineering*, vol. 131, pp. 1–8, 1967.

[25] H. Muguruma and S. Morita, "Fundamental study on bond between steel and concrete-part 1: basic laws of bond stress distribution II," *Journal of Structural Construction Engineering*, vol. 132, pp. 1–6, 1967.

[26] S. Morita and S. Fujii, "Bond-slip models in finite element analysis," in *Proceedings of the Japan-US Seminar on Finite Element Analysis of Reinforced Concrete Structures*, pp. 348–363, Tokyo, Japan, May 1985.

[27] G. Rehm, *Kriterien zur beurteilung von bewehrungsstäben mit hochwertigem Verbund*, Ernst & Sohn, Berlin, German, 1969.

13

Chloride Penetration through Cracks in High-Performance Concrete and Surface Treatment System for Crack Healing

In-Seok Yoon

Department of Construction Information Engineering, Induk University, Seoul 139-749, Republic of Korea

Correspondence should be addressed to In-Seok Yoon, isyoon@induk.ac.kr

Academic Editor: Jin Ping Lu

For enhancing the service life of concrete structures, it is very important to minimize crack at surface. Even if these cracks are very small, the problem is to which extend these cracks may jeopardize the durability of these decks. It was proposed that crack depth corresponding with critical crack width from the surface is a crucial factor in view of durability design of concrete structures. It was necessary to deal with chloride penetration through microcracks characterized with the mixing features of concrete. This study is devoted to examine the effect of high strength concrete and reinforcement of steel fiber on chloride penetration through cracks. High strength concrete is regarded as an excellent barrier to resist chloride penetration. However, durability performance of cracked high strength concrete was reduced seriously up to that of ordinary cracked concrete. Steel fiber reinforcement is effective to reduce chloride penetration through cracks because steel fiber reinforcement can lead to reduce crack depth significantly. Meanwhile, surface treatment systems are put on the surface of the concrete in order to seal the concrete. The key-issue is to which extend a sealing is able to ensure that chloride-induced corrosion can be prevented. As a result, penetrant cannot cure cracks, however, coating and combined treatment can prevent chloride from flowing in concrete with maximum crack width of 0.06 mm and 0.08 mm, respectively.

1. Introduction

Over the last decades, many improvements have been presented in both measuring techniques and modeling of ionic flows in the field of chloride penetration. However, the majority of this research has been performed on sound uncracked concrete, even though many in situ concrete structures have more or less micro-cracks. Although these microcracks may not degrade the structural safety immediately, the question is to which extent these cracks may jeopardize the durability of cracked concrete by permitting penetration of aggressive substances into concrete easily. To begin with, moisture penetration through cracks can defeat the purpose of the surface treatment by allowing localized reinforcement corrosion [1]. Especially, penetration of chlorides into concrete through the micro-cracks can make a significant harmful effect on reinforcement corrosion [2]. There are several different reasons for not taking the effect of crack on chloride penetration into account. The most important ones are that it is difficult to make accurate measurements on the flow through cracks and that there are many affecting factors and complexities. Nevertheless, the existent cracks are a reality and it is necessary to deal with the effect of cracks on chloride penetration.

In particular, high-performance concrete (high-strength concrete) can be regarded as a durable material to insure a long service life, as long as there are no cracks. Barrier function to protect high-strength concrete is severely broken, however, if a crack is generated. Since high-strength concrete is more brittle than ordinary concrete, durability design must be accomplished for greater safety. Nevertheless, it is very rare to deal with the issue in the study of high-performance concrete.

Meanwhile, surface treatment system for small cracks and epoxy injection for medium or big cracks have been recommended for crack repair [3]. That is, if the size of the crack is small, surface treatment system can be considered as one of the best options to heal the crack in terms of cost effectiveness versus durability performance and to extend the service life of concrete structures. However, there seems

FIGURE 1: Outline of experimental procedures: from making concrete samples to getting data.

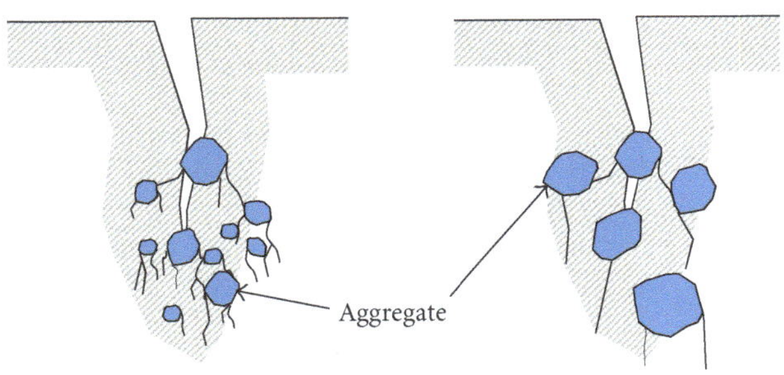

(a) Small size of aggregate (b) Large size of aggregate

FIGURE 2: Effect of size of coarse aggregate on microcracking.

TABLE 1: Material properties of penetrant (PT).

Items	Value
Main ingredient	Silane
Turbidity	below 1.0
Ratio of water sorptivity (KSF 4930)	0.05
Color	Colorless
pH	8.3
Type	Liquid

to be no research to explain technical limitation of surface treatment system, in particular, critical crack width which can be healed by the system [4, 5]. ACI [6], JCI [7], and KISTEC [8] just suggested that low solids and low-viscosity resin-based systems are useful to cure fine cracks. However, the limitation of surface treatment system as a crack repair method should be understood. That is, it is necessary to define the scale of the crack that can be healed by the method, in order to make sure of the reliability of the crack repair method.

The purpose of this study is to examine the effect of microcracks on chloride penetration in cracked HPC and study the efficiency of certain surface treatments for the crack healing. Accelerating experiment was carried out to investigate the effect of surface treatment systems to enhance service life of cracked concrete. For plain concrete and surface treated concrete, the critical crack size that can be healed by surface treatment systems is examined. Cracks smaller than this critical crack size are considered not to have a significant influence on the rate of chloride transport inwards, while chloride penetration does proceed faster above the critical crack size.

2. Experiment Program

2.1. Program I: Chloride Penetration through Cracks in HPC

2.1.1. Preparation of Concrete Samples and Cracking. All steps for experiment program are depicted in Figure 1. Mixing proportion is given in Table 1. For comparison with high-performance concrete (HPC), OPC samples were made with ordinary Portland cement, water/cement ratio of 0.50. OPC samples had two types with maximum size of coarse aggregate: 8 mm and 16 mm. It is well known that interfacial transition zone (ITZ) between cement paste and aggregate forms a more porous framework than bulk cement paste. Since ITZ is considered as the weakest link of the chain, cracks are preferentially propagated in concrete under compressive or tensile stress. The crack propagation depends on the size of coarse aggregate, as shown in Figure 2. Cracking is very complicated in concrete with the small size of coarse aggregate and this is intended in two types of OPC concrete samples.

HPC samples were mixed with fly ash and silica fume and its 28-day strength was equivalent to 100.1 Mpa. HPFRC samples mean high-strength concrete reinforced by steel fiber. Steel fiber can lead to complicated cracks and cracks bridging in the concrete. This was intended for investigation on chloride penetration through the complicated cracks in HPFRC samples.

The concrete was cast into 150 × 150 × 150 mm cube moulds, which were moist-cured at 20°C for 48 hours, followed by air exposure at a relative humidity of 65% and 20°C. Sample series and preparation were set up in such a way that the specimens could be tested for chloride penetration at an age of 28 days. In these specimens, a notch was made of 5 mm wide and 5 mm depth. Next to the notch steel plates were glued to introduce the tensile forces to create a fracture.

To control the crack formation, two LVDTs were fixed on each side of the specimen, perpendicular to the loading direction, to monitor the crack mouth opening displacement (CMOD). The tests were performed in deformation control mode up to the target CMOD. Then the specimen was unloaded and taken out of the machine. Time, force, stroke, and CMOD were recorded during the test. The effective CMOD value (i.e., the unloaded value) was used in the analysis and graphical plots. After unloading but before performing the RCM test the notch had been cut off, to arrive at a sample thickness of 50 mm.

2.1.2. Chloride Penetration Experiment. Rapid chloride migration (RCM) test, standardized in NordTest NT-Build 492 [9], has been used more and more to investigate pene-tration of chloride into concrete. This is set up as a non-steady-state migration test. These are then used to help design concrete materials for long-lasting structures. In such design process ordinary crack-free concrete was considered; however, this experiment was effective to quantify the effect of cracks on chloride penetration. One side of the sample was exposed to a salt solution, while the other side of the sample was exposed to a 0.3 N NaOH solution. Using stainless steel plate electrodes on both sides of the sample, a DC voltage was applied to force the negatively charged chloride ions through the sample. After the test the samples were split and sprayed with 0.01 N $AgNO_3$ solution to visualize the ingress of chloride into the sample.

Since it is the purpose of this research to investigate the influence of micro-cracks on the penetration of chloride, the RCM test has been performed on fractures samples after 28 days. The introduction and formation of cracks are described later in this paper. The expected end result of an RCM test is given in Figure 3. Chloride penetration depth in the unfractured zone (d_1) and the fractured zone (d_T) was measured and the values were used for chloride diffusion coefficient in the unfractured zone (D_1) and the fractured zone (D_T).

2.2. Program II: Crack Healing to Prevent Chloride Penetration. Surface treatment systems can be classified into three

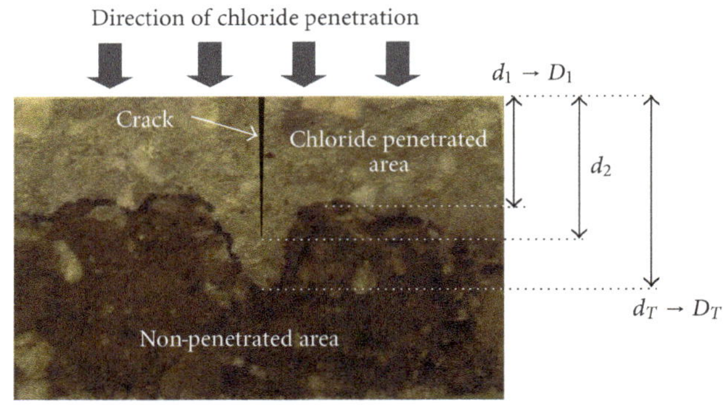

Figure 3: Schematic end result of the RCM test on fractured specimen, with different notation codes.

Table 2: Material properties of coating material (CT).

Items	Value
Main ingredient	Polymer
Bonding strength (Mpa)	2.12
Color	Light grey
Type	powder + water

types: (a) penetrant, whereby surface near zone is impregnated and dense layer with pore blocker is formed, (b) coating, whereby in most cases a thin or thick continuous membrane polymer film is applied, and (c) sealer, whereby properties of both the penetrant and coating can be expected. Figure 4 illustrates these three groups of protection.

After generating cracks in OPC samples, surface treatment system was applied in the surface. In the experiment, penetrant and coating material were used and materials properties are shown in Tables 2 and 3. For penetrant, the achievable penetration depth in concrete mainly depends on four factors: the type of hydrophobic agent applied, the W/C of the concrete substrate, the initial moisture content, and the surface preparation of the concrete substrate [10].

Thickness of surface coating was measured by a microscope and the value was about 70 um, as shown in Figure 5. Samples treated with penetrant were split and sprayed with water. Then the penetration depth was measured and the value was 2.8 mm.

3. Result and Discussion

3.1. Chloride Penetration through Cracks. Figure 6 represents the relative diffusivity (D_T/D_1) of concrete with various sizes of coarse aggregate. The relative diffusivity of concrete with G_{max} 16 mm tends to be higher than that of concrete with G_{max} 8 mm, however, the trend is not significant. Although micro-cracks are concentrated at the surface of coarse aggregate, the cracks do not influence chloride penetration. Since the micro-cracks are smaller than 0.012 mm, which is a critical crack width in accordance with a previous research of author, the crack is thought not to be effective for chloride penetration [11]. According to the study, chloride ions cannot immediately penetrate all the way up to the

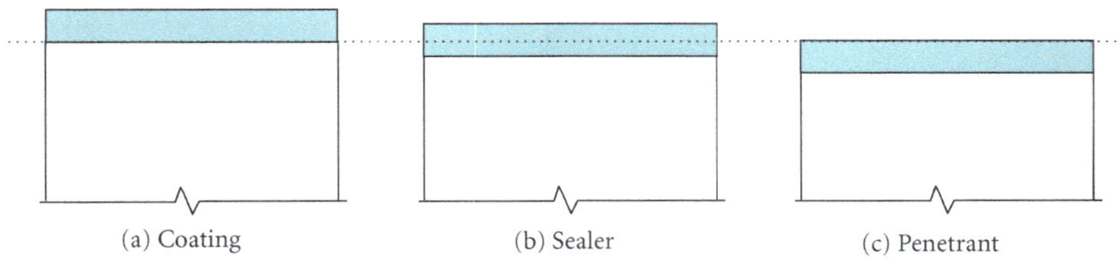

FIGURE 4: Type of surface treatment systems.

TABLE 3: Mixing proportion of concrete.

Mix symbol	28-day strength (MPa)	W/C	G_{max} (mm)	Unit weight (kg/m^3)				
				Water	Cement	Mineral Adx.	Sand	Gravel
OPC	28.5	0.50	16	185	370	—	720	1021
OPC II	28.5	0.50	8	185	370	—	720	1021
HPC	100.1	0.20	8	128	440	Fly ash 132 Silica fume 64	1045	485
HPFRC*	125.5	0.33	—	145	439	—	1912	—

*Volume of steel fiber: 175 kg/m^3 (length: 16 mm, diameter: 0.13 mm).

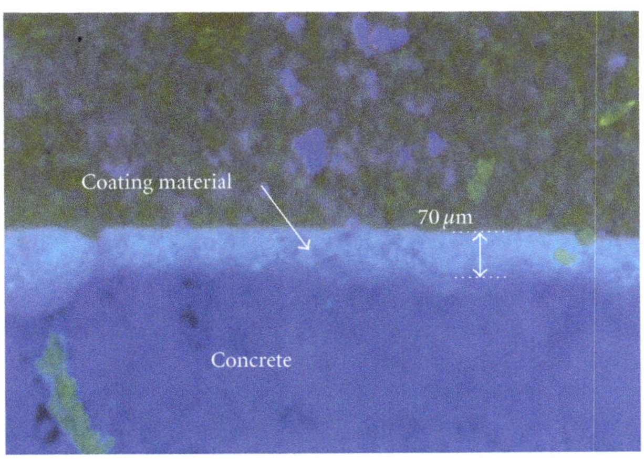

FIGURE 5: Thickness of coating material on surface concrete.

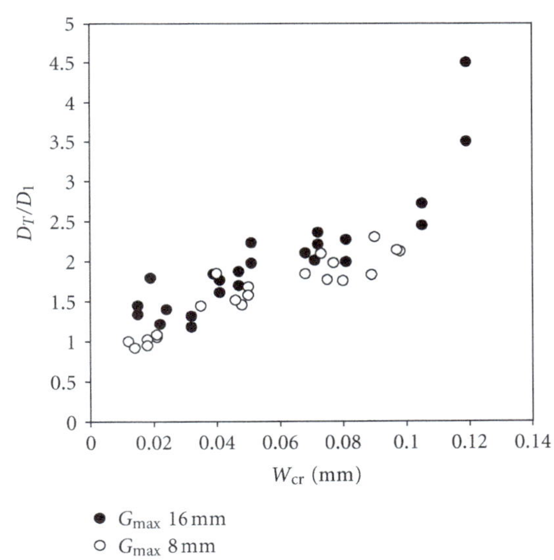

FIGURE 6: Relative diffusivity versus G_{max}.

crack tip when the cracked concrete is faced with chloride solution. The rate of penetration decreases with increasing crack depth. This rate of penetration is influenced by the geometrical properties of the cracks. With increasing the depth of the crack, the crack width decreases and, more importantly, the crack is tortuous and has more chance to be unconnected closer to the crack tip, because of crack branching and bridging. Based on the findings in this research, it is necessary to make a strong argument to consider the crack width versus chloride penetration under various tension conditions.

Figure 7 depicts the relative diffusivity of various concrete samples. Compared with OPC, HPC and HPFRC have a high relative diffusivity as if the durability performance of HPC/HPFRC is worse than that of OPC. The result is far from previous researches because HPC is recognized as a-high-durability concrete. The reason can be explained by Figures 8 and 10.

Figure 8 shows the relationship between chloride penetration depth and crack width. HPC and OPC showed the same chloride penetration depth if the crack width is higher than 0.03 mm. Overall, a little chloride penetration is examined at HPC and HPFRC. This is because high strength leads to dense micro-structure of concrete and steel fiber inducing

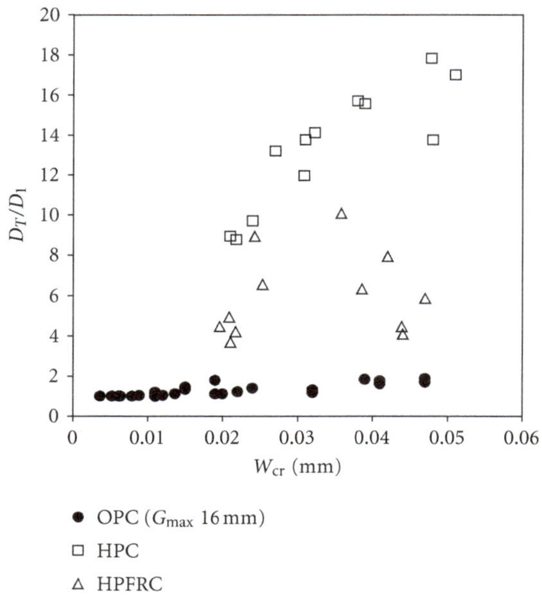

FIGURE 7: Relative diffusivity versus concrete type.

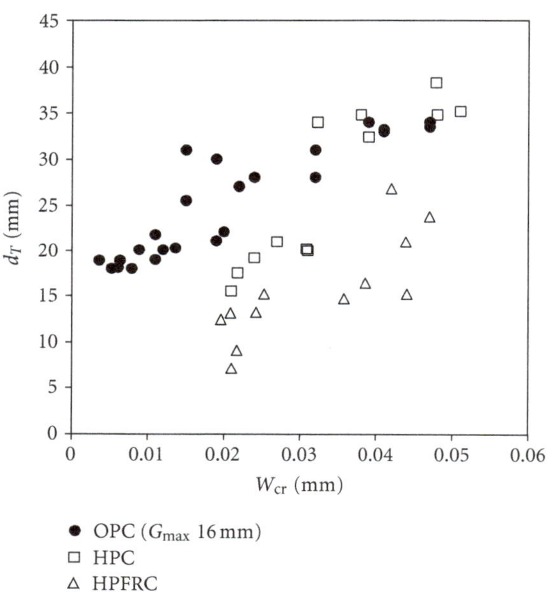

FIGURE 8: Chloride penetration depth versus crack width.

delay strain induced crack propagation. That is to say, steel fiber has a function to increase crack depth, $d_2 - d_e$ (see Figure 9), corresponding with critical crack width 0.012 mm, and to decrease effective crack depth d_e.

Figure 10 shows a photo of chloride profile in concrete contaminated by chloride ions. Compared with OPC, HPC does not allow chloride penetration. Once cracks are generated, however, HPC loses its high protection performance significantly. This is why relative diffusivity (D_T/D_1) of HPC is much higher than that of OPC. HPC without cracks has a very high resistance to deterioration due to chloride penetration; however, HPC is prone to be cracked. If there are

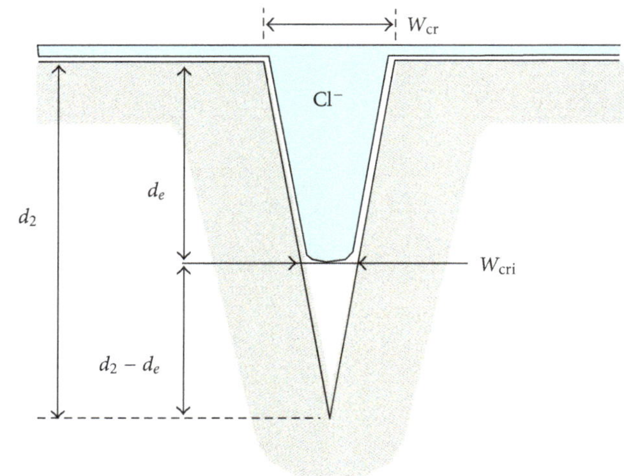

FIGURE 9: Schematic representation of effective crack depth according to critical crack depth.

no cracks in HPC, its service life can be overestimated too much in durability designing. Therefore, the penetration of harmful substances through cracks must be considered in establishing the maintenance strategy of HPC structures under severe conditions.

3.2. Crack Healing Effect. Figure 11 shows chloride diffusion coefficient for penetrant-treated concrete. As expected, the chloride diffusion coefficient of cracked concrete is higher than that of uncracked concrete. From CMOD 0.024 mm, chloride penetration through crack is found. CMOD 0.024 mm is too small, and, thus, it can be suggested that penetrant is difficult to expect crack healing.

Figure 12 represents chloride diffusion coefficient for CT-treated concrete. Compared with PT-treated concrete, CT-treated concrete showed low diffusion coefficient. This means that the performance of surface coating is better than that of penetrant for chloride resistance. Above all, the experimental result shows obviously that when the crack width was large, beyond around 0.060 mm, the rate of increase of chloride penetration increased. In other words, increasing chloride diffusivity is responsive to 0.060 mm of CMOD. This means the critical crack opening that can be healed with coating system is 0.060 mm, and this is higher than 0.024 mm, which is a limitation of crack width healed with penetrant. Therefore, it is concluded that coating also does not have a satisfactory performance although coating is better than penetrant for crack healing.

Figure 13 illustrates chloride diffusion coefficient for PCT-treated concrete. Already below a value of CMOD of 0.08 mm the influence of the crack on the chloride transport starts to become noticeable. The penetrant can lead to dense fracture; however, space between cracks should still exist. If penetrant-treated concrete is coated, membrane coating film is formed and this is effective to protect the concrete from chloride penetration. The double protective system can be believed to cure crack with 0.08 mm maximum width. This

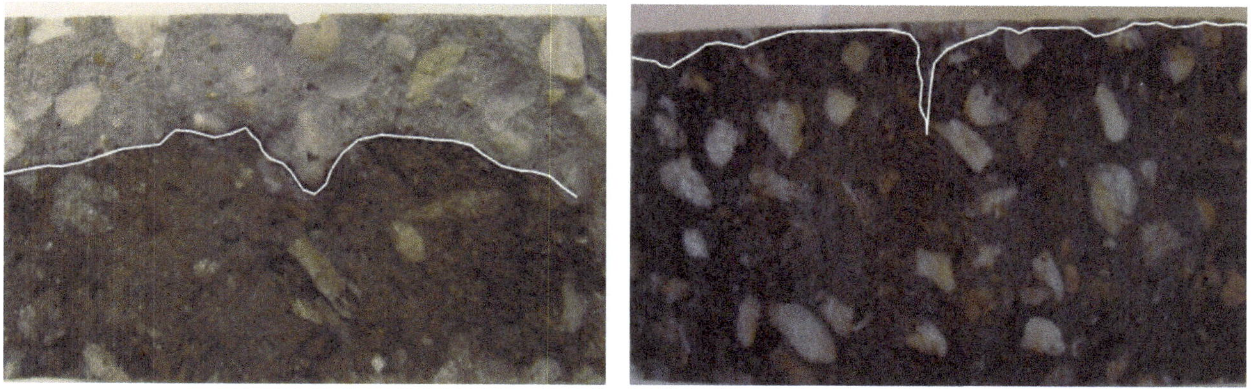

(a) OPC concrete

(b) HPC concrete

FIGURE 10: Comparison of chloride profile of OPC concrete with that of HPC concrete.

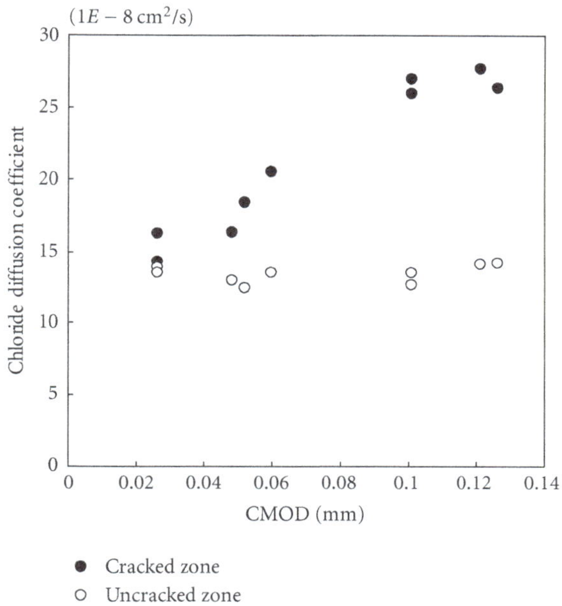

FIGURE 11: Chloride diffusion coefficient for PT-treated concrete.

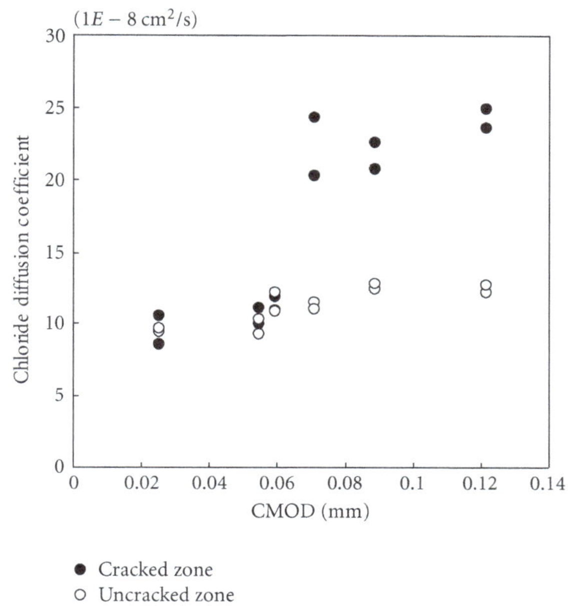

FIGURE 12: Chloride diffusion coefficient for CT-treated concrete.

is also not a satisfactory performance for healing. Of course, the critical crack width depends on type of surface treatment system, quality, the number of treatments, number of layers, and so on.

Figure 14 represents the effect of type of surface treatment on the prevention of extra chloride penetration through cracks. There is no doubt that the penetrant cannot cure cracks. This is because the penetrant cannot fill space between crack fractures and does not have significant crack-bridging capabilities, although the hydrophobic nature of the material may reduce the intrusion of moisture into narrow cracks. The experimental result shows obviously, however, that the coating system is effective for healing cracks up to maximum CMOD 0.06 mm, the so-called critical crack width that can be healed by surface treatment system. For combined application of the penetrant and coating system, when the CMOD was large, beyond around 0.08 mm, the rate of increase of chloride penetration increased. This means that there is no extra chloride penetration through the crack if the crack width is smaller than 0.08 mm. As a matter of course, the value depends on material properties of surface treatment; however, the value is much smaller than we have expected. The value should lead to extremely conservative criterion. It is difficult to apply to estimation the integrity of in situ concrete structures with a value. JCI [7] and KISTEC [8] recommended the surface treatment method including the penetrant to keep the performance of watertightness for concrete with maximum crack width 0.20 mm. Obviously 0.20 mm is much higher than the critical crack width of 0.06 ∼ 0.08 mm. It is necessary to examine its effective application.

Based on the results of this study, it is concluded that penetrant is not suitable for crack healing and the coating system should be applied after understanding the limitation

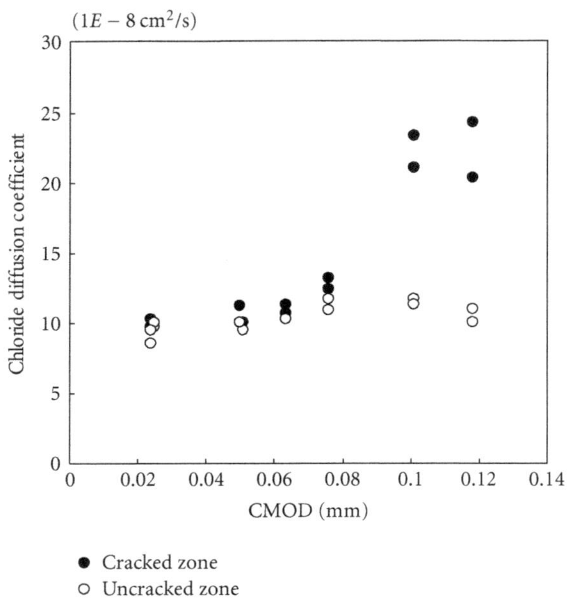

Figure 13: Chloride diffusion coefficient for PCT treated concrete.

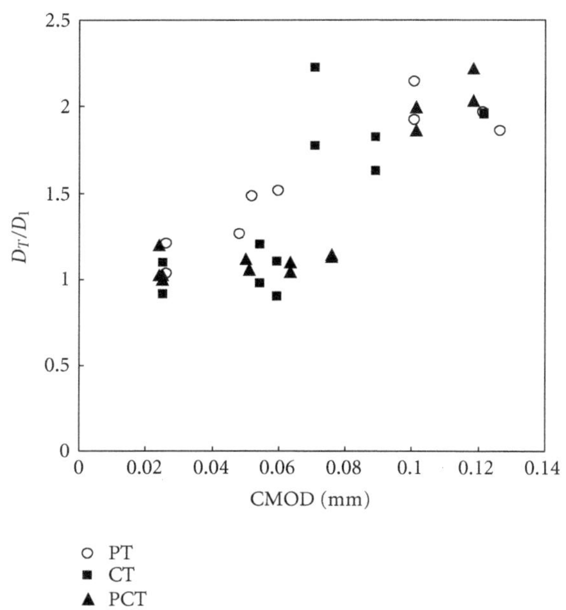

Figure 14: Relative chloride diffusion coefficient (D_T/D_I) for surface treated concrete.

of its effectiveness. For verification, the crack healing effect and critical crack width should be examined with the outcome of chloride immersion testing for long term.

This study is expected to provide us with insight into developing design criteria for a durable concrete and in understanding the effect of surface treatment system to repair cracked concrete structure. It is expected that the outcome of this study could be an engineering ground to choose the type of surface treatment system and to judge the durability performance of surface treated concrete. For more research, the capacity to bridge and heal cracks should be carefully studied depending on the type of the surface treatment system and number of coating layers.

4. Conclusions

(1) Experiment recipe was designed to investigate chloride penetration through cracks in OPC and HPC concrete. Compared with OPC, HPC did not allow chloride penetration. However, HPC with cracks lost its high protection performance seriously. That is, the relative diffusivity of HPC is much higher than that of OPC. This means that chloride penetration through cracks must be considered in HPC structures exposed to sea water. Steel fiber had a function to increase crack depth and to decrease effective crack depth.

(2) Penetrant cannot cure cracks. However, it was noticed that concrete with maximum crack width 0.06 mm was healed by coating system. Combination of penetrant and coating system made a positive contribution to crack healing up to maximum crack width 0.08 mm.

Acknowledgment

This study was sponsored by Induk University, Seoul, Republic of Korea.

References

[1] ACI Committee 546, *Concrete Repair Guide, ACI Manual of Concrete Practice—Part 6*, ACI International, 2002.

[2] R. François and G. Arliguie, "Effect of microcracking and cracking on the development of corrosion in reinforced concrete members," *Magazine of Concrete Research*, vol. 51, no. 2, pp. 143–150, 1999.

[3] Concrete Society, *Non-Structural Cracks in Concrete, Concrete Society Technical Report No.22, Concrete Repair Manual*, vol. 1, ACI, BRE, Concrete Society, International Concrete Repair Institute, London, UK, 2nd edition, 1992.

[4] E. Nolan, P. A. M. Basheer, and A. E. Long, "Effects of three durability enhancing products on some physical properties of near surface concrete," *Construction and Building Materials*, vol. 9, no. 5, pp. 267–272, 1995.

[5] S. O. Nwaubani and J. Dumbelton, "A practical approach to in-situ evaluation of surface-treated structures," *Construction and Building Materials*, vol. 15, no. 4, pp. 199–212, 2001.

[6] ACI Committee 224, *Causes, Evaluation and Repair of Cracks in Concrete Structures, ACI Manual of Concrete Practice—Part 2*, ACI International, 2002.

[7] JCI, *Investigation, Evaluation and Repair of Cracks in Concrete Structures*, Japanese Concrete Institute, 2003 (in Japanese).

[8] KISTEC, *Specification on Repair of Cracks in Concrete Structures*, Korea Infrastructure Safety & Technology Corporation, 1999 (in Korean).

[9] NT Build 492, *Concrete, Mortar and Cement-Based Repair Materials: Chloride Migration Coefficient from Non-Steady-State Migration Experiment*, Nordtest, Espoo, Finland, 1999.

[10] J. G. Dai, Y. Akira, F. H. Wittmann, H. Yokota, and P. Zhang, "Water repellent surface impregnation for extension of service

life of reinforced concrete structures in marine environments: the role of cracks," *Cement and Concrete Composites*, vol. 32, no. 2, pp. 101–109, 2010.

[11] I. S. Yoon and E. Schlangen, "Long/short term experimental study on chloride penetration in cracked concrete," *Key Engineering Materials*, vol. 417-418, pp. 765–768, 2010.

14

Viscoelastic-Viscoplastic Modelling of the Scratch Response of PMMA

G. Kermouche,[1] N. Aleksy,[2] and J. M. Bergheau[2]

[1] *Ecole Nationale Supérieure des Mines de Saint-Etienne, SMS Division, LGF UMR CNRS 5307, 158 Cours Fauriel 42023 Saint-Etienne Cedex, France*
[2] *Université de Lyon, ENISE, LTDS UMR CNRS 5513, 58 Rue Jean Parot, 42023 Saint-Etienne Cedex 2, France*

Correspondence should be addressed to G. Kermouche; kermouche@emse.fr

Academic Editor: Pavel Lejcek

This paper aims at understanding how to model the time-dependent behavior of PMMA during a scratch loading at a constant speed and at middle strain levels. A brief experimental study is first presented, consisting of the analysis of microscratches carried out at various scratching velocities and normal loads. The loading conditions have been chosen in such a way that neither (visco)elasticity nor (visco)plasticity of the PMMA may be neglected a priori. The main analyzed parameter is the tip penetration depth measured during the steady state. Then, a finite element model is used to investigate the potential of classical elastic-viscoplastic constitutive models to reproduce these experimental results. It is mainly shown that these models lead to unsatisfying results. More specifically, it is pointed out here that the time-independent Young modulus used in such models is not suitable. To take into account this feature, a viscoelastic-viscoplastic model based on the connection in series of a viscoelastic part with a viscoplastic part is proposed. It is shown that it leads to more acceptable results, which points out the importance of viscoelasticity in the scratch behavior of solid polymers.

1. Introduction

The scratch test is one of the most efficient tests to investigate the mechanical resistance of coated and uncoated surfaces [1–3]. In a typical single pass scratch test, a hard indenter slides along the surface of a relatively soft material (see Figure 1). Depending on the materials and on the scratch conditions, crack propagations [4, 5] and/or elastic-plastic deformations are observed [6, 7]. The scratch test on amorphous polymers has been widely investigated during the past ten years [8–14], because of the large use of amorphous polymer films as surface protection in many industrial sectors. It is now well known that the scratch resistance of amorphous polymers depends on the scratching speed and on the load applied on the tip [15, 16]. It points out their complex mechanical behavior which depends strongly on strain and strain rate levels and on the temperature. Three kinds of constitutive models are commonly used: viscoelastic models [17] for low strain levels, elastic-viscoplastic models [18] for large strain levels, and viscoelastic-viscoplastic models [19] for middle strain levels. These different strain levels are illustrated in Figure 2. In the case of scratch loadings, it is difficult to state on the appropriate model to use [20, 21]. Indeed, depending on the scratch conditions, the strain levels can be very different, and thus different phenomena have to be taken into account to really understand the scratch behavior of such materials. For instance, one issue addresses the necessity to take into account the viscoelasticity at small strains. One other issue could concern the influence of the strain softening or the effect of the hyperelastic hardening on the scratch results. The present paper focuses on the time-dependent properties (viscoelasticity, viscoplasticity) of the PMMA and their relation to the scratch response. The aim of this paper is to determine which is the best model to reproduce the influence of the scratching speed induced by a rigid spherical indenter on a PMMA.

Such investigations cannot only be based on experimental measurements, because viscoelasticity and viscoplasticity may have a similar effect on the measured penetration depth, contact pressure, and friction coefficient [22] at given loads and scratching speeds. Finite element calculations appear as a great help for such an investigation because it allows to

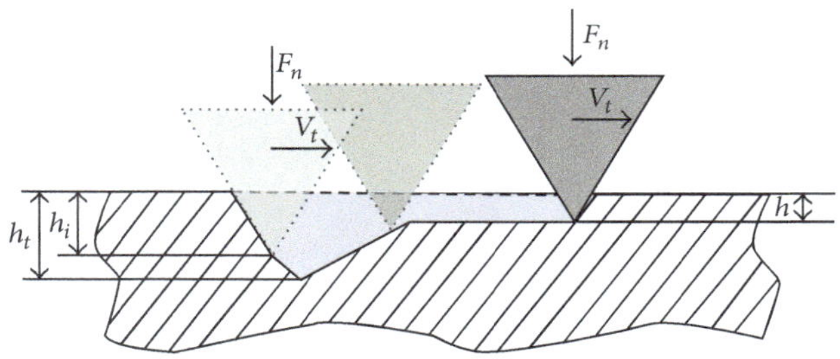

FIGURE 1: Penetration during load controlled scratch test.

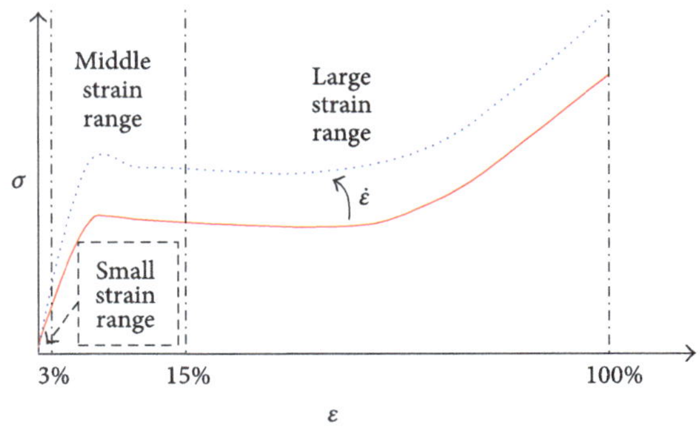

FIGURE 2: Stress/strain curve of amorphous glassy polymers in tensile test: the small strain level corresponds to linear viscoelasticity, the large strain level can be well described by an elastic-viscoplastic model, and the middle strain level should account for both viscoelasticity and viscoplasticity.

understand the own influence of each material property on the scratch response. The finite element method has often been used in the past to understand phenomena occurring during scratch tests on semi-infinite homogeneous solids [6, 7, 13, 20, 23–25] or on coating over substrate systems [3, 26, 27]. Here, we propose a coupled experimental and numerical investigation to address this issue.

Let us now describe the scratch test realized. At fixed normal load, penetration of the tip increases during the indentation stage up to the static mechanical balance. When the tip begins its tangential displacement, the contact area decreases by the rear face of the tip. That leads to a new increase of the penetration depth to satisfy the static mechanical balance. Then, a frontal ridge is created and grows until the steady state is reached, which leads also to a decrease of the penetration depth (Figure 1). The analysis of this transient regime is very difficult because it depends both on the kinematics of the indentation stage and on the scratch stage. Moreover, the static adhesion should also have a strong influence. Consequently, in this paper, only the results in the steady state are investigated.

This paper is splitted into a brief experimental part and a modelling part. The experimental approach consists in the analysis of microscratches carried out on PMMA samples at different scratching velocities and at different normal loads. The loading conditions have been chosen in such a way that neither the (visco)elasticity nor (visco)plasticity of the PMMA may be neglected a priori. The main analyzed parameter is the penetration depth of the tip measured during the steady state. The modeling part aims at determining the constitutive model that can reproduce the experimental results with enough accuracy. First, the standard elastic-viscoplastic model developed by Argon and later improved by Arruda and Boyce (AB model in the sequel) is used. It is shown that this model is not accurate enough. More specifically, it is pointed out here that the time-independent Young modulus used in such models is not suitable. To take into account this feature, a viscoelastic-viscoplastic model based on the connection in series [22] of the classical viscoelastic model and the AB viscoplastic model is proposed. It is shown that this model leads to more acceptable results. Finally, we conclude on the importance to take into account viscoelasticity to model the scratch behavior of solid polymers.

2. Experimental Study

2.1. Scratch Experiments. The scratch tests are performed with the nanoindenter XP from Agilent. This device is set on an air table, and the whole system is installed in a thermal room. The temperature, the normal load, and the scratch velocity are controlled [28]. The penetration depth of the tip is measured. Two samples of PMMA from two different suppliers are scratched three times in the same conditions in order to ensure the stability and the reproducibility of the results. Nine series of microscratch are carried out for three normal loads, $F_n = 20\,\text{mN}$, $F_n = 40\,\text{mN}$, and $F_n = 80\,\text{mN}$, and three tip velocities $V = 0.1\,\mu\text{m}{\cdot}\text{s}^{-1}$, $V = 1\,\mu\text{m}{\cdot}\text{s}^{-1}$, and $V = 10\,\mu\text{m}{\cdot}\text{s}^{-1}$ with a spherical tip (radius $R = 35\,\mu\text{m}$). The whole experiments are performed at the room temperature ($T = 25^{\circ}\text{C}$) with the scratch distance of $100\,\mu\text{m}$. The steady state has been reached for each scratch. The results are summarized in Table 1. The measured friction coefficient is about 0.15 for each scratch.

2.2. Analysis. As shown in Table 1, the penetration depth is maximum for the highest normal load and the lowest scratching speed used. According to many authors [20, 29], the average strain level induced by a scratch test can be estimated using the same formula as for indentation test

$$\varepsilon_r = \kappa \frac{a}{R}, \tag{1}$$

where a is the contact radius and κ is a coefficient. For indentation test, κ is about 0.2 whereas for scratch test κ is about 0.5–1. Neglecting piling up or sinking-in phenomena, the contact radius a equals to the geometrical contact radius a_c which may be estimated directly from penetration depth h by the following equation:

$$a_c = \sqrt{h\,(2R - h)}. \tag{2}$$

Here, the ratio a_c/R is about 0.25–0.35. According to Pelletier et al. [29], it corresponds to the middle strain level for PMMA, where both viscoelasticity and viscoplasticity may affect the scratch response of the PMMA. Let us also note that at such a strain level, the hyperelastic hardening has a very weak

TABLE 1: Penetration depth measured for different scratching speeds and different load applied.

F_n in mN	$V = 0.1\,\mu$m/s	$V = 1\,\mu$m/s	$V = 10\,\mu$m/s
20	870 ± 10	800 ± 8	730 ± 9
40	1550 ± 10	1420 ± 8	1280 ± 8
80	2780 ± 12	2540 ± 10	2240 ± 11

influence on the PMMA mechanical response contrary to viscoelasticity.

Let us now focus on the effect of the scratch speed: the results summarized in Table 1 point out that whatever the normal load is, h decreases when V increases. Consequently, the contact area should also decrease when V increases, which indicates that the mean contact pressure—usually called hardness—increases with the tip velocity. As shown in previous papers [3, 22], the increase in mean contact pressure with scratching speed is a clue that the scratched material exhibits a time-dependent behavior. This could be explained with the help of the scratch representative parameters as shown by Aleksy et al. [22]. Similarly to indentation testing [30, 31], the representative stress is proportional to the mean contact pressure. According to Briscoe et al. [32], the representative strain rate is given by

$$\dot{\varepsilon}_r \propto \frac{V}{a_c}. \quad (3)$$

For a given normal load, the increase in V leads to a decrease in a_c and, according to (3), an increase in $\dot{\varepsilon}_r$. As the mean contact pressure also increases when V increases, it points out that the representative stress depends on the representative strain rate. Consequently, the PMMA exhibits a time-dependent behavior during such a scratch test. As shown by Aleksy et al. [22], both viscoelasticity and viscoplasticity may lead to the same result. The aim of the following section is to state whether viscoelasticity and/or viscoplasticity are the the most suitable models to describe the scratch behavior of PMMA at such middle strain levels.

As a preliminary conclusion, this experimental study points out that the mechanical behavior of the PMMA depends on the representative strain rate during the scratch test which is in good agreement with the literature on this subject [8, 32] and also with the tensile and compressive behavior of this well-known time-dependent material.

3. Numerical Investigation

The aim of the numerical investigation is to determine the best model which reproduces the experimental variation of steady state penetration depth h with the tip velocity. Experiments have highlighted the time-dependent behavior of the PMMA. As microscratches at such strain levels involve residual grooves [29], therefore the simplest model to be employed is an elastic-viscoplastic one. However, the Young modulus of the glassy amorphous polymers is well known to depend on the strain rate, which is an evidence of a viscoelastic behaviour. In this section, we try to adress the following

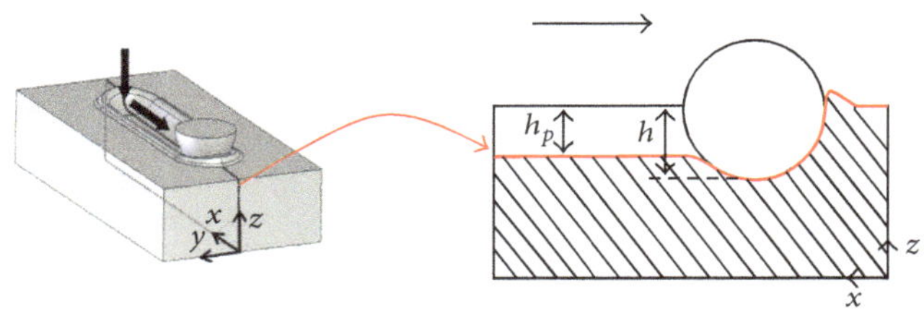

FIGURE 3: Representation of the zx symmetry cut plane.

issue: *Is the elastic-viscoplastic model accurate enough or is it necessary to use a combined viscoelastic-viscoplastic model?*

3.1. Numerical Model. The simulation of a scratch test on a polymeric material is nontrivial because of two major numerical problems: contact with friction under large displacement and large deformations which could induce severe mesh distortions. All of these problems have been treated in previous articles [3, 7, 22], and this is the reason why they are not described in details in the present paper, to focus on the main goal of this work: the mechanical behaviour of PMMA under a scratch loading.

The scratch test presents a symmetry plane, $y = 0$, defined by the indentation z-axis and the scratching x-axis (Figure 3). Consequently, the numerical study can be carried out on half of the domain. The finite element domain is a right angled parallelepiped.

The dimensions of this domain have been chosen high enough to avoid boundary effects. The mesh is constituted of 8-node-brick isoparametric element with selective-reduced integration scheme to ensure plastic incompressibility. The numerical simulations are performed using a large displacement/large strain option (updated Lagrangian formulation). To increase the computation accuracy and to limit the computation time, the mesh is refined in the vicinity of the scratch. The contact and friction inequalities are solved using the penalty method [33]. For very low levels of scratching conditions (small penetration compared to tip radius, low friction coefficient), numerical simulations can be performed without remeshing algorithm. Let us note that an automatic remeshing procedure was implemented in the finite element software Systus to study more severe scratching conditions [3].

Similarly to the experimental study, the tip radius is $R = 35\,\mu$m, the temperature is $T = 25^\circ$C, the tip velocities are $V = 0.1\,\mu\text{m}\cdot\text{s}^{-1}$, $V = 1\,\mu\text{m}\cdot\text{s}^{-1}$, and $V = 10\,\mu\text{m}\cdot\text{s}^{-1}$, and the normal load is constant and identical to the experimental one ($F_n = 20$ mN, $F_n = 40$ mN and $F_n = 80$ mN). However, the numerical loading is achieved by controlling the quasistatic displacement of the indenter. A specific algorithm has thus been developed in the finite element software (Systus) to ensure the control of the load. It adjusts the vertical displacement to maintain the normal load constant, at each time step. Let us note that the adhesion friction coefficient (Coulomb) is adapted at each velocity and each normal load to exactly reproduce the friction coefficient measured during the experimental microscratches ($\mu_{\text{measured}} = \mu_{\text{adhesion}} + \mu_{\text{ploughing}}$ [2, 34]).

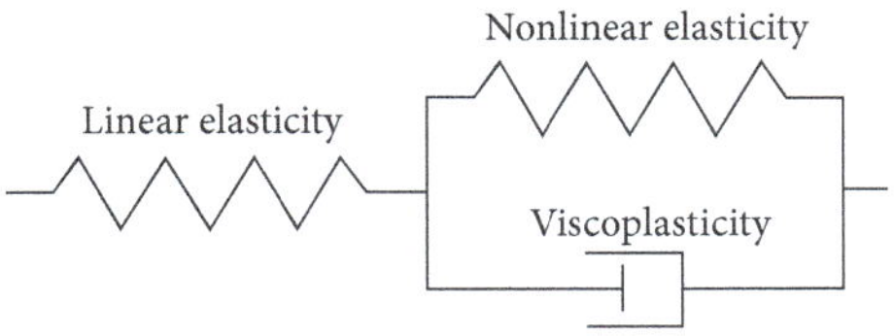

FIGURE 4: Standard rheology used to describe the large strain range of glassy polymers [18].

FIGURE 5: Arruda-Boyce model without the strain hardening element.

3.2. Elastic-Viscoplastic Modeling.

Typical tensile stress/strain curves obtained for amorphous glassy polymers are represented in Figure 2. This mechanical behavior can be easily described with an elastic-viscoplastic formulation [35–37]. Such model can be constituted by three rheological elements connected as in Figure 4. The first spring represents the linear elasticity referring to the behavior at small strains. The dashpot, here assumed as a viscoplastic element, represents the time-dependence yield stress and the strain softening of such materials in plastic regime. The second spring refers to the nonlinear elastic strain hardening of amorphous polymers, which is similar to the hyperelastic behavior of elastomeric materials [18, 38, 39].

3.2.1. Constitutive Model.

The PMMA elastic-viscoplastic behavior is here modeled with the modified Arruda-Boyce model [18, 40, 41]. It is composed of an Argon viscosity [42], in parallel with a Langevin rubbery hyperelasticity. This element, used to model the strain hardening, is not accounted for in the present paper because we have observed that for a scratch severity lower than $a_c/R = 0.4$, the strain level is not high enough to activate this element. Finally, a linear elasticity is added in, series in order to model the behavior at small strains (Figure 5). Scalar equation of the Argon viscous element is written:

$$\dot{\gamma}^p = \dot{\gamma}_0 \exp\left[\frac{A}{kT}(\tau - s)\right], \quad (4)$$

where $\dot{\gamma}^p$ is the plastic shear strain rate, $\dot{\gamma}_0 = 2.8 \cdot 10^7\ \mathrm{s}^{-1}$ and $A = 1.39 \cdot 10^{-27}$ are material constants, τ is the effective equivalent shear strength, k is Boltzmann's constant, T is absolute temperature, and s is the athermal shear strength. This strength s varies during strain softening according to

$$\dot{s} = \xi\left(1 - \frac{s}{s_{ss}}\right)\dot{\gamma}^p, \quad (5)$$

where $\xi = 315$ MPa is the softening slope, $s_{max} = 0.077\,G/(1-\nu)$ is the athermal shear strength without strain softening, and s_{ss} is the preferred state of the material ($s_{ss}/s_{max} = 0.87$). G is the elastic shear modulus, and the Poisson ratio is assumed to be constant and equal to 0.33. Different values of the Young modulus have been used in order to highlight its effect on the scratch response. More specifically, two different values have been used at each normal load. They have been chosen in order to fit the experiments for the lowest and highest scratch speeds. The numerical results constitute the upper bound and lower bound of the experiments. These values are consistent with the classical value of the Young modulus of the PMMA [43–45].

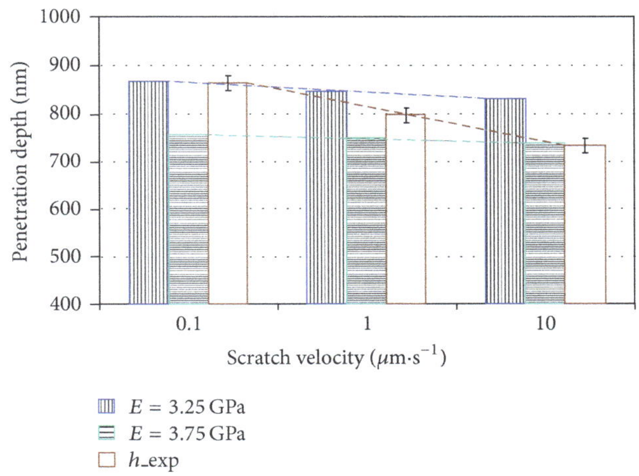

FIGURE 6: Comparison of experimental results (h_exp) and numerical results for F_n = 20 mN. The data denoted by E = 3.25 GPa (resp., E = 3.75 GPa) are those calculated with the elastic-viscoplastic model in using a Young modulus of 3.25 GPa (resp., 3.75 GPa).

3.2.2. Numerical Results.

The variation of the penetration depth with the scratch speed is shown in Figure 6 for F_n = 20 mN. The data denoted that h_exp corresponds to the experimental data whereas E = 3.25 GPa (resp., E = 3.75 GPa) are those calculated with the FEM model in using a Young modulus of 3.25 GPa (resp., 3.75 GPa).

One can firstly observe that the slope of the decrease of the penetration depth calculated with the FEM is less pronounced than the experimental one. It means that the time-dependent behavior of the PMMA is not well reproduced with this elastic-viscoplastic model in these conditions (F_n and V). One can also observe that the Young modulus has to be adapted to the tip velocity (i.e., to say the representative strain rate) to fit the experimental data. Figures 7 and 8 show the penetration depth in different tip velocities for F_n = 40 mN and F_n = 80 mN. Here again, the same conclusions can be drawn.

The major assumption in this study is that the Young modulus remains time independent. However, this parameter is well known to depend on the strain rate for most polymers in particular for the PMMA [45, 46]. This aspect could explain the bad agreement between experimental and numerical results. Such phenomenon is often modeled by using a viscoelastic model [47].

It can also be observed that the Young modulus which fits the experimental variation depends also on the value of the normal load used. For instance, a Young modulus of 3.75 GPa is required to fit the penetration depth corresponding to

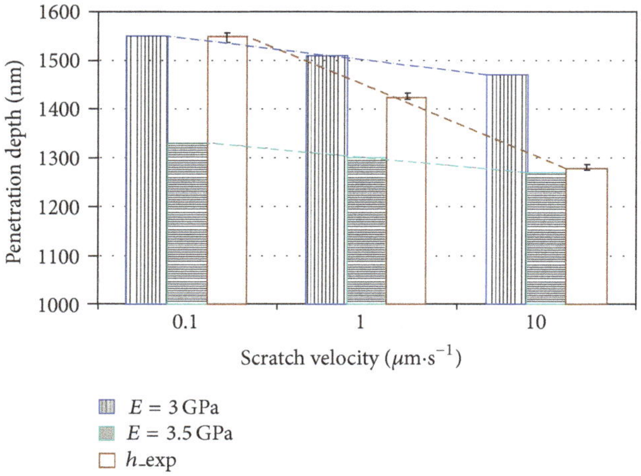

FIGURE 7: Comparison of experimental results ($h_$exp) and numerical results for F_n = 40 mN. The data denoted by E = 3.0 GPa (resp., E = 3.5 GPa) are those calculated with the elastic-viscoplastic model in using a Young modulus of 3.0 GPa (resp., 3.5 GPa).

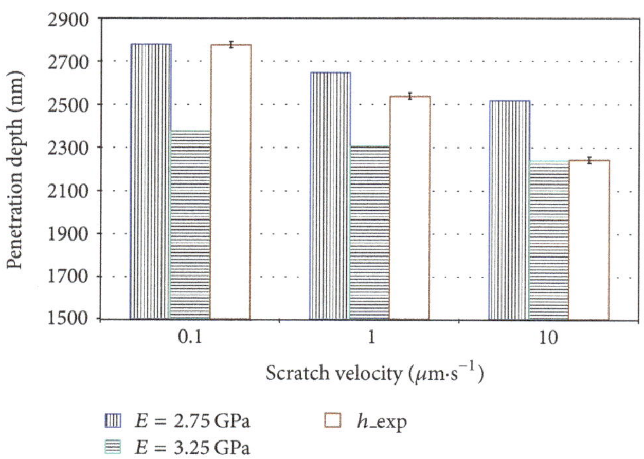

FIGURE 8: Comparison of experimental results ($h_$exp) and numerical results for F_n = 80 mN. The data denoted by E = 2.75 GPa (resp., E = 3.25 GPa) are those calculated with the elastic-viscoplastic model in using a Young modulus of 2.75 GPa (resp., 3.25 GPa).

$V = 10\,\mu\text{m}\cdot\text{s}^{-1}$ and $F_n = 20$ mN, whereas a Young modulus of 3.25 GPa is required to fit the penetration depth corresponding to $V = 10\,\mu\text{m}\cdot\text{s}^{-1}$ and $F_n = 80$ mN. Here again, this result is consistent with a viscoelastic behavior, since the elastic modulus depends on the strain rate [22]. Indeed, the higher the load, the higher the contact radius and according to (3), the lower the representative strain rate and thus the lower the elastic modulus. However, prior to start a more complex viscoelastic-viscoplastic investigation, we have studied the influence of the viscoplastic parameters on the penetration depth. Indeed, the set of the parameters used here has been identified by Arruda and Boyce [18] from macroscopic compression testing and thus, it could be not appropriate to model the response of our PMMA samples at such a microscopic scale.

3.2.3. A New Set of Viscoplastic Parameters. In this section, we attempt to fit the experimental data by modifying the viscoplastic parameters of Arruda-Boyce model [18]. The strategy consists in determining the values of theses parameters leading to the closest results to $h_$exp for $F_n = 20$ mN. Then, this set of parameters is used to simulate micro-scratches at $F_n = 40$ mN and $F_n = 80$ mN, and the results are compared with experiments.

The first step deals with the determination of the Young modulus. The original value provided by Arruda and Boyce [18] is $E = 3.2$ GPa. By applying the elastic theory of Hertz given by (6), the penetration depth in the case of a purely linear elastic contact should be equal to $h = 800$ nm at $F_n = 20$ mN:

$$h_{\text{elastic}} = \left(\frac{3}{4}\frac{F}{E^\star}\sqrt{R}\right)^{2/3} \quad \text{with } E^\star = \frac{E}{1-\nu^2}. \tag{6}$$

The elastic behavior constitutes the limit state of the elastic-viscoplastic framework for extremely high strain rate. In our study that means that elastic simulations lead necessary to less deep penetration than elastic-viscoplastic simulations, for identical Young modulus. However, elastic theory provides $h = 800$ nm (with $E = 3.2$ GPa) which is deeper than experimental penetration. It can thus be concluded that $E = 3.2$ GPa cannot be applied. $E = 3.75$ GPa appears to be the most accurate value (for $F_n = 20$ mN).

The second step consists in determining the most suitable set of parameters of the viscoplastic element. This element is governed by (4) rewritten as

$$\tau = s + \frac{k}{A} T \ln\left(\frac{\dot{\gamma}^p}{\dot{\gamma}_0}\right). \tag{7}$$

The modification of parameters A and $\dot{\gamma}_0$ alters the viscoplastic response. A modifies the sensitivity to the strain rate, and $\dot{\gamma}_0$ is the strain rate corresponding to the athermal shear stress s. Several scratch simulations have been carried out for $F_n = 20$ mN with various values of A and $\dot{\gamma}_0$. The best results according to experiments are shown in Figure 9. They have been obtained with: $A = 1.956 \cdot 10^{-27}\,\text{m}^3$ and $\dot{\gamma}_0 = 300\,\text{s}^{-1}$. One can observe the penetration depth in different tip velocities. A very good agreement is obtained between numerical and experimental results. Before concluding about the accuracy of this elastic-viscoplastic model, this result needs to be confirmed for the other normal loads.

Two series of simulations have been thus carried out at $F_n = 40$ mN and $F_n = 80$ mN with this particular set of parameters ($E = 3.75$ GPa, $A = 1.956 \cdot 10^{-27}\,\text{m}^3$, and $\dot{\gamma}_0 = 300\,\text{s}^{-1}$). Figures 10 and 11 show the penetration depth obtained in this way compared to the experimental one. Firstly, the numerical penetration at each tip velocity and each normal load is significantly different to the experimental one. Secondly, the variation of the numerical penetration depth with the tip velocity at each normal load does not reproduce the experimental one. As a conclusion, this elastic-viscoplastic model is not appropriate even after the adjustment of its parameters. In this context, we propose to investigate the addition of a viscoelastic model to this viscoplastic one.

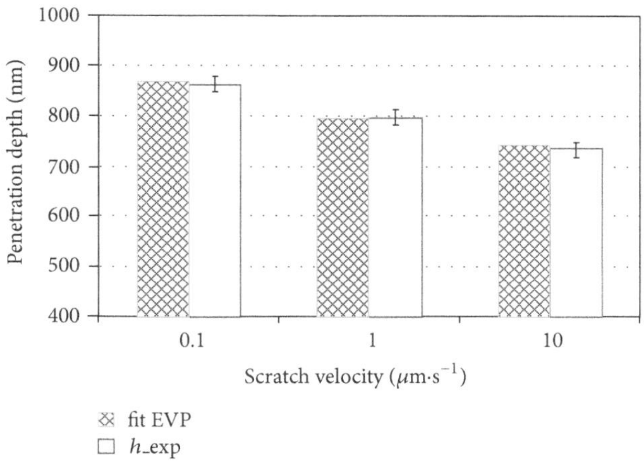

FIGURE 9: Comparison of experimental results ($h\_\exp$) and numerical results (fit_EVP) for $F_n = 20$ mN. The numerical results have been obtained with the elastic-viscoplastic model in using a Young modulus of 3.75 GPa. The set of parameters have been identified from experimental scratch results at 20 mN.

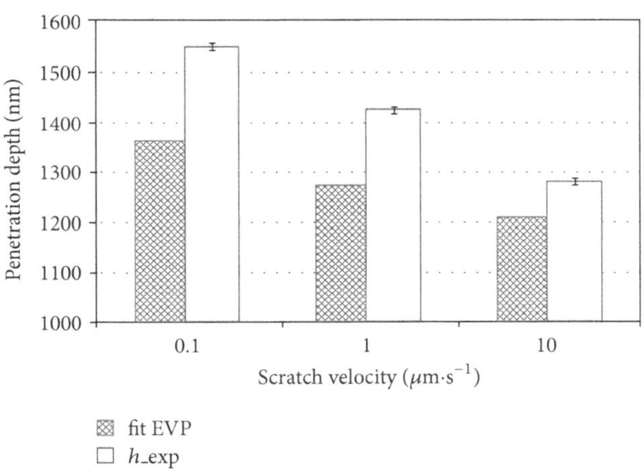

FIGURE 10: Comparison of experimental results ($h\_\exp$) and numerical results (fit_EVP) for $F_n = 40$ mN. The numerical results have been obtained with the elastic-viscoplastic model in using a Young modulus of 3.75 GPa. The set of parameters have been identified from experimental scratch results at 20 mN.

3.3. *Viscoelastic-Viscoplastic Model with the "Series Model".*

To account for both viscoelasticity and viscoplasticity, we have developed a specific algorithm [22], named "series model" in the finite element code Systus. It makes it possible to connect in series two models of any kind (Figure 12). Such algorithm is helpful to combine very quickly whatever models, if their constitutive equations are available in the finite element code in the same framework (updated lagrangian, total lagrangian, etc.). The algorithm uses an iterative process which calculates the dispatching of the strain rate to ensure the equality of the stress tensors. This algorithm can be summarized as in Figure 13. Our case study needs to connect a viscoelastic model with a viscoplastic one.

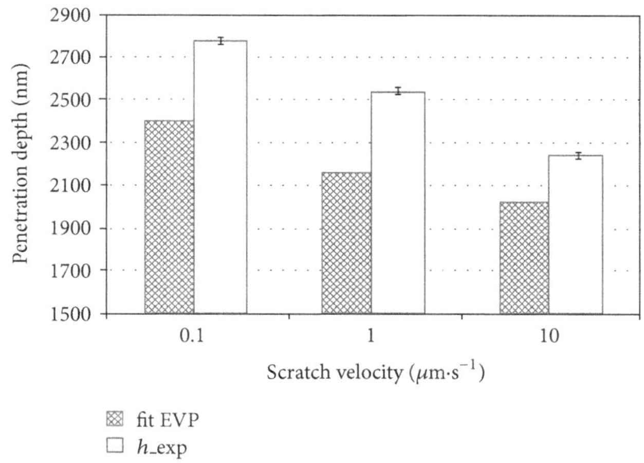

FIGURE 11: Comparison of experimental results ($h\_\exp$) and numerical results (fit_EVP) for $F_n = 80$ mN. The numerical results have been obtained with the elastic-viscoplastic model in using a Young modulus of 3.75 GPa. The set of parameters have been identified from experimental scratch results at 20 mN.

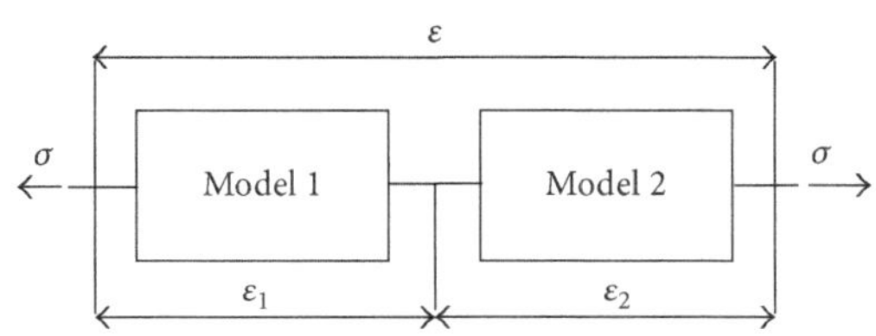

FIGURE 12: Connection in series of two rheological models.

3.3.1. *The Viscoelastic-Viscoplastic Model.*

In our case study, model 1 refers to the Kelvin-Voigt model and model 2 corresponds to the Arruda-Boyce model previously presented in Figure 14. However, the viscoelastic response of the whole model is the consequence of the connection in series of the Kelvin-Voigt model and the elastic element of Arruda-Boyce model. That constitutes the Poynting model having a viscoelastic transition wide enough to cover the sharp experimental velocities range (only two decades) and the parameters used in the viscoelastic model derived from the elastic-viscoplastic simulations. More precisely, the values of the Young modulus chosen in the previous section can be considered as the instantaneous and delayed modulus of the viscoelastic model (Figure 14). The viscous parameter η is identified by choosing the best fit to the experimental results at each normal load. The parameters are given in Table 2. One can observe that η increases with the normal load F_n. This is meaningful because E_1 and E_2 also depend on the normal load as observed in the previous section. At a fixed tip velocity, the representative strain rate decreases when F_n increases. Consequently, the range of strain rate shifts toward low values when F_n increases, and thus, a higher viscous parameter is required.

3.3.2. *Numerical Results.*

Figure 15 shows the penetration depth in different scratching velocities at $F_n = 20$ mN. First, the variation of the viscoelastic-viscoplastic penetration

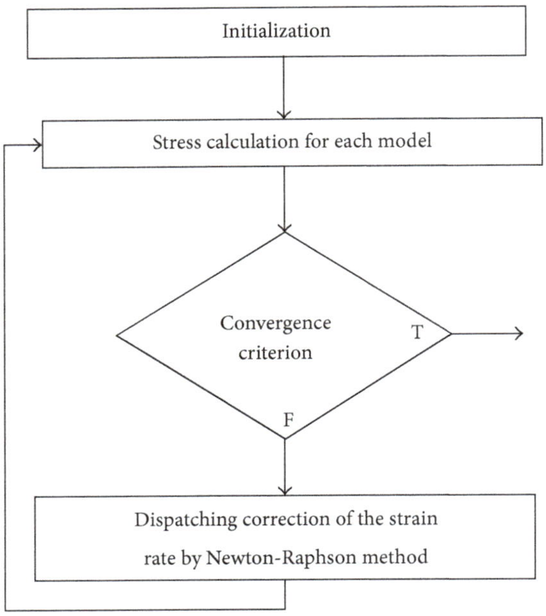

FIGURE 13: Series model algorithm.

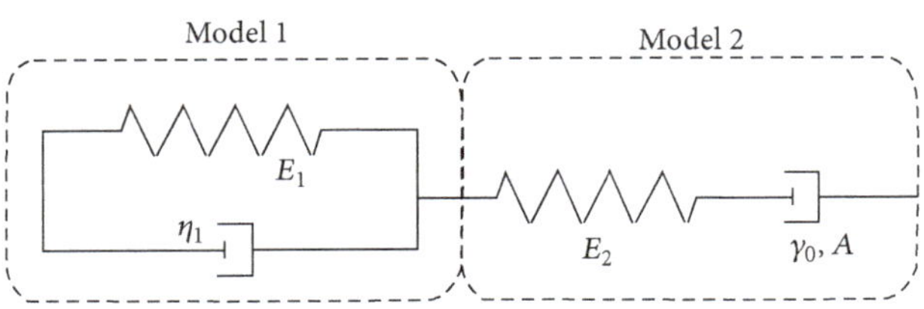

FIGURE 14: Proposed viscoelastic-viscoplastic model.

TABLE 2: Parameters of the Kelvin-Voigt viscoelastic model.

F_n in mN	E_1 in MPa	E_2 in MPa	η in GPa · s^{-1}
20	3250	3750	300
40	3000	3500	500
80	2750	3250	700

VEVP is almost the same as the experimental one. Moreover, the viscoelastic-viscoplastic penetration VEVP at each tip velocity is very close to the experimental one. That means that the time-dependent behavior of the PMMA in scratch test is better reproduced with this viscoelastic-viscoplastic model. Figures 16 and 17 show the penetration depth in different scratching velocities at F_n = 40 mN and at F_n = 80 mN. The same conclusions can be drawn: this viscoelastic-viscoplastic model is a pertinent solution to model the time-dependent scratch response of the PMMA. But how does this model modify the response to conventional testing? To adress this question, the shear stress-strain curves found by numerical simulation is plotted in Figure 18 using the set of parameters identified. The strain rate dependence of the yield stress level appears to be much pronounced than the strain rate dependence of the elastic modulus. Therefore, it could explain why

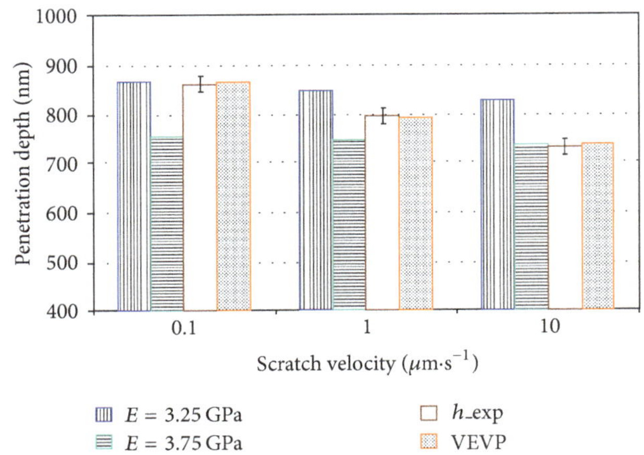

FIGURE 15: Comparison of experimental results (h_exp) and numerical results (elastic-viscoplastic and viscoelastic-viscoplastic) for F_n = 20 mN. The data denoted by E = 3.25 GPa (resp., E = 3.75 GPa) are those calculated with the elastic-viscoplastic model in using a Young modulus of 3.25 GPa (resp., 3.75 GPa). The data denoted VEVP corresponds to the viscoelastic-viscoplastic model.

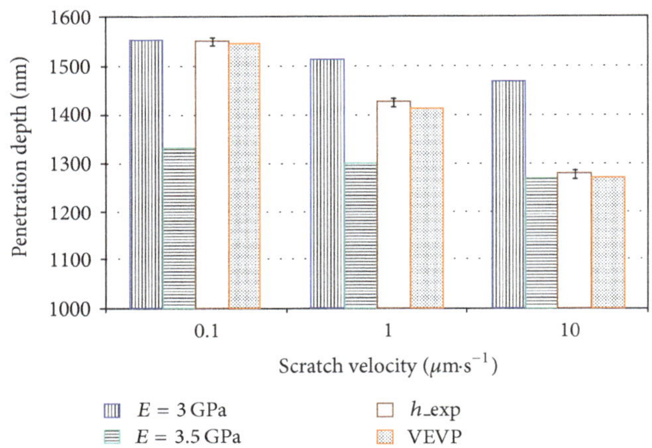

FIGURE 16: Comparison of experimental results (h_exp) and numerical results (elastic-viscoplastic and viscoelastic-viscoplastic) for F_n = 40 mN. The data denoted by E = 3.0 GPa (resp., E = 3.5 GPa) are those calculated with the elastic-viscoplastic model in using a Young modulus of 3.0 GPa (resp., 3.5 GPa). The data denoted VEVP corresponds to the viscoelastic-viscoplastic model.

viscoelasticity has been often neglected in the past for investigating the scratch response of PMMA.

Let us pay attention that the viscoelastic model chosen here is too simplistic to be used in a more global context. In practice, it should be calibrated with the help of DMA testing, which should lead to the connection in series of another Kelvin-Voigt models. The aim of this paper was not to propose the right model of PMMA in scratch test, but to highlight that the understanding of the scratch behavior of PMMA at middle strain levels requires a viscoelastic-viscoplastic model.

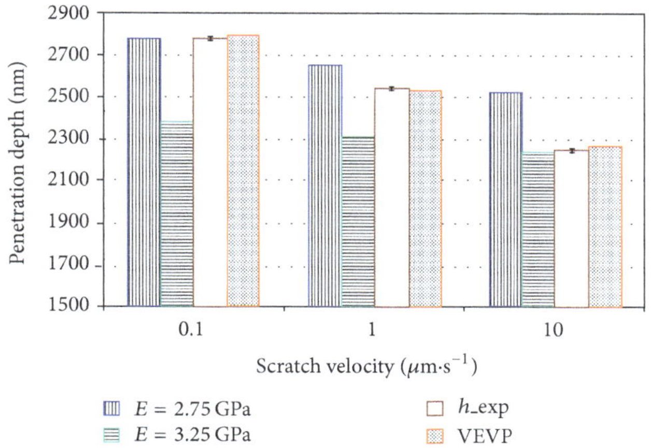

FIGURE 17: Comparison of experimental results (h_exp) and numerical results (elastic-viscoplastic and viscoelastic-viscoplastic) for $F_n = 80$ mN. The data denoted by $E = 2.75$ GPa (resp., $E = 3.25$ GPa) are those calculated with the elastic-viscoplastic model in using a Young modulus of 2.75 GPa (resp., 3.25 GPa). The data denoted VEVP corresponds to the viscoelastic-viscoplastic model.

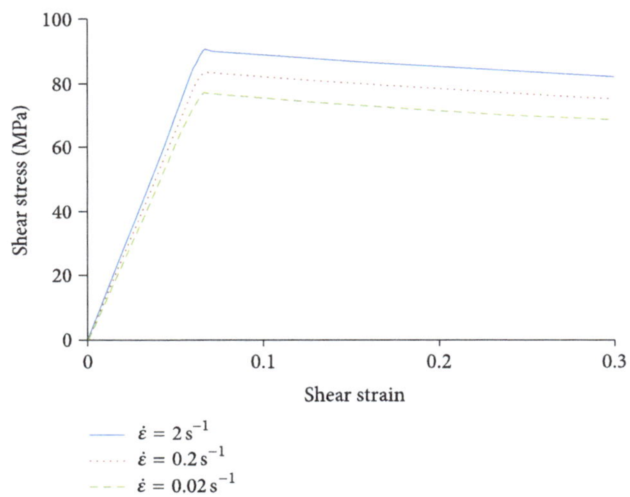

FIGURE 18: Shear stress-strain curves found by numerical simulation using the viscoelastic-viscoplastic model.

4. Conclusions

The investigation presented in this paper aims at understanding which is the best model to reproduce the time-dependent scratch behavior of PMMA at middle strain levels. A brief experimental study is first presented. It consists in the analysis of microscratches carried out on PMMA samples at different scratching velocities and at different normal loads. The loading conditions have been chosen in such a way that neitheir the (visco)elasticity nor (visco)plasticity of the PMMA may be neglected a priori. The main analyzed parameter is the penetration depth of the tip measured during the steady state. This experimental part is then followed by a modelling part which aims at determining the best constitutive model, elastic-viscoplatic or viscoelastic-viscoplastic, which allows to reproduce the experimental results. The standard elastic-viscoplastic model developed by Argon and later improved by Arruda and Boyce is used. It is mainly shown that this model leads to non satisfying results. More specifically, it is pointed out here that the time-independent Young modulus used in such models is not suitable. To take into account this feature, a viscoelastic-viscoplastic model based on the connection in series of the classical viscoelastic model and the AB viscoplatic model is proposed. It is shown that this model leads to more acceptable results, which points out the importance of viscoelasticity in the scratch behavior of solid polymers. Nevertheless, the viscoelastic model chosen here is too simplistic to be used in a more global context, because it only covers two decades of scratching speed. The goal of a forthcoming work could be to calibrate the viscoelastic parameters on a larger range in using a spectral approach based on DMA testing for instance.

The aim of this paper was not to propose the right model of PMMA in scratch test. It was to highlight the need to take into account both viscoelasticity and viscoplasticity to understand the scratch behavior of PMMA at middle strain levels. This observation is of primary importance in regards to the mechanical resistance of amorphous polymer films because the effect of viscoelasticity and viscoplasticity on the scratch results are not similar. This methodology can obviously be applied to other amorphous polymers such as PC or PS. One important perspective of this work is the possibility to use the scratch loading to identify the surface mechanical properties of solid polymers. Nevertheless, such investigations require to state on the influence of the mechanical properties affecting the response to a scratch loading to optimize the set of parameters to identify. Up to date, there is only a few papers dealing with this subject, mainly in reason of the difficulties related to the modeling of the scratch loading [3, 7, 13, 25, 29]. The present work can thus be viewed as a first step for this important issue.

Acknowledgments

The authors thank Professors J. L. Loubet and S. Pavan from Ecole Centrale de Lyon, France, for their help in the experimental part of this paper.

References

[1] V. Jardret, H. Zahouani, J. L. Loubet, and T. G. Mathia, "Understanding and quantification of elastic and plastic deformation during a scratch test," *Wear*, vol. 218, no. 1, pp. 8–14, 1998.

[2] D. Tabor, "Friction and wear-developments over the last fifty years," *Mechanical Engineering Publications*, vol. 1, pp. 157–172, 1989.

[3] G. Kermouche, N. Aleksy, J. L. Loubet, and J. M. Bergheau, "Finite element modeling of the scratch response of a coated time-dependent solid," *Wear*, vol. 267, no. 11, pp. 1945–1953, 2009.

[4] M. C. Baietto, J. Rannou, A. Gravouil, H. Pelletier, C. Gauthier, and R. Schirrer, "3D crack network analysis during a scratch test of a polymer: a combined experimental and multigrid x-fem

based numerical approach," *Tribology International*, vol. 44, no. 11, pp. 1320–1328, 2011.

[5] I. Demirci, C. Gauthier, and R. Schirrer, "Mechanical analysis of the damage of a thin polymeric coating during scratching: role of the ratio of the coating thickness to the roughness of a scratching tip," *Thin Solid Films*, vol. 479, no. 1-2, pp. 207–215, 2005.

[6] M. Barge, G. Kermouche, P. Gilles, and J. M. Bergheau, "Experimental and numerical study of the ploughing part of abrasive wear," *Wear*, vol. 255, no. 1-6, pp. 30–37, 2003.

[7] J. L. Bucaille, E. Felder, and G. Hochstetter, "Mechanical analysis of the scratch test on elastic perfectly plastic materials with the three-dimensional finite element modeling," *Wear*, vol. 249, no. 5-6, pp. 422–432, 2001.

[8] C. Gauthier, S. Lafaye, and R. Schirrer, "Elastic recovery of a scratch in a polymeric surface: experiments and analysis," *Tribology International*, vol. 34, no. 7, pp. 469–479, 2001.

[9] H. Pelletier Herve, A.-L. Durier, C. Gauthier, and R. Schirrer, "Viscoelastic and elastic-plastic behaviors of amorphous polymeric surfaces during scratch," *Tribology International*, vol. 41, no. 11, pp. 975–984, 2008.

[10] H. Pelletier, C. Gauthier, and R. Schirrer, "Influence of local friction coefficient and strain hardening on the scratch resistance of polymeric surfaces investigated by finite element modeling," *Procedia Engineering*, vol. 10, pp. 1772–1778, 2011.

[11] P. Bertrand-Lambotte, J. L. Loubet, C. Verpy, and S. Pavan, "Nano-indentation, scratching and atomic force microscopy for evaluating the mar resistance of automotive clearcoats: study of the ductile scratches," *Thin Solid Films*, vol. 398-399, pp. 306–312, 2001.

[12] N. Aleksy, G. Kermouche, J. M. Bergheau, J. L. Loubet, and S. Pavan, "Mechanical investigation of the healing phenomenon of the PMMA," *International Journal of Material Forming*, vol. 3, no. 1, pp. 575–578, 2010.

[13] L. C. A. van Breemen, L. E. Govaert, and H. E. H. Meijer, "Scratching polycarbonate: a quantitative model," *Wear*, vol. 274, pp. 238–247, 2012.

[14] V. Jardret and P. Morel, "Viscoelastic effects on the scratch resistance of polymers: relationship between mechanical properties and scratch properties at various temperatures," *Progress in Organic Coatings*, vol. 48, no. 2–4, pp. 322–331, 2003.

[15] B. J. Briscoe, P. D. Evans, S. K. Biswas, and S. K. Sinha, "The hardnesses of poly(methylmethacrylate)," *Tribology International*, vol. 29, no. 2, pp. 93–104, 1996.

[16] P. Kurkcu, L. Andena, and A. Pavan, "An experimental investigation of the scratch behaviour of polymers: 1. influence of rate-dependent bulk mechanical properties," *Wear*, vol. 290-291, pp. 86–93, 2012.

[17] A. Hernández-Jiménez, J. Hernández-Santiago, A. Macias-García, and J. Sánchez-González, "Relaxation modulus in PMMA and PTFE fitting by fractional Maxwell model," *Polymer Testing*, vol. 21, no. 3, pp. 325–331, 2002.

[18] E. M. Arruda and M. C. Boyce, "Evolution of plastic anisotropy in amorphous polymers during finite straining," *International Journal of Plasticity*, vol. 9, no. 6, pp. 697–720, 1993.

[19] F. Zaïri, K. Woznica, and M. Naït-Abdelaziz, "Phenomenological nonlinear modelling of glassy polymers," *Comptes Rendus Mecanique*, vol. 333, no. 4, pp. 359–364, 2005.

[20] E. Felder and J. L. Bucaille, "Mechanical analysis of the scratching of metals and polymers with conical indenters at moderate and large strains," *Tribology International*, vol. 39, no. 2, pp. 70–87, 2006.

[21] J. L. Bucaille, C. Gauthier, E. Felder, and R. Schirrer, "The influence of strain hardening of polymers on the piling-up phenomenon in scratch tests: experiments and numerical modelling," *Wear*, vol. 260, no. 7-8, pp. 803–814, 2006.

[22] N. Aleksy, G. Kermouche, A. Vautrin, and J. M. Bergheau, "Numerical study of scratch velocity effect on recovery of viscoelastic-viscoplastic solids," *International Journal of Mechanical Sciences*, vol. 52, no. 3, pp. 455–463, 2010.

[23] G. Kermouche, J. Rech, H. Hamdi, and J. M. Bergheau, "On the residual stress field induced by a scratching round abrasive grain," *Wear*, vol. 269, no. 1-2, pp. 86–92, 2010.

[24] J. Rech, G. Kermouche, W. Grzesik, C. García-Rosales, A. Khellouki, and V. García-Navas, "Characterization and modelling of the residual stresses induced by belt finishing on a AISI52100 hardened steel," *Journal of Materials Processing Technology*, vol. 208, no. 1–3, pp. 187–195, 2008.

[25] S. C. Bellemare, M. Dao, and S. Suresh, "Effects of mechanical properties and surface friction on elasto-plastic sliding contact," *Mechanics of Materials*, vol. 40, no. 4-5, pp. 206–219, 2008.

[26] G. Kermouche, A. L. Kaiser, P. Gilles, and J. M. Bergheau, "Combined numerical and experimental approach of the impact-sliding wear of a stainless steel in a nuclear reactor," *Wear*, vol. 263, no. 7–12, pp. 1551–1555, 2007.

[27] K. Holmberg, A. Laukkanen, H. Ronkainen, K. Wallin, S. Varjus, and J. Koskinen, "Tribological contact analysis of a rigid ball sliding on a hard coated surface. Part I: modelling stresses and strains," *Surface and Coatings Technology*, vol. 200, no. 12-13, pp. 3793–3809, 2006.

[28] P. Bertrand-Lambotte, J. L. Loubet, C. Verpy, and S. Pavan, "Understanding of automotive clearcoats scratch resistance," *Thin Solid Films*, vol. 420-421, pp. 281–286, 2002.

[29] H. Pelletier, C. Gauthier, and R. Schirrer, "Relationship between contact geometry and average plastic strain during scratch tests on amorphous polymers," *Tribology International*, vol. 43, no. 4, pp. 796–809, 2010.

[30] G. Kermouche, J. L. Loubet, and J. M. Bergheau, "Cone indentation of time-dependent materials: the effects of the indentation strain rate," *Mechanics of Materials*, vol. 39, no. 1, pp. 24–38, 2007.

[31] G. Kermouche, J. L. Loubet, and J. M. Bergheau, "Extraction of stress-strain curves of elastic-viscoplastic solids using conical/pyramidal indentation testing with application to polymers," *Mechanics of Materials*, vol. 40, no. 4-5, pp. 271–283, 2008.

[32] B. J. Briscoe, P. D. Evans, E. Pelillo, and S. K. Sinha, "Scratching maps for polymers," *Wear*, vol. 200, no. 1-2, pp. 137–147, 1996.

[33] D. Chamoret, P. Saillard, A. Rassineux, and J. M. Bergheau, "New smoothing procedures in contact mechanics," *Journal of Computational and Applied Mathematics*, vol. 168, no. 1-2, pp. 107–116, 2004.

[34] H. Pelletier, C. Gauthier, and R. Schirrer, "Influence of the friction coefficient on the contact geometry during scratch onto amorphous polymers," *Wear*, vol. 268, no. 9-10, pp. 1157–1169, 2010.

[35] J. Sweeney and I. M. Ward, "A constitutive law for large deformations of polymers at high temperatures," *Journal of the Mechanics and Physics of Solids*, vol. 44, no. 7, pp. 1033–1049, 1996.

[36] Y. Tomita, "Constitutive modelling of deformation behavior of glassy polymers and applications," *International Journal of Mechanical Sciences*, vol. 42, no. 8, pp. 1455–1469, 2000.

[37] B. P. Gearing and L. Anand, "On modeling the deformation and fracture response of glassy polymers due to shear-yielding and

crazing," *International Journal of Solids and Structures*, vol. 41, no. 11-12, pp. 3125–3150, 2004.

[38] J. S. Bergström, S. M. Kurtz, C. M. Rimnac, and A. A. Edidin, "Constitutive modeling of ultra-high molecular weight polyethylene under large-deformation and cyclic loading conditions," *Biomaterials*, vol. 23, no. 11, pp. 2329–2343, 2002.

[39] J. Richeton, S. Ahzi, K. S. Vecchio, F. C. Jiang, and A. Makradi, "Modeling and validation of the large deformation inelastic response of amorphous polymers over a wide range of temperatures and strain rates," *International Journal of Solids and Structures*, vol. 44, no. 24, pp. 7938–7954, 2007.

[40] O. A. Hasan and M. C. Boyce, "A constitutive model for the nonlinear viscoelastic viscoplastic behavior of glassy polymers," *Polymer Engineering and Science*, vol. 35, no. 4, pp. 331–344, 1995.

[41] E. M. Arruda, M. C. Boyce, and H. Quintus-Bosz, "Effects of initial anisotropy on the finite strain deformation behavior of glassy polymers," *International Journal of Plasticity*, vol. 9, no. 7, pp. 783–811, 1993.

[42] A. S. Argon, "A theory for the low temperature plastic deformation of glassy polymers," *Philosophical Magazine*, vol. 28, no. 4, pp. 839–865, 1973.

[43] A. Vautrin, "Etude en traction du processus de relaxation béta du polyméthacrylate de méthyle," *Comptes Rendus de l'Academie des Sciences de Paris B*, vol. 279, pp. 629–632, 1974.

[44] A. Vautrin, *Etude du comportement viscoélastique des solides par la méthode de traction modulée: relaxation secondaire du PMMA [Ph.D. thesis]*, Université de Nancy I, 1976.

[45] J. Richeton, G. Schlatter, K. S. Vecchio, Y. Rémond, and S. Ahzi, "A unified model for stiffness modulus of amorphous polymers across transition temperatures and strain rates," *Polymer*, vol. 46, no. 19, pp. 8194–8201, 2005.

[46] M. F. Ashby and D. R. H. Jones, *Matériaux : 2. Microstructures et Mise en Oeuvre*, Dunod, 1991.

[47] C. G'Sell and J. M. Haudin, *Introduction à la Mécanique des Polymères*, Institut National Polytechnique de Lorraine, 1995.

Study on Thixojoining Process Using Partial Remelting Method

M. N. Mohammed, M. Z. Omar, M. S. Salleh, and K. S. Alhawari

Department of Mechanical and Materials Engineering, Faculty of Engineering and Built Environment, Universiti Kebangsaan Malaysia, 43600 Selangor, Malaysia

Correspondence should be addressed to M. Z. Omar; zaidi5886@gmail.com

Academic Editor: Rui Vilar

Cold-work tool steel is considered to be a nonweldable metal due to its high percentage content of carbon and alloy elements. The application of a new process of the semisolid joining of two dissimilar metals is proposed. AISI D2 cold-work tool steel was thixojoined to 304 stainless steel by using a partial remelting method. After thixojoining, microstructural examination including metallographic analysis, energy dispersive spectroscopy (EDS), and Vickers hardness tests was performed. From the results, metallographic analyses along the joint interface between semisolid AISI D2 and stainless steel showed a smooth transition from one to another and neither oxides nor microcracking was observed. Hardness values obtained from the points in the diffusion zone were much higher than those in the 304 stainless steel but lower than those in the AISI D2 tool steel. The study revealed that a new type of nonequilibrium diffusion interfacial structure was constructed at the interface of the two different types of steel. The current work successfully confirmed that avoidance of a dendritic microstructure in the semisolid joined zone and high bonding quality components can be achieved without the need for force or complex equipment when compared to conventional welding processes.

1. Introduction

Cold-work tool steel has many attractive properties such as high wear resistance, high compressive strength, high hardness after hardening, and good dimension stability during heat treatment. Due to the growing demand for cold-work tool steel in various industrial applications, it is crucial to improve the welding technique for joining cold-work tool steel. However, welding this type of material is very difficult and it is almost impossible to achieve a good result using conventional processes. There have been a number of reported studies on using a feedstock material with a thixotropic characteristic (globular microstructure) to weld by employing different methods such as laser welding and friction stir welding, but problems arise with these methods including porosities, loss of alloying elements, bad geometry, and softening of the heat-affected zone resulting in a nonhomogenous microstructure in the heat-affected zone caused by the remelting temperature [1–3]. Clearly this last effect is particularly undesirable from a metallurgic point of view due to the lack of control of the microstructure. Moreover, the high temperatures which are necessary to achieve the complete melting of the various parts of the substrate induce a heat-affected zone. In this zone, deterioration of the fracture toughness, corrosion resistance, and yield strength is often attributed to the material undergoing a thermal history, which, in turn, results in making its properties different, and generally worse, than those of the rest of the bulk material in the substrate [4].

Joining in the thixotropic state is considered to be an effective alternative joining method in which the material is processed when it is in a semisolid state between solidus and liquidus temperatures, that is, when the material shows thixotropic behaviour. Metal in a thixotropic state has many features such as high softness, good flowability, superb workability, and excellent join ability [5, 6]. Therefore, the key potential advantage of the semisolid joining of alloys is that it can avoid many of the above-mentioned problems because of the fact that the solidification and heat transfer processes are basically different from those of typical welding. However, as reported by other studies, there are three methods that can be used to produce semisolid joining. The first method involves the use of a thixotropic metal as a filler applied into the joint groove for the purpose of joining materials [4]. The second method uses semisolid properties in combination with forming in a single process [7, 8]. The third method

TABLE 1: Chemical composition (wt%) of AISI D2 tool steel and AISI 304 stainless steel by XRF technique.

	C	Si	Mn	Cr	Ni	Mo	V	P	S
AISI D2	1.46	0.258	0.239	11.2	0.197	0.769	0.711	—	—
AISI 304	0.08	1	2	18	8	—	—	0.04	0.03

uses the properties of two different metals in their semisolid state [9]. It should be mentioned that these variants of semisolid joining technology are new processes that have been proposed fairly recently and thus far, researchers have not focused their attention on this particular area of work. Therefore, it is necessary to carry out fundamental investigations for the purpose of obtaining additional information about the general characteristics of this technology. In this study, thixojoining was applied to join AISI D2 cold-work tool steel to a 304 stainless steel. The joining process for this welded material was carried out by using a partial remelting method. The main aim of this paper is to provide globular joining structure in semisolid state. The study revealed that a new type of nonequilibrium diffusion interfacial structure was constructed at the interface of the two different types of steel.

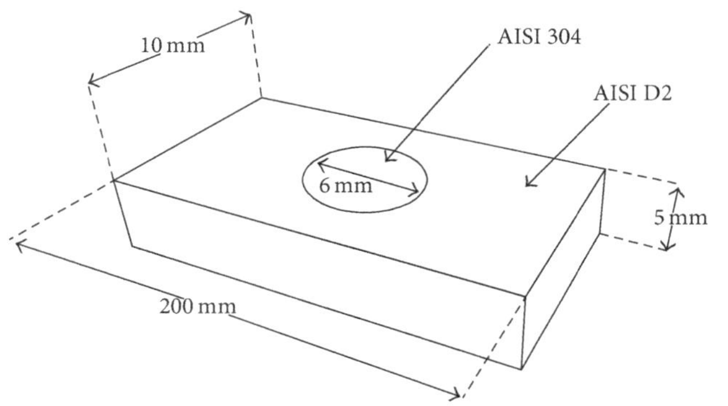

FIGURE 1: Diagram of the combination of the two different samples.

2. Experimental Procedure

2.1. Materials. The investigated materials were AISI D2 (ENX160CrMo12-1) cold-work tool steel and AISI 304 stainless steel. The chemical composition of the starting materials is given in Table 1.

2.2. JMatPro Simulation. JMatPro (Java-based material properties) is software that was developed to augment thermodynamic calculations by incorporating diverse theoretical material models and a properties database that allows quantitative calculation of the requisite material properties to be made within a larger software structure [10–12]. Here, JMatPro software is used to estimate solidus and liquidus temperatures as well as the liquid fraction profile within the semisolid zone for both metals.

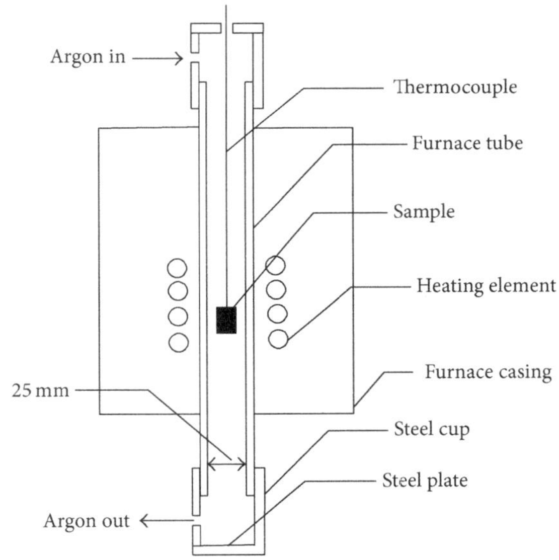

FIGURE 2: Schematic of the furnace setup for remelting experiment.

2.3. Sample Preparation. The first step involved cutting the base metal (AISI D2) into samples with a size of 20 × 10 × 5 mm and cutting the insert metal (AISI 304 stainless steel) into samples with a diameter of Ø6 × 5 mm, as shown in Figure 1. In the second step each sample was immersed into a pickling solution with a concentration of 40 mL HCL (37%) + 60 mL H_2O for 5 min to get rid of rust and descale the metal to obtain a clean surface. In the third and final preparation step the sample was washed in acetone followed by rinsing with distilled water and then dried in air.

2.4. Direct Partial Remelting Experiment. Partial remelting is considered one of the most effective solid state routes to produce a nondendritic microstructure especially for high melting point metals when the metal is directly heated to a temperature between solidus and liquidus [13]. A direct partial remelting experiment was performed using a vertical, high temperature carbolite furnace with protective atmosphere of argon gas, as shown in Figure 2. When the furnace had reached the predefined temperature, the sample (base and insert) was lowered into the hottest place (1320°C) of the furnace by using chromel wire and held there for 30 min after which it was subjected to air cooling to room temperature. This approach ensures a rapid heating of the sample which normally enables it to reach the predefined temperature in about 3 min, as shown in Figure 3.

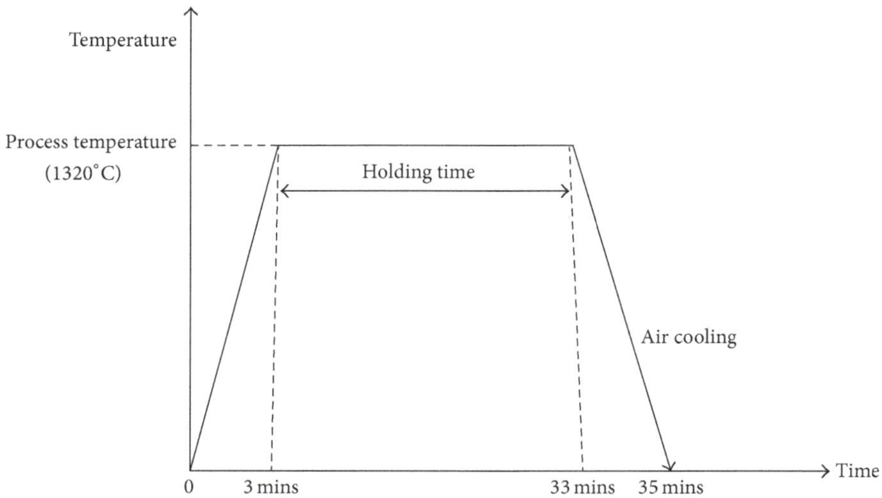

FIGURE 3: Temperature profile of sample inside vertical tube furnace.

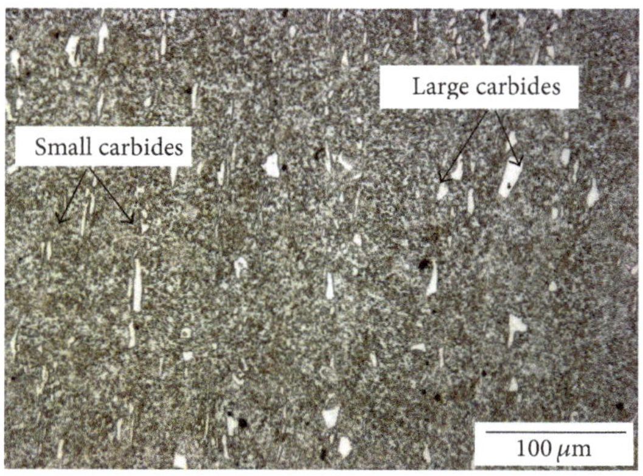

FIGURE 4: Optical micrographs of as-received AISI D2 tool steel.

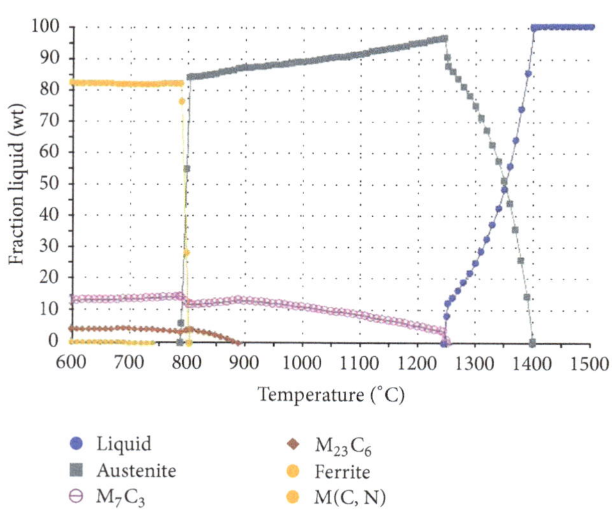

FIGURE 5: Phase equilibrium diagram of AISI D2 tool steel obtained from JMatPro simulation.

2.5. Vickers Hardness Testing. Vickers hardness (HV) tests were performed on the transverse sections. The hardness measurement profiles were obtained through the sample's horizontal cross-section using a load of 2 kg applied for 15 sec.

2.6. Metallography and Image Analysis. Samples from the direct partial remelting experiments were carefully ground and polished to obtain mirror surfaces. Microstructural characterization was carried out by using a BX-51 Olympus optical microscope, a Hitachi S3400N scanning electron microscope (SEM), and an energy dispersive X-ray (EDX) equipped with energy (EDS dispersive spectroscopy). All samples were etched using Villela reagent (1 g picric acid, 5 mL hydrochloric acid, and 95 mL ethyl alcohol) to reveal their microstructures. Average grain sizes were calculated by using the mean lineal intercept method (after the ASTM E112-96 standard). The ideal grain size for semisolid forming should not exceed 100 μm [14].

3. Results and Discussion

3.1. Starting Material. The as-received material contains surplus arrays of coarse large carbides with smaller carbides distributed homogeneously in a ferrite matrix parallel to the working direction, as shown in Figure 4. This structure has been identified in annealed cold-work tool steel [15, 16] and thus confirms that soft annealing treatment had been carried out on the as-received material as described by the supplier. From Figure 5, the phase stability diagrams obtained from the JMatPro simulation for as-received AISI D2 show that, at room temperature, the carbides consisted of M_7C_3, MC, and $M_{23}C_6$ type carbides, and these have been experimentally verified by Omar et al. [11].

3.2. Microstructure and Joint Quality. According to the graphs in Figure 6, the liquid fraction profile curve at 1320°C for AISI D2 was around 35% while for AISI 304 stainless steel it was around 2.5%. That is to say, there is a huge difference in the liquid fraction percentage at the same temperature between the base metal and the insert.

The microstructure of the welded joints revealed by the optical microscope is shown in Figure 7, while the microstructure of the welded joints observed by SEM is shown in Figure 8. The microstructure has three distinct features, the regions of which can be determined by the difference in colour and grain size. It can also be seen that there is a fine boundary between the base blank and insert with a smooth transition of the diffusion zone from one metal to the other. At the transition there is no evidence of microcracking or porosity, and the shape of the joint appears to be very smooth and not corrugated. A full penetration welded joint can also be clearly seen, as denoted by grain boundary migration and the observation of different colours in one grain. The grain size of AISI D2 is 85 μm, while the grains adjacent to the diffusion zone interface are around 140 μm. In addition, the base metal grains have changed through the semisolid process to form new columnar crystal growth near the convergence zone (diffusion zone). Nonequilibrium diffusion has occurred due to the difference in chemical composition of the two metals and the huge difference in the liquid fraction percentage of the two metals at the same temperature. As for the interfacial structure, both metals exhibit different reaction and diffusion abilities between solidus and liquidus whereby the base metal is connected to the insert metal along the bonding boundary to achieve what appears to be a perfect joining at the interfaces of both metals.

Table 2: EDX analysis and composition of the points marked in Figure 10.

Element	AISI D2	Diffusion zone	AISI 304
C	1.89	1.61	1.07
Si	0.21	0.28	0.36
P	—	0.033	0.037
Mo	0.79	0.39	—
S	—	0.04	0.05
V	0.45	0.38	—
Cr	7.2	9.89	11.57
Mn	0.34	0.49	0.77
Ni	0.49	1.97	6.1
Fe	85.57	84.917	80.043

Scanning electron microscopy and EDX analysis combined with EDS analysis were carried out in order to scrutinize the phases that occurred during the joining process at the interface region. The line scan analysis was performed along the marked line as shown in Figure 9. From this figure it is observed that nickel and chromium diffuse towards the AISI D2 tool steel from the AISI 304 stainless steel and the diffusion of iron and carbon from tool steel side towards stainless steel. Figure 10 illustrates the EDX analysis results from the SEM image taken from points 1, 2, and 3 representing the AISI-D2, the diffusion zone, and 304 stainless steel in the AISI D2-AISI 304 joint, respectively, while distribution of elements within the determined location is shown in Table 2. It can be clearly seen that elements P and S begin to transfer from the AISI 304 side in the interface region and grain boundaries, while Mo, V, and Cu of AISI D2 side diffuse into the convergence zone. The EDS results confirm the formation of congregation compounds of these two different metals. The formation of the new phase compounds has a tendency to depend on the difference in the chemical composition of the two metals, which leads to the diffusion of different elements through the interface region (diffusion zone).

3.3. Vickers Hardness Testing. The variation in the hardness values in the transverse cross-section of the joint is shown in Figure 11. It can be seen that the hardness values obtained from the points in the base metal are much higher than those in the insert, while in the diffusion zone the hardness value is higher than that in the insert but lower than that in the base metal. These differences are probably due to the diffusion of the elements from both metals at the interface zone of the

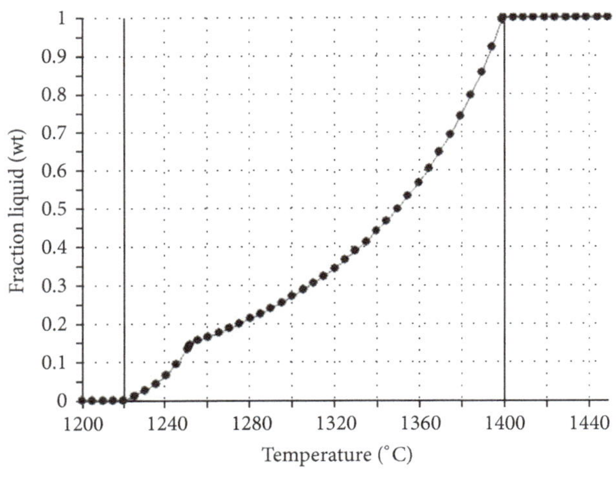

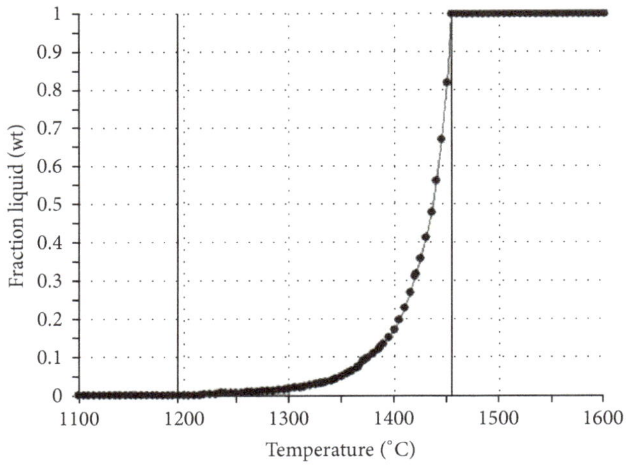

FIGURE 6: Liquid fraction profile for chemical composition of (a) AISI D2 tool steel and (b) AISI 304 stainless steel obtained from JMatPro simulation.

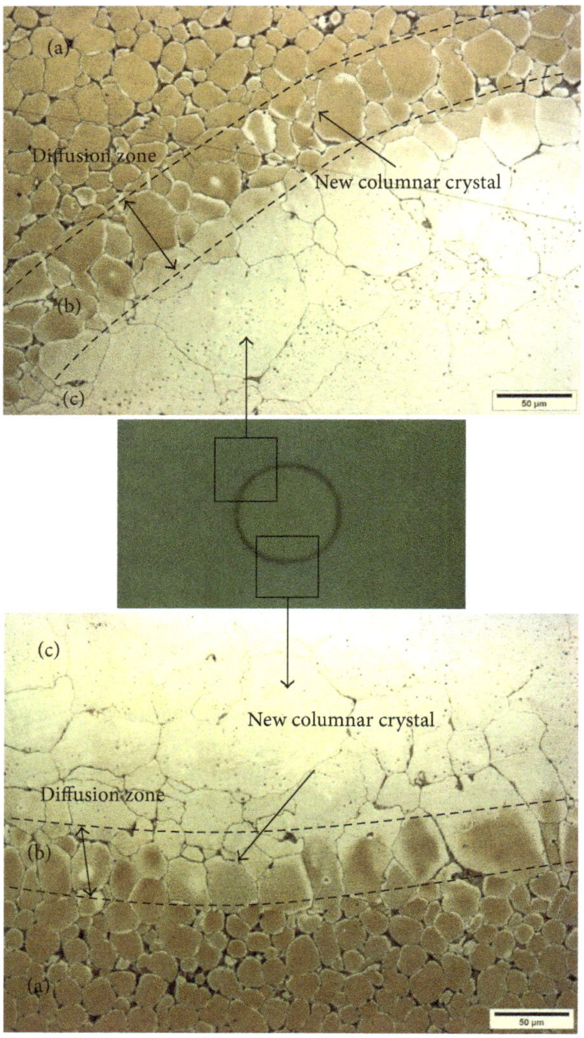

FIGURE 7: Microstructures of thixojoint of (a) AISI D2, (b) diffusion zone, and (c) AISI 304 after application of 1320°C for 30 minutes.

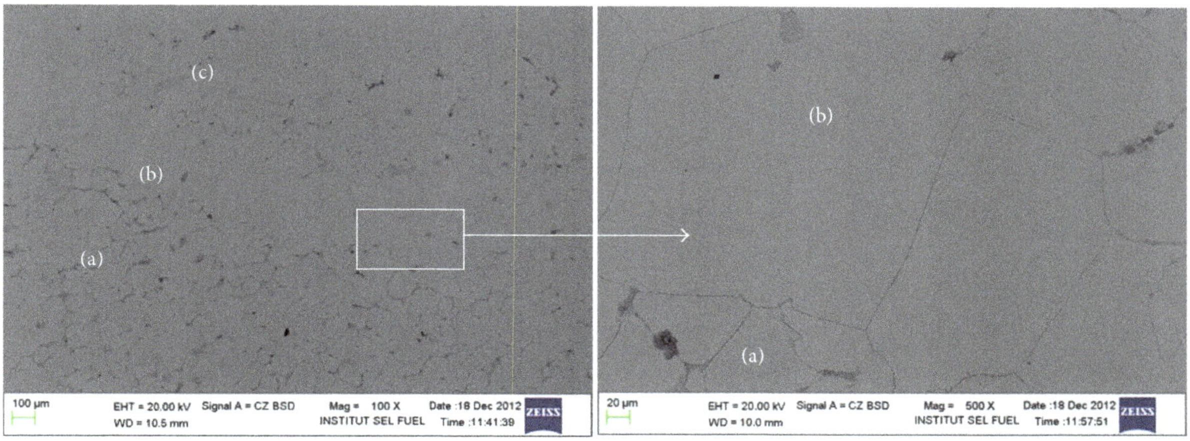

FIGURE 8: Image of microstructures of (a) AISI D2, (b) diffusion zone, and (c) AISI 304.

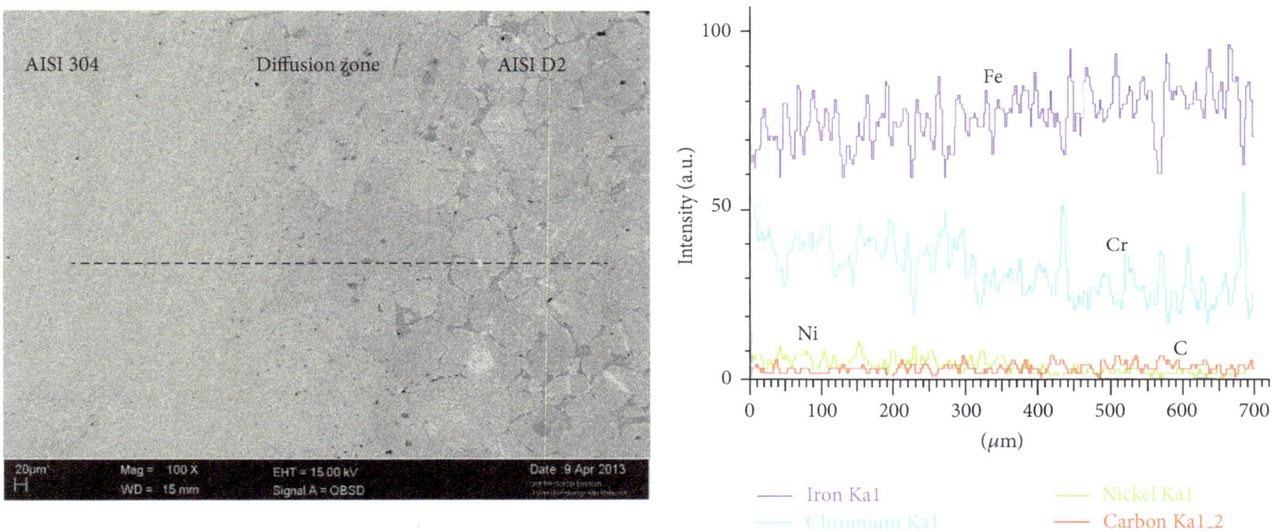

(a) (b)

FIGURE 9: Elemental distribution across the joint centre in AISI D2-AISI 304.

joint. Then, there is a decrease towards the side of the insert part on the horizontal distance of the joints, as shown in Figure 11. This hardness distribution data for the transverse cross-section of joints can be attributed to the fine transition structure between the base metal and the insert as well as to the natural diffusion between these different metals. It can be seen that the hardness of the cross-section area is almost uniform throughout the whole length of the sample.

4. Conclusion

A new type of thixojoining process for AISI D2 with AISI 304 stainless steel using a partial remelting method was proposed. This process is not based on conventional methods (adhesive, nails, and screws) but can produce homogeneous properties with high surface quality and avoids the creation of a dendritic microstructure at the join zone. In addition, this technique has two key advantages when compared with conventional joining methods, namely, the possibility of producing functional components with multiple materials and minimizing the defects associated with conventional welding processes. In addition, by using this type of thixojoining process, there is no need for force or complex equipment when compared to conventional welding processes such as friction welding. Based on the results of the experiments conducted in this study, the use of this thixojoining technique can create a fine interfacial diffusion along the bonding boundary between the base blank and insert. Furthermore, a smooth transition from one metal to the other is achieved through the formation of new columnar crystal growth near the convergence zone. The transition occurs without evidence of microcracking or porosity.

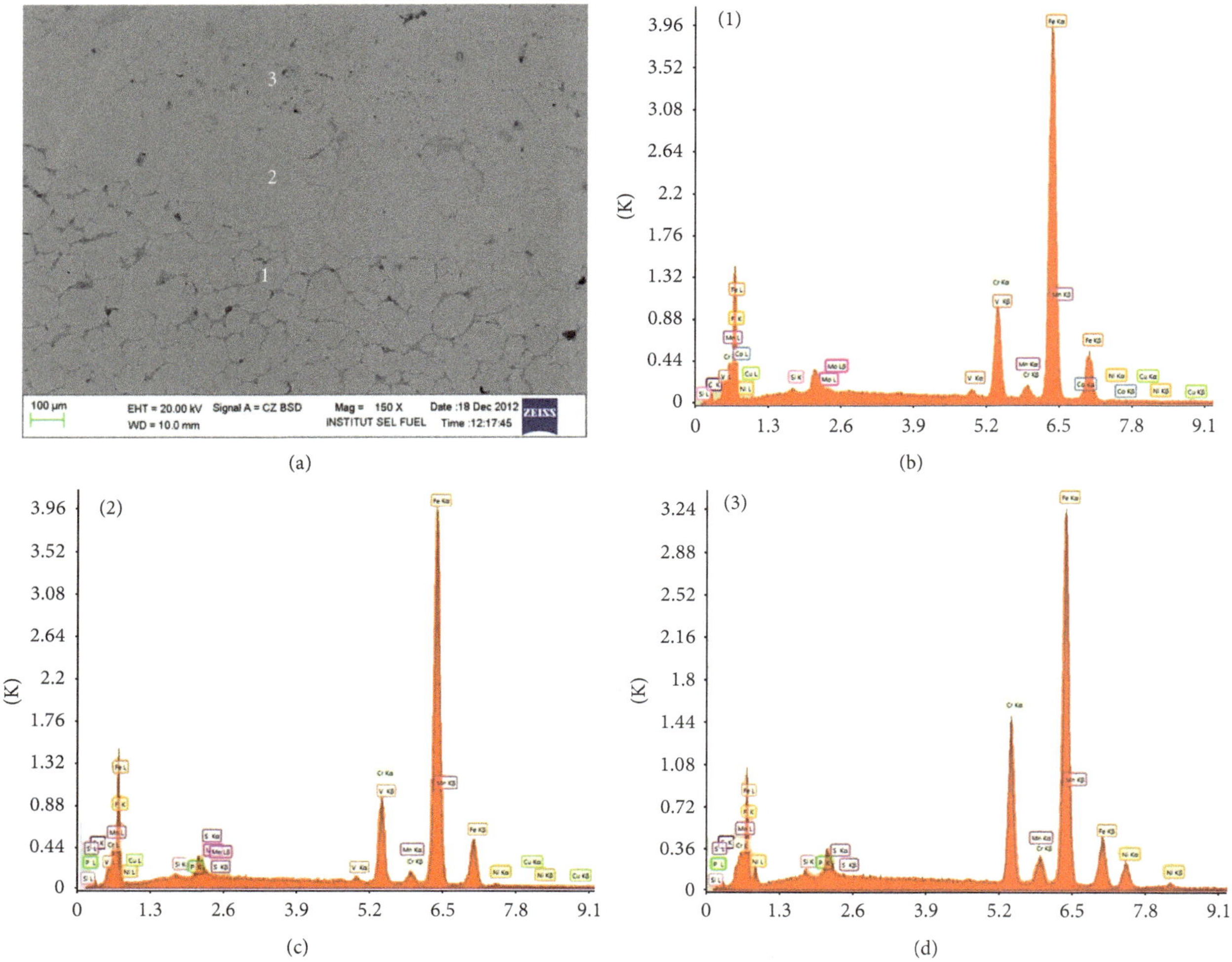

FIGURE 10: SEM micrograph of joining compound and EDX analysis at (point 1) AISI D2, (point 2) diffusion zone, and (point 3) 304 stainless steel.

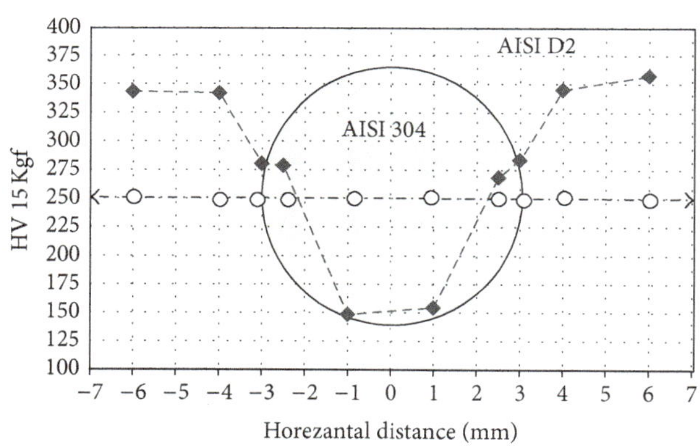

FIGURE 11: Vickers hardness profile.

Acknowledgments

The authors would like to thank Universiti Kebangsaan Malaysia (UKM) and the Ministry of Science, Technology and Innovation (MOSTI), Malaysia for the financial support under research Grants GUP-2012-040 and AP-2012-014.

References

[1] G. Govender, L. Ivanchev, H. Burger, R. Knutsen, and G. Kunene, "Weldability of SSM rheo processed aluminum alloy A356," *Solid State Phenomena*, vol. 141–143, pp. 773–778, 2008.

[2] S. Zhang, X. Guo, and G. Ma, "Investigation of semisolid welding characteristics of 2A14 aluminum alloy," *Advanced Materials Research*, vol. 154-155, pp. 639–642, 2011.

[3] A. Narimannezhad, H. Aashuri, A. H. Kokabi, A. Khosravani, M. Kiani, and A. Foroughi, "Semisolid joining of zinc AG40A alloy by partial remelting and mechanical stirring," *Solid State Phenomena*, vol. 141–143, pp. 225–230, 2008.

[4] P. F. Mendez, C. S. Rice, and S. B. Brown, "Joining using semi-solid metals," *Welding Journal*, vol. 81, no. 9, pp. 181–187, 2002.

[5] M. Kiuchi, J. Yanagimoto, and S. Sugiyama, "Mashy-state joining, a new process for joining materials together," in *Proceedings of the 5th International Conference on Semi-Solid Processing of Alloys and Composites*, pp. 123–130, Golden, Colo, USA, 1998.

[6] M. Kiuchi, J. Yanagimoto, and S. Sugiyama, "Application of mushy/semi-solid joining part-2," in *Proceedings of the 7th International Conference on Semi-Solid Processing of Alloys and Composites*, pp. 707–712, Tsukuba, Japan, 2002.

[7] R. Baadjou, H. Shimahara, and G. Hirt, "Automated semi-solid forging of steel components by means of thixojoining," *Solid State Phenomena*, vol. 116-117, pp. 383–386, 2006.

[8] R. Kopp, J. Kallweit, T. Moller, and I. Seidl, "Forming and joining of commercial steel grades in the semi-solid state," *Journal of Materials Processing Technology*, vol. 130-131, pp. 562–568, 2002.

[9] M. N. Mohammed, M. Z. Omar, M. S. Salleh, M. A. Zailani, and K. S. Alhawari, "Joining two metals via partial remelting method," *Journal of Asian Scientific Research*, vol. 2, pp. 724–730, 2012.

[10] N. Saunders, Z. Guo, X. Li, A. P. Miodownik, and J. P. Schillé, "Using JMatPro to model materials properties and behavior," *JOM*, vol. 55, no. 12, pp. 60–65, 2003.

[11] M. Z. Omar, A. Alfan, J. Syarif, and H. V. Atkinson, "Microstructural investigations of XW-42 and M2 tool steels in semi-solid zones via direct partial remelting route," *Journal of Materials Science*, vol. 46, no. 24, pp. 7696–7705, 2011.

[12] M. S. Salleh, M. Z. Omar, J. Syarif, and M. N. Mohammed, "Thermodynamic modelling on the mutual effect of copper, manganese and iron addition in Al-Si-Cu for semisolid processing," *Journal of Asian Scientific Research*, vol. 2, pp. 614–619, 2012.

[13] M. Z. Omar, H. V. Atkinson, and P. Kapranos, "Thixotropy in semisolid steel slurries under rapid compression," *Metallurgical and Materials Transactions A*, vol. 42, no. 9, pp. 2807–2819, 2011.

[14] G. Hirt, L. Khizhnyakova, R. Baadjou, F. Knauf, and R. Kopp, "Semi-solid forming of aluminium and steel-introduction and overview," in *Thixoforming-Semi-Solid Metal Processing*, G. Hirt and R. Kopp, Eds., pp. 1–27, Willey-VCH GmbH, Weinheim, Germany, 2009.

[15] G. A. Roberts, J. C. Hamaker Jr., and A. R. Johnson, *Tool Steels*, American Society for Metals, Materials Park, Ohio, USA, 3rd edition, 1971.

[16] A. Alfan, M. Z. Omar, J. Syarif, and Z. Sajuri, "Direct partial remelting of XW-42 steel in semi-solid zone," *Journal of Applied Sciences*, vol. 10, no. 13, pp. 1255–1262, 2010.

Investigations of Physical and Rheological Properties of Aged Rubberised Bitumen

Asim Hassan Ali, Nuha S. Mashaan, and Mohamed Rehan Karim

Center for Transportation Research, Faculty of Engineering, University of Malaya, 50603 Kuala Lumpur, Malaysia

Correspondence should be addressed to Asim Hassan Ali; asim@siswa.um.edu.my

Academic Editor: Dachamir Hotza

Several road pavement distresses are related to rheological bitumen properties. Rutting and fatigue cracking are the major distresses that lead to permanent failures in pavement construction. Influence of crumb rubber modifier (CRM) on rheological properties of bitumen binder such as improvement of high and intermediate temperatures is investigated in the binder's fatigue and rutting resistance through physical-rheological changes in this research. The bitumen binders were aged by rolling thin film oven (RTFOT) to simulate short-term aging and pressure aging vessel (PAV) to simulate long-term aging. The effects of aging on the rheological and physical properties of bitumen binders were studied conducting dynamic shear rheometer test (DSR), Brookfield viscometer test, softening point test, and penetration test. The results showed that the use of rubberised bitumen binder reduces the aging effect on physical and rheological properties of the bitumen binder as illustrated through lower aging index of viscosity, lower aging index of $G^*/\sin\delta$, and an increase in Tan δ with crumb rubber modifier content increasing, indicating that the crumb rubber might improve the aging resistance of rubberised bitumen binder. In addition, the results showed that the softening point increment (ΔS) and penetration aging ratio (PAR) of the rubberised bitumen binder decreased significantly due to crumb rubber modification. Furthermore, the higher crumb rubber content, the lower $G^*\sin\delta$ after PAV aging, which led to higher resistance to fatigue cracking bitumen.

1. Introduction

Several road pavement distresses are related to bitumen properties. Rutting and fatigue cracking are the major distresses that lead to permanent failures in pavement construction. Bitumen is a viscoelastic material; its rheological properties are very sensitive to temperature and rate of loading. In general, road pavement performance properties are mainly affected by the bitumen binder properties; it is well known that the rheological properties and durability of conventional bitumen are not sufficient to resist pavement distresses. A conventional bitumen 80/100 penetration grade is commonly used in Malaysia, and moreover, it is subjected to the high traffic load and hot climate. The use of crumb rubber (waste tyre) in bitumen modification is considered as a sustainable technology which transforms unwanted residue into a new bituminous mixture. Aging is known as a factor influencing the performance and characteristics of bitumen binder. Many factors might contribute to this hardening of the bitumen such as oxidation, volatilisation, polymerization, and thixotropy [1]. Field studies have shown that crumb rubber can improve the bitumen pavement performance [2–5].

Mohamed et al. [6] illustrated the influence of low CRM content on rheological properties of aged bitumen, reporting that the rheological changes in bitumen binder led to improving the mechanical properties of rubberised bitumen binder. In a joint experiment, Abdelaziz and Karim [7] conducted a study on rheological evaluation of aged rubberised bitumen. The results showed that CRM has an obvious significant effect on bitumen rheology by increasing complex shear modulus G^* and decreasing phase angle δ. Also, the RTFOT aging results of this study indicated that G^* was increased by 1 to 1.5 times for unmodified bitumen and 2.5 times for rubberised bitumen. The PAV test indicated an increase in complex shear modulus which was about 2 and 3 times for unmodified and rubberised bitumen, respectively.

The applications of rubberised bitumen binders have reported many advantages; for instance improved bitumen resistance to rutting due to high viscosity, high softening point and better resilience, reduction in road pavement maintenance cost, lower fatigue/reflection cracking, and less temperature susceptibility [8]. In general, the application of crumb rubber modified in bitumen binder can improve binder properties by lowering the inherent temperature susceptibility of binders [9, 10]. The use of crumb rubber modifier with bitumen binder led to enhancing the fatigue resistance [11, 12]. The improved performance of bitumen rubber pavements compared with conventional bitumen pavements is primarily a result of improved rheological properties of the rubberised bitumen binder.

A study by Aflaki and Memarzadeh [13] investigated the effect of rheological properties of crumb rubber on fatigue cracking at low and intermediate temperatures using different shear methods. The results of this study showed that the high shear blending indicated more influence on improvement of low-temperature properties of rubberised binder than the low shear blending. The mechanism by which binder properties may change at low temperature was presented. This mechanism is referred to as physical hardening occurring at temperatures next to or lower than the glass transition temperature and causes significant hardening of the bitumen binder. Aging influenced the bitumen chemistry and rheology significantly; additionally, the use of the two aging conditions, that is, thin film oven test (TFOT) and rolling thin file oven test (RTFOT) showed a strong correlation and similar severity [14]. The ideal use of bitumen binders as paving materials is dependent on its resistance to the physical change across the range of temperatures encountered in a conventional pavement. Aging significantly changed both the chemical and physical properties of asphalts. Physical properties such as penetration, ductility, viscosity, and stiffness are all affected to some extent by aging [15].

Aging index measurement is one of the most popular procedures used to determine aging susceptibility and provides ranking of different bitumen binders. Aging index is defined as the ratio of a given binder property after aging to that property before aging as shown in (1):

$$\text{Aging index} = \frac{\text{Bitumen property after aging}}{\text{same property before aging}}. \quad (1)$$

1.1. Research Objective and Scope. The primary objective of this research is to evaluate and investigate the influence of crumb rubber content on the rubberised binder's rheological and physical changes subsequent to aging. The secondary objective is to identify the degree of aging in rheological and physical properties of rubberised bitumen and its effect on high-temperature properties (rutting) as well as intermediate temperature (fatigue cracking) through short-term aging test (rolling thin film oven test (RTFOT)) and long-term aging test (pressure aging vessel (PAV)).

TABLE 1: Physical and rheological properties of 80/100 PG bitumen.

Test properties	Results
Viscosity at 135°C (Pa · s)	0.65
$G^*/\sin\delta$ at 64°C (kPa)	1.35
Ductility at 25°C (cm)	>100
Softening point (°C)	47
Penetration at 25°C (d-mm)	88
Ductility after RTFOT at 25°C (cm)	>100
RTFOT aged $G^*/\sin\delta$ at 64°C (kPa)	6.022
PAV aged $G^*\sin\delta$ at 25°C (kPa)	3122.5

TABLE 2: The chemical components of the original bitumen before and after ageing.

	Unaged	After RTFOT	After PAV
Saturates (%)	15.1	13.88	12.47
Aromatics (%)	48.55	44.19	38.55
Resins (%)	26.22	28.94	31.21
Asphaltenes (%)	10.13	12.99	17.77

TABLE 3: (CRM gradation).

Sieve n (size, mm)	Particle size distribution
No. 10 (2.00)	100
No. 16 (1.180)	100
No. 20 (0.850)	100
No. 30 (0.600)	97.7
No. 40 (0.425)	61.1
No. 50 (0.300)	33.9
No. 80 (0.180)	12.5
No. 100 (0.150)	7.5
No. 200 (0.075)	0.0

2. Experimental Program and Materials

The base 80/100 penetration grade bitumen was used in this study. Tables 1 and 2 show the properties of base bitumen and the chemical components of original bitumen (before and after aging), respectively. Rubberised bitumen binder was prepared by mixing bitumen 80/100 penetration grade with CRM (8% and 16% by binder weight) passing the 30 mesh sieve as shown in Table 3, the propeller mixer used to mix the base bitumen, and the crumb rubber at constant blending conditions of 180°C temperature, time of 60 minutes, and velocity of 200 rpm.

The samples were aged using the standard rolling thin file oven test (RTFOT) (ASTM D-2872) at a temperature of 163°C for 85 min as short-term aging and pressure aging vessel (PAV) (AASHTO PP1) at 100°C and an air pressure of 2.1 Mpa for 20 hours to simulate the long-term aging. To investigate the influence of oxidative aging on rheological changes of rubberised bitumen, the age hardening is evaluated by measuring Brookfield viscosity (135°C) ASTM D4402, penetration test (25°C) ASTM D5, softening point test (25°C) ASTM D36, and dynamic shear (DSR) ASTM D-4 Proposal

TABLE 4: DSR condition test used in the current study.

Rheological parameter	Plate Dia. (mm)	Gap (mm)	Sample	T (°C)
$G^*/\sin\delta$	25	1	RTFOT	76
$G^*\sin\delta$	8	2	PAV	31

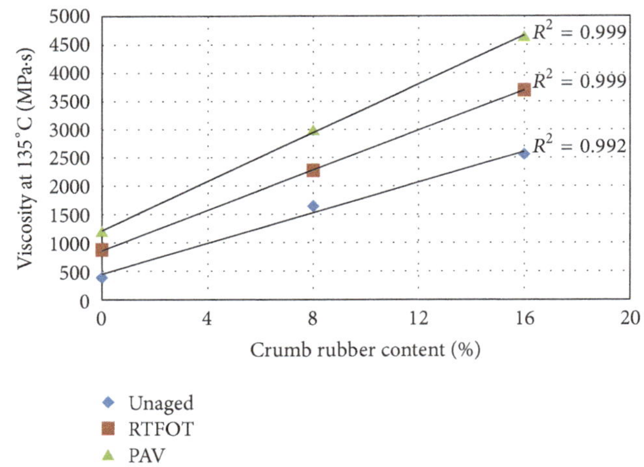

FIGURE 1: Brookfield viscosity results versus crumb rubber content.

P246. In this study, the condition test of DSR test of parallel plate (diameter and gaps), frequency sweep, and the testing temperature for samples during RTFOT and PAV aging are listed in Table 4.

3. Results and Discussion

3.1. Performance Properties at High Temperature

3.1.1. Brookfield Viscosity at 135°C. The viscosity of bitumen binders at high temperatures is an important property as it is a good indicator of the binder's ability of pumping through bitumen plant [16]. Figures 1 and 2 show the viscosity results of the rubberised bitumen tested in this study for various rubber contents and aging phases/conditions. As expected, the increase in crumb rubber content led to high viscosity of the rubberised binder. Additionally, the increase in viscosity of aged rubberised bitumen was about 36% and 44% for rubber contents of 8% and 16%, respectively, under RTFOT aging conditions, while the increase in the viscosity under PAV aging values was about 78% and 82% for rubber content of 8% and 16%, respectively. Thus, the increase in crumb rubber content has an obvious effect on aged rubberised bitumen viscosity compared to aged unmodified bitumen with correlation coefficient $R^2 = 0.997$. The reason is the conversion to asphaltenes (the highest molecular component fraction) of the other lower molecular weight asphalt components through oxidation. This is clearly evidenced by molecular weight analysis of asphalts before and after aging as shown in Table 2.

Figure 3 shows VAI of the original bitumen and the rubberised binder after RTFOT and PAV aging. VAI is the

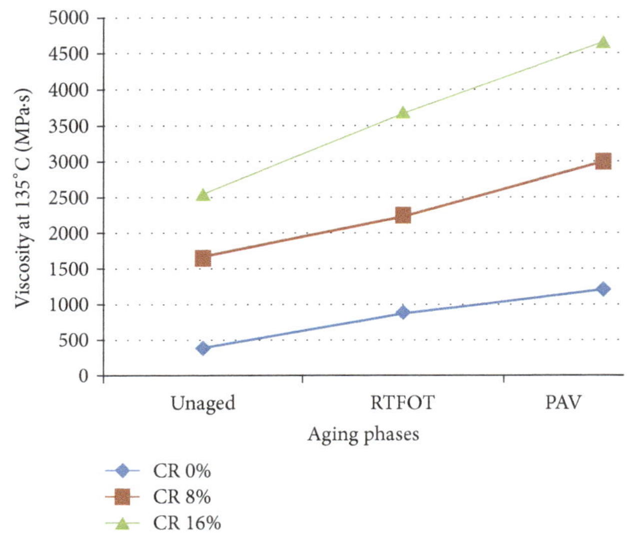

FIGURE 2: Brookfield viscosity results of CRM binders versus aging conditions.

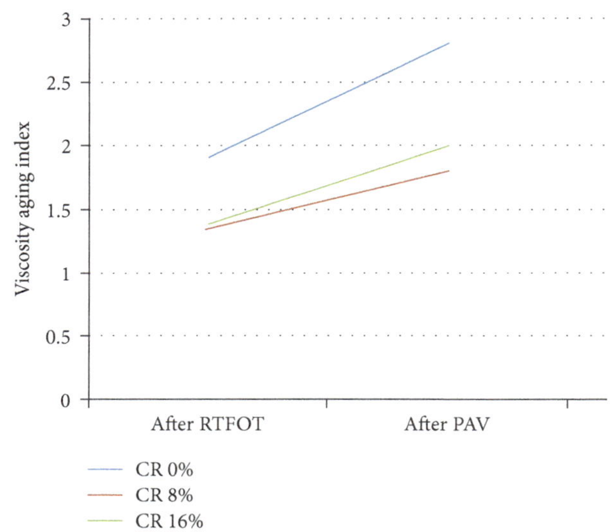

FIGURE 3: VAI of original and rubberised modified bitumen after RTFOT and PAV.

ratio of aged viscosity of the binder to the unaged viscosity of same binder as illustrated in (2) [17]:

$$\text{VAI} = \frac{V_{\text{aged}}}{V_{\text{unaged}}}, \quad (2)$$

where VAI is the viscosity aging index, V_{unaged} and V_{aged} are the viscosities of unaged and aged values of the binder, respectively. This exhibits that viscosity aging index of rubber modified binder is significantly lower than that of unmodified bitumen, thus indicating that the rubberised bitumen would improve the thermooxidative aging properties of bitumen. This result can be explained by the chemical components change of the binder during the aging that led to the increase

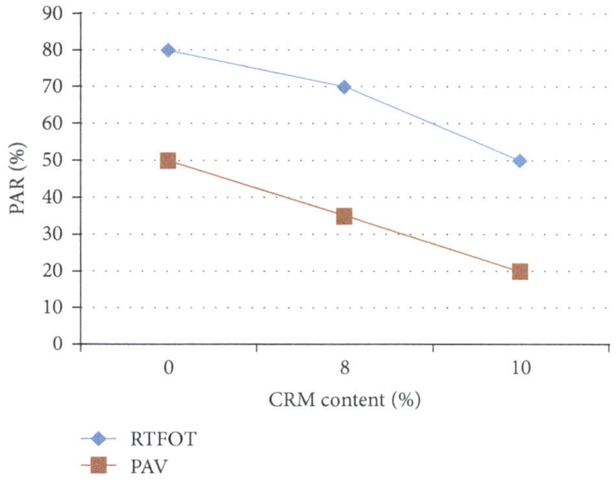

FIGURE 4: Effect of aging on penetration of bitumen binder.

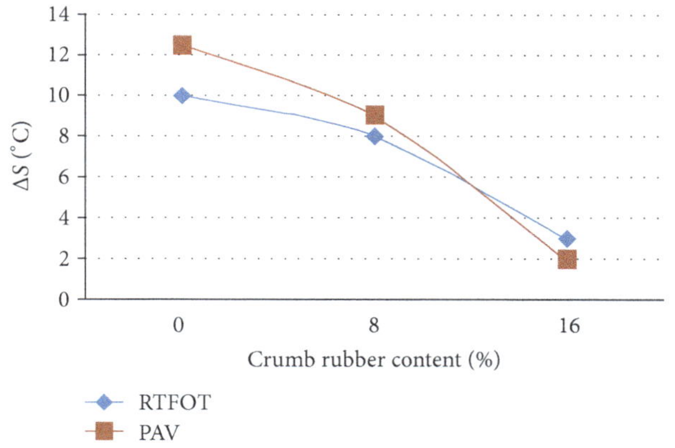

FIGURE 5: Effect of aging on softening point increment of bitumen binder.

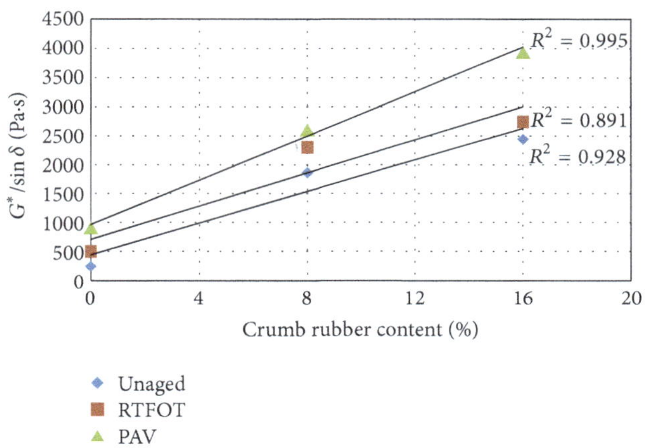

FIGURE 6: $G^*/\sin\delta$ results versus crumb rubber content.

of the asphaltene content which was responsible for increasing the viscosity flow of the bitumen binder. The viscosity aging index reflected better resistance of crumb rubber modified bitumen to aging as it showed lower value after short-term aging, RTFOT, and long-term aging, PAV, respectively. Thus, the use of viscosity aging index determines the effect of aging of the original bitumen and modified bitumen which provides a much clearer effect of the hardening (aging) rate [17].

3.1.2. Penetration Aging Ratio Results. The penetration aging ratio is an alternative option to show the effects of aging on the physical properties of the bitumen binder. Penetration aging ratio (PAR) can be defined as the ratio of aged penetration value to unaged penetration as illustrated in (3):

$$\mathrm{PAR} = \left(\frac{\text{penetration}_{\text{aged}}}{\text{penetration}_{\text{unaged}}}\right) \times 100. \quad (3)$$

Figure 4 shows the penetration aging ratio of original and rubberised bitumen after rolling thin film oven test and pressure aging vessel test. The results showed that the penetration of unmodified bitumen and rubberised bitumen binders decreased after the two different aging. In addition, the higher crumb rubber content, the lower penetration aging ratio which led to reducing the degree of aging of rubberised bitumen binder. Thus, the crumb rubber addition led to improving the binder resistance to oxidative aging.

3.1.3. Softening Point Increment (ΔS). The softening point increment can be used as a good indicator to measure the degree of aging, and it can be defined by ΔS and calculated as shown in (4):

$$\Delta S = \mathrm{S.P}_{\text{aged}} - \mathrm{S.P}_{\text{unaged}}, \quad (4)$$

where ΔS is the softening point increment, $\mathrm{S.P}_{\text{aged}}$ is aged softening point value, and $\mathrm{S.P}_{\text{unaged}}$ is the unaged softening point value. Figure 5 shows the effect of crumb rubber modified binder bitumen on softening point changes after RTFOT aging and PAV aging. The softening point results of unmodified bitumen and rubberised bitumen showed an increase after both aging conditions of RTFOT and PAV, respectively, as the rubber content increased, indicating the hardening level of the materials during aging. In comparing the softening point increment (ΔS) of modified binder with bitumen without crumb rubber modifier, the crumb rubber modified bitumen showed lower (ΔS) as rubber crumb content was increased. This can be explained by the structural change of molecules in asphalt of fraction content during aging, which led to an increase in the softening point [17, 18]. Thus, the results of using rubberised bitumen binder have a significant effect on improving the aging resistance and show lower hardening aging which can reduce the oxidative aging resistance of the bitumen.

3.1.4. Influence of Aging on Rheological Properties at 76°C. The dynamic shear rheometer test was conducted to select the rutting resistance parameter, $G^*/\sin\delta$, of the unmodified binder and rubberised bitumen binder were mastered at 76°C, and the results are shown in Figure 6. Generally, the higher rubber content led to high value of rutting factor, $G^*/\sin\delta$,

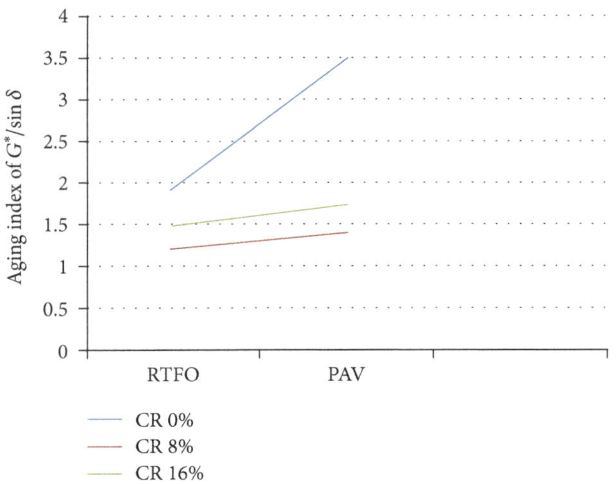

FIGURE 7: Aging index of $G^*/\sin\delta$ of bitumen binder versus CRM content.

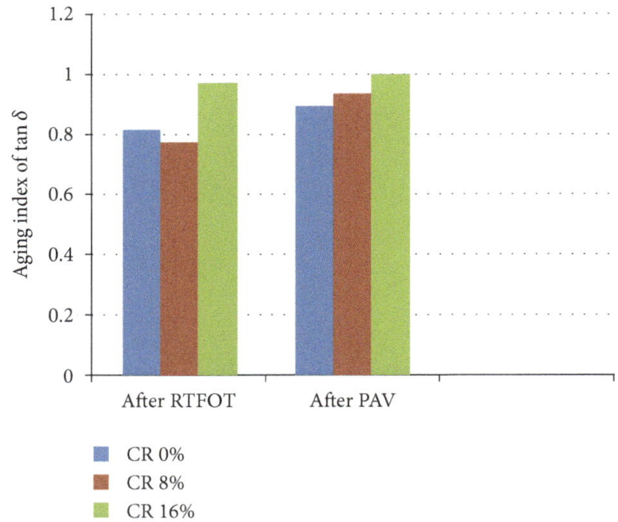

FIGURE 8: Effect of aging on Tan δ of bitumen binder.

for all combinations tested. Additionally, $G^*/\sin(\delta)$ of aged rubberised bitumen was about 11% and 21% for rubber contents of 8% and 16% compared to unaged base binder using RTFOT aging conditions. The increase in the $G^*/\sin(\delta)$ was about 38% and 60% for samples of 8% and 16% rubber content, respectively, compared to unaged base binder using PAV aging conditions. Thus, the increase in crumb rubber content has an obvious effect on rutting factor, $G^*/\sin(\delta)$, with correlation coefficient $R^2 = 0.995$. In general, the results of rubberised binder are in tandem with the superpave binder specification of a minimum $G^*/\sin\delta$ requirement using RTFOT (2.2 Pa) which showed that the binder with higher $G^*/\sin\delta$ value is less susceptible to rutting deformation at high pavement temperatures [10, 19]. Figure 7 shows the aging index of rutting factor, $G^*/\sin\delta$, of original and crumb rubber modified bitumen binder at 76°C as calculated by (5) as follows:

$$\text{AIRF} = \frac{(G^*/\sin\delta)_{\text{aged}}}{(G^*/\sin\delta)_{\text{unaged}}}. \tag{5}$$

In the case of AIRF as the aging index of rutting factor $(G^*/\sin\delta)_{\text{aged}}$ and $(G^*/\sin\delta)_{\text{unaged}}$ is the $G^*/\sin\delta$ value of bitumen binder after and before aging, respectively. The results showed that $G^*/\sin\delta$ of bitumen binder with different content of crumb rubber increased after different aging, obviously after PAV aging. Thus, crumb rubber modified bitumen binder showed better resistance to oxidation aging. As displayed in Figure 7, the unmodified binder was of higher AIRF which led the binder to being more susceptible to aging, while modified bitumen shows lower AIFR as crumb rubber increased in the specimen, leading to better aging resistance of modified binder.

Figure 8 shows the effect of aging on loss tangent of bitumen binder after RTFOT aging and PVA aging. The results displayed that the higher the crumb rubber content, the higher aging index of loss tangent, Tan δ, indicating better aging resistance. The aging ratio of loss tangent is defined as the ratio of aged Tan δ to unaged Tan δ of the bitumen binder as shown in (6):

$$\text{Tan}\,\delta\text{ avg.} = \left[\frac{(\text{Tan}\,\delta)_{\text{aged}}}{(\text{Tan}\,\delta)_{\text{unaged}}}\right] \times 100, \tag{6}$$

where Tan δ avg. is the average of Tan δ. $(\text{Tan}\,\delta)_{\text{aged}}$ and/ $(\text{Tan}\,\delta)_{\text{unaged}}$ is the (Tan δ) value of bitumen binder after and before aging, respectively. The higher value of aging index Tan δ is related to higher elastic behaviour (high in storage modulus G') since Tan $\delta = G''/G'$; thus the high-loss tangent of rubberised bitumen led to more elastic behaviour and more solid-like material. This suggests that the rubberised bitumen binder can improve the resistance to oxidative aging.

3.2. Fatigue Cracking Properties at Intermediate Temperature. The Strategic Highway Research Program (SHRP) had a maximum value of 5000 kPa for $G^*\sin(\delta)$, and low values of these parameters are considered good indicators of fatigue cracking resistance [16]. In this study, the fatigue resistance parameter, $G^*\sin(\delta)$ values, of the unaged bitumen and rubberised binders after pressure aging vessel (PAV), were measured using the dynamic shear rheometer (DSR) at 31°C, and the results are illustrated in Figure 9. In general, the high crumb rubber content led to lower $G^*\sin(\delta)$ values of the rubberised bitumen binders. The decrease in the $G^*\sin(\delta)$ was about 13% and 25% for rubber content of 8% and 16%, respectively, after pressure aging vessel (PAV), aging conditions, respectively. Thus, the increase in crumb rubber content has an obvious effect on aged rubberised bitumen fatigue parameter, $G^*\sin\delta$, with correlation coefficient $R^2 = 0.816$ after PAV aging. Owing to the aging, the results showed that the lower the $G^*\sin\delta$ value, the less shearing energy loss and the better ability of the fatigue resistance. Also, $G^*\sin\delta$ of the modified bitumen is lower than the unmodified bitumen at temperature of 31°C, indicating that rubberised bitumen

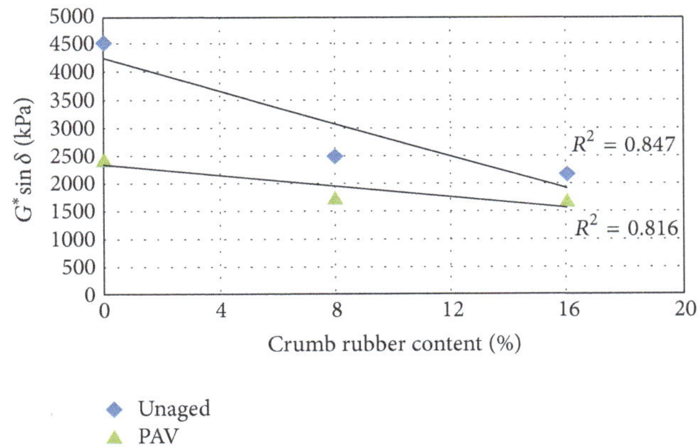

FIGURE 9: G^* sin δ results versus crumb rubber content.

binder would improve the fatigue resistance of bitumen as a result of introduction of crumb rubber modifier (CRM).

4. Conclusion

After the introduction to the problem, a review of the literature, and analysis and discussion of the findings, this section intends to summarise the overall conclusions achieved through this research. Therefore, the significant findings of the current study are as follows.

(1) The results showed that the use of rubberised bitumen binder reduced the aging effect on physical and rheological properties of the bitumen binder as illustrated through lower aging index of viscosity, lower aging index of $G^*/\sin\delta$ at 76°C, and an increase in Tan δ with increased crumb rubber modifier content, indicating that the antioxidant released in asphalt from the tire rubber particles is a significant factor in reducing the aging effect as correctly observed by the data results.

(2) In addition, the results showed that the softening point increment (ΔS) and penetration aging ratio (PRA) of the rubberised bitumen binder decreased significantly due to crumb rubber modification.

(3) The aging index of viscosity reflected better resistance of crumb rubber modified bitumen to aging as it showed lower value after short-term aging, RTFOT, and long-term aging (PAV), respectively. Thus, the use of viscosity index eliminated the variability caused by difference in viscosity of the original bitumen and modified bitumen which provided a more reliable picture of the hardening (aging) rate.

(4) Furthermore, the higher crumb rubber content, the lower G^* sin δ at 31°C after PAV aging, which led to higher resistance to fatigue cracking.

(5) Crumb rubber modifier increased the stiffness of the bitumen binder improving its resistance to rutting.

References

[1] J. C. Petersen, "Chemical composition of asphalt as related to asphalt durability—state of the art," *Transportation Research Record*, vol. 999, pp. 13–30, 1984.

[2] D. R. Brown, D. Jared, C. Jones, and D. Watson, "Georgia's experience with crumb rubber in hot-mix asphalt," *Journal of Transportation Research Record*, vol. 1583, pp. 45–51, 1997.

[3] G. W. Maupin Jr., "Hot mix asphalt rubber applications in Virginia," *Journal of Transportation Research Record*, vol. 1530, pp. 18–24, 1996.

[4] E. Charania, J. O. Cano, and R. H. Schnormeier, "Twenty-year study of asphalt rubber pavement in Phoenix, Arizona," *Transportation Research Board*, vol. 1307, pp. 29–38, 1991.

[5] M. Stroup-Gardiner, B. Chadbourn, and D. E. Newcomb, "Babbitt, Minnesota: case study of pretreated crumb rubber modified asphalt concrete," *Transportation Research Record*, vol. 1530, pp. 34–42, 1996.

[6] A. A. Mohamed, O. Husaini, M. O. Hamzah, and H. Ismail, "Rheological properties of crumb rubber modified bitumen containing antioxidant," *The Arabian Journal For Science and Engineering*, vol. 34, article 1B, 2009.

[7] M. Abdelaziz and M. R. Karim, "Rheological evaluation of ageing properties of rubber crumb modified bitumen," *Journal of the Eastern Asia Society for Transportation Studies*, vol. 5, pp. 820–833, 2003.

[8] S. Liu, W. Cao, J. Fang, and S. Shang, "Variance analysis and performance evaluation of different crumb rubber modified (CRM) asphalt," *Construction and Building Materials*, vol. 23, no. 7, pp. 2701–2708, 2009.

[9] S. J. Lee, C. K. Akisetty, and S. N. Amirkhanian, "The effect of crumb rubber modifier (CRM) on the performance properties of rubberized binders in HMA pavements," *Construction and Building Materials*, vol. 22, no. 7, pp. 1368–1376, 2008.

[10] K. D. Jeong, S. J. Lee, S. N. Amirkhanian, and K. W. Kim, "Interaction effects of crumb rubber modified asphalt binders," *Construction and Building Materials*, vol. 24, no. 5, pp. 824–831, 2010.

[11] L. Raad and S. Saboundjian, "Fatigue behavior of rubber-modified pavements," *Transportation Research Record*, vol. 1639, pp. 73–82, 1998.

[12] H. R. Soleymani, H. Zhai, and H. Bahia, "Role of modified binders in rheology and damage resistance behavior of asphalt mixtures," *Transportation Research Record*, vol. 1875, pp. 70–79, 2004.

[13] S. Aflaki and M. Memarzadeh, "Using two-way ANOVA and hypothesis test in evaluating crumb rubber modification (CRM) agitation effects on rheological properties of bitumen," *Construction and Building Materials*, vol. 25, no. 4, pp. 2094–2106, 2011.

[14] H. U. Bahia and D. A. Anderson, "Glass transition behavior and physical hardening of asphalt binders," *Journal of the Association of Asphalt Paving Technologists*, vol. 62, pp. 93–129, 1993.

[15] X. Lu and U. Isacsson, "Effect of ageing on bitumen chemistry and rheology," *Construction and Building Materials*, vol. 16, no. 1, pp. 15–22, 2002.

[16] The Asphalt Institute, *Performance Graded Asphalt Binder Specification and Testing*, Superpave Series (SP-1), Lexington, Lexington, Ky, USA, 2003.

[17] P. Cong, S. Chen, J. Yu, and S. Wu, "Effects of aging on the properties of modified asphalt binder with flame retardants,"

Construction and Building Materials, vol. 24, no. 12, pp. 2554–2558, 2010.

[18] M. N. Siddiqui and M. F. Ali, "Studies on the aging behavior of the Arabian asphalts," *Fuel*, vol. 78, no. 9, pp. 1005–1015, 1999.

[19] A. K. Apeagyei, "Laboratory evaluation of antioxidants for asphalt binders," *Construction and Building Materials*, vol. 25, no. 1, pp. 47–53, 2011.

17

Stress Corrosion Cracking of Steel and Aluminum in Sodium Hydroxide: Field Failure and Laboratory Test

Y. Prawoto,[1] K. Sumeru,[1,2] and W. B. Wan Nik[3]

[1] *Faculty of Mechanical Engineering, Universiti Teknologi Malaysia, Skudai, 81310 Johor, Malaysia*
[2] *Department of Refrigeration and Air Conditioning, Politeknik Negeri Bandung, Bandung 40012, Indonesia*
[3] *Faculty of Maritime Studies and Marine Science, University Malaysia Terengganu, 21030 Kuala Terengganu, Malaysia*

Correspondence should be addressed to Y. Prawoto, yunan.prawoto@gmail.com

Academic Editor: Pavel Lejcek

Through an investigation of the field failure analysis and laboratory experiment, a study on (stress corrosion cracking) SCC behavior of steel and aluminum was performed. All samples were extracted from known operating conditions from the field failures. Similar but accelerated laboratory test was subsequently conducted in such a way as to mimic the field failures. The crack depth and behavior of the SCC were then analyzed after the laboratory test and the mechanism of stress corrosion cracking was studied. The results show that for the same given stress relative to ultimate tensile strength, the susceptibility to SCC is greatly influenced by heat treatment. Furthermore, it was also concluded that when expressed relative to the (ultimate tensile strength) UTS, aluminum has similar level of SCC susceptibility to that of steel, although with respect to the same absolute value of applied stress, aluminum is more susceptible to SCC in sodium hydroxide environment than steel.

1. Introduction

Engineers in the field often assume that within the same group of materials, for example, group of ferrous or group of nonferrous, only the chemical composition is relevant when selecting materials for corrosive environments. Not so many of them consider the effect of heat treatment, let alone the microstructure. On the other hand, it is widely known that petroleum and natural gas systems can be contaminated with solutions that are sometimes very aggressive to engineering materials, such as steels and aluminums, which are used in the transport and processing of petroleum products. It has been reported from decades ago that up to a quarter of equipment failures in the petroleum refining industry are in some way associated with (stress corrosion cracking) SCC and hydrogen damage [1]. Although SCC of austenitic steel is the most common, steel with other microstructures can also undergo SCC. Therefore, SCC is not only unique to austenitic steel but also problematical in ferrite-pearlite and martensitic steels and aluminum alloys as well. This also has been known for long time [2]. Publications on the studies of SCC on ferrite-pearlite in the environment of NaOH were started quite a while ago, and it remains actively studied up to now [3, 4]. The SCC behavior is highly dependant on the microstructure. El Sayed et al. studied SCC at various stress levels ranging from 10% to 100% of the (ultimate tensile strength) UTS of the materials [5]. However, they focused on the depth of the crack and did not analyze a morphology of the crack.

Similarly, SCC failure is also reported in aluminum and its derivative alloys [6–8]. It is also commonly understood that aluminum alloys that contain considerable amounts of soluble alloying elements, especially copper, magnesium, silicon, and zinc, are susceptible to SCC. They can fail by cracking along grain boundaries when simultaneously exposed to specific environments and sufficient magnitude of stresses. Well-known environments that can trigger SCC include but are not limited to water vapor, aqueous solutions, acids (such as acetic acid), organic liquids, liquid metals, and salt solutions. Stresses sufficient for crack initiation and crack propagation can be far below the stresses required for gross yielding. Unlike for high-strength materials such as steels, papers discussing about SCC of aluminum are rather scarce [7–10]. While majority of papers focus on

Metallographic appearance | Metallographic appearance

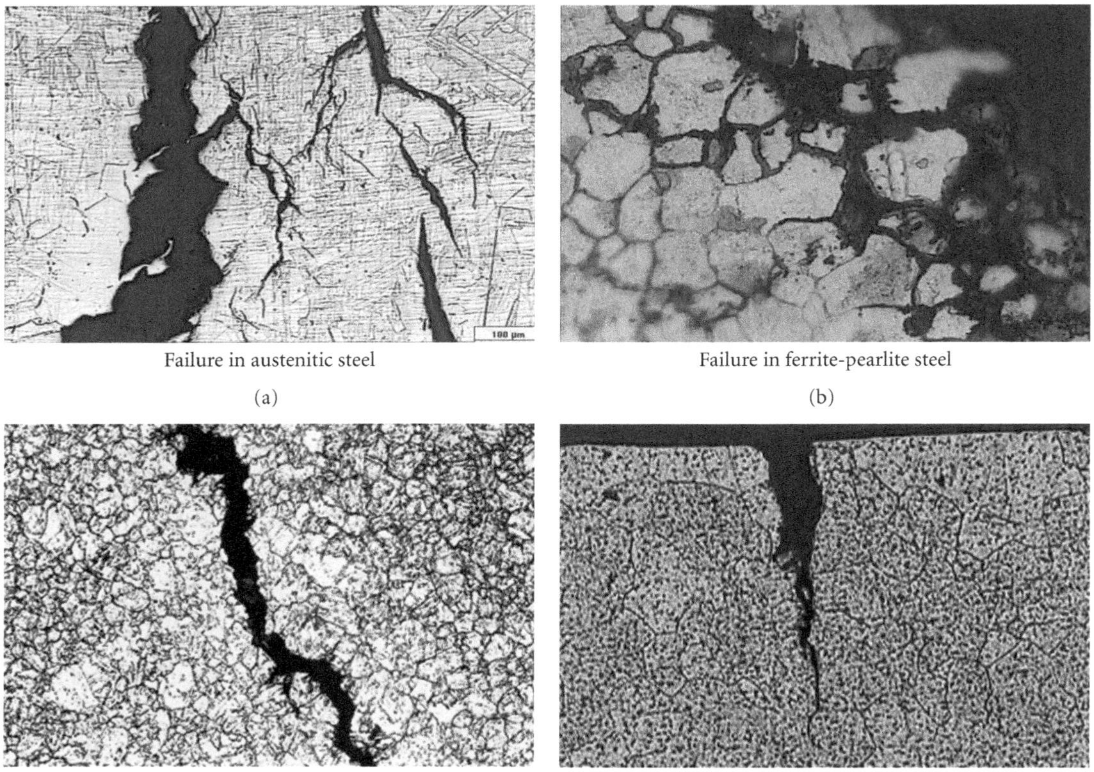

Failure in austenitic steel (a)

Failure in ferrite-pearlite steel (b)

Failure in martensitic steel (c)

Failure in aluminum alloy (d)

FIGURE 1: Appearance of the microstructures of samples extracted from field failure [12].

the chemical composition of aluminum alloys in various corrosive solutions, few of them focus on the effect of heat treatments or microstructures [11].

This paper is a combination of analysis of failed field specimens and laboratory experiments using three different microstructures: austenite, martensite, and ferrite-pearlite for steels and two different heat treatments for aluminum. The environment used was sodium hydroxide. The laboratory test was an accelerated version of what happens in the field.

2. Case Study

Figure 1 shows metallography of the samples from the field failures of both aluminum and the steel samples failed in similar environments. Although the environments in which all actual parts used were similar, the degree of the stress corrosion cracking is clearly different. For the steel, austenite was the most sensitive to SCC, followed by martensite, and finally ferrite-pearlite. All the samples were obtained from an oil company. While further detailed information is confidential, the environments in which they were being used contained some sodium hydroxide (NaOH). All the pipes had been in service for more than 8 years. Further discussion only on the steel part of this project has been reported along with the basic general mechanism of the SCC [12], while the brief of the computational principle of this is available in the computational journal [13]. For the aluminum, the shape of the crack tip is somehow similar to that of fatigue crack tip, which is wedge-shaped crack morphology typically associated with thermal and/or mechanical fatigue cracking [14].

Because of the difference in concentrations used, stress levels applied, and temperatures involved, direct comparison of these cases cannot be made without laboratory accelerated experiment. Therefore, similar but an accelerated version of the simulation was created in the laboratory.

3. Brief on Stress Corrosion Cracking

SCC is fundamentally different from other types of corrosion. It needs a combination of mechanical-and-corrosive environment, and the failure due to SCC only occurs under specific conditions. It is highly chemically specific in that certain alloys are likely to undergo cracking only when exposed to a small number of chemical environments. Similar to the nature of the fatigue failure, in industries, SCC progress is very slow and difficult to detect until sudden failure. Typically, outside appearance of the SCC does not reflect the condition inside. This fact often deceives engineers

TABLE 1: (a) Chemical composition analysis results of steel materials [12]. (b) Chemical composition analysis results of aluminum alloys.

(a)

% wt.	Field failure case study			Ferrite-pearlite tube		
	Ferrite-pearlite tube	Martensite tube	Austenite tube	Ferrite-pearlite tube	Martensite tube	Austenite tube
C	0.274	0.494	0.164	0.382		0.093
Si	0.178	0.167	0.044	0.149		0.437
Mn	0.462	0.379	0.896	0.263		0.672
P	0.016	0.019	0.032	0.011		0.041
S	0.022	0.030	0.021	0.028		0.028
Cr	Negligible	negligible	12.128	negligible		13.427
Fe	Balance	Balance	Balance	Balance		Balance

(b)

% wt.	Field failure sample	Laboratory sample
Si	0.692	0.674
Fe	0.223	0.280
Cu	0.044	0.041
Cr	0.033	0.031
Ni	0.009	0.007
Al	Balance	Balance

working on the fields. Stress corrosion cracking of carbon steels and aluminum is relatively common in many parts, especially in the oil and gas sectors as well as in chemical manufacture.

Although there is much published research into SCC problems, complete elimination of the SCC problem is difficult, especially in oil and gas industry or even in general engineering applications. Therefore, suitable approaches are needed not only to eliminate this problem and guarantee the safety of human life but also to avoid the consequences of the malfunction. Common SCC avoidance programs such as materials selection, corrosion inhibitors, protective coatings and cathodic protection are usually preferred method to overcome the corrosion predicament.

Numerous papers on materials with different microstructures are available individually but studies where different microstructures have been examined under similar conditions and compared directly are very rare, if it is available at all. Some used martensite tube and immersion in NaOH, K_2CO_3, and their combination. The effect of variation of microstructure on SCC is also shown in other researches as well [15, 16], which reported on the susceptibility to SCC in Na_2S of steels for various samples with assorted heat treatments and clarified the influence of the heat treatment parameters on the susceptibility to SCC.

Mechanisms of Stress Corrosion Cracking. The mechanisms of SCC are relatively widely discussed and therefore will be omitted here [17, 18]. Furthermore, both stages, Crack Initiation and Crack Propagation, produce a unique crack shape depending upon their mechanism. In short, the mechanisms for crack initiation include mechanical features, local galvanic cells initiating dissolution, pitting type crack initiation, and initiation at a stress-induced phenomenon. The crack growth mechanisms include film rupture mechanism, film cleavage mechanism, adsorption induced cleavage, adsorption induced plasticity, atomic surface mobility, and corrosion tunnel model.

4. Experiment

4.1. Field Failure Case. The samples extracted from field failure were total of four samples, three steel samples, and one aluminum sample. Figure 1 shows representatives of metallography of the field failure cases of four samples received. All of them came from comparable usage conditions. The comprehensive data are not disclosed here due to the agreement with the owner. All materials were retrieved from sites that had alkalinity of approximately 1.1% and pH of 9.2. The main source of this environment was NaOH.

The chemical composition analysis results for steel and aluminum samples are shown in Tables 1(a) and 1(b), respectively. They are austenitic stainless steel, low carbon steel, and aluminum 6XXX series. Glow discharge spectroscopy was used to obtain the data. For the accelerated simulation, similar materials available in the market were used.

4.2. Laboratory Experiment. Following the ASTM standard [19], both the steel and aluminum samples were shaped into C-ring samples.

Specimen Preparation. The low carbon steel and austenitic stainless steel pipes were used in standard size with NPS

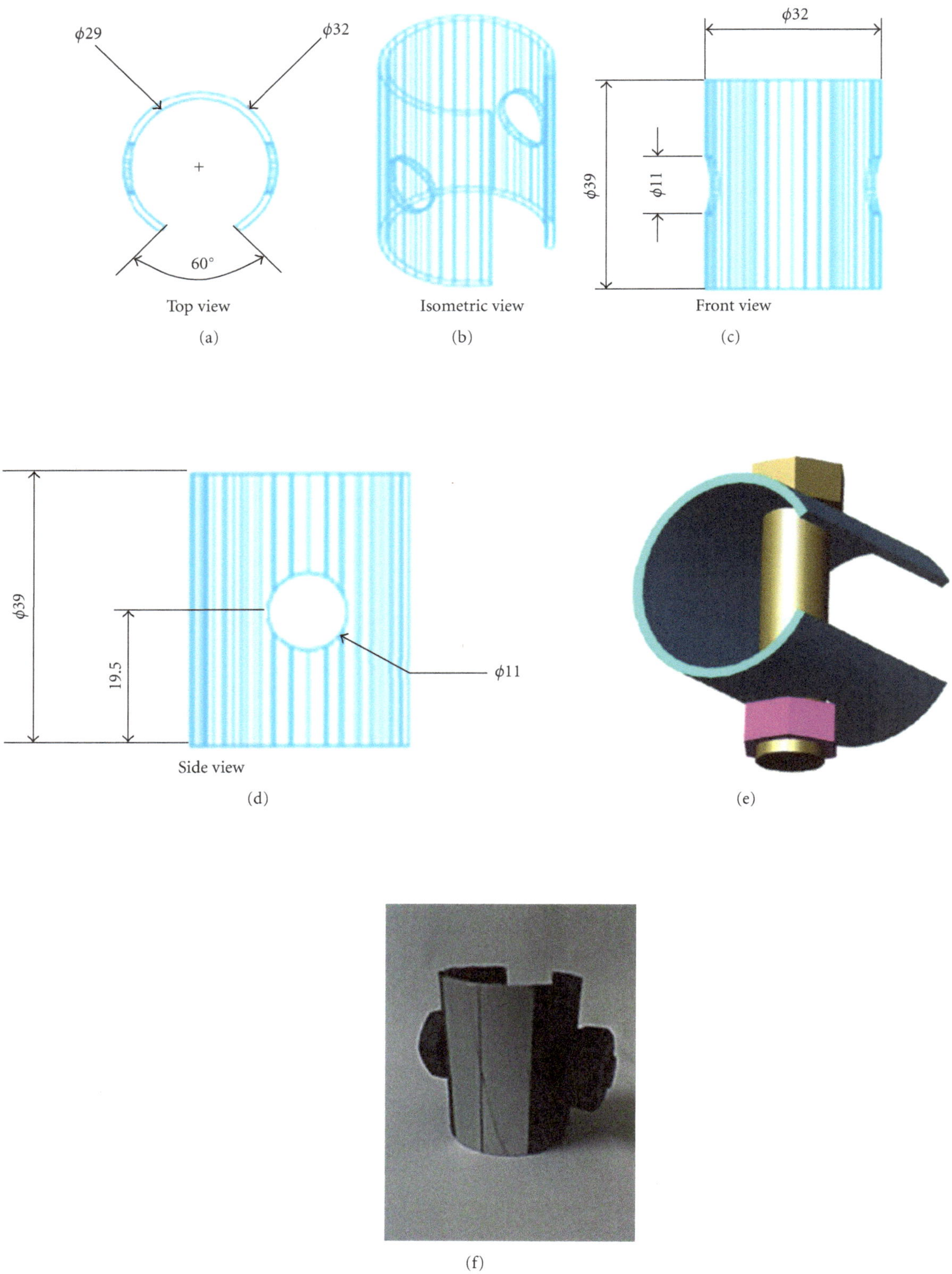

FIGURE 2: Appearance of C-ring specimen. Away from the areas of interest, coating was applied to avoid corrosion.

designator number 1.5. Its outer diameter is 48.3 mm (1.9 in.) with thickness 3.68 mm (0.145 in.). Figure 2 shows the appearance of the C-ring specimen that was used. This specimen was machined with a Computerized Numerical Control (CNC) machine to the designed dimensions.

The low carbon steel specimens were then undergone heat treating to obtain ferrite/pearlite (FP) and martensite (M), while the stainless steel was austenite (A). For the aluminum 6XXX series sample, two different heat treatments, namely, HT1 (precipitation heat treatment) and HT2 (annealing), was performed. Bolt and nuts was utilized to apply particular stress levels according to the ASTM standard. Three different stress values were applied, 40%, 75%, and 95%, of the material's UTS. The UTS values of the samples were obtained from tensile tests. This is the best choice of stress value because changes of behavior in SCC can be evaluated easily. The following equation was used to calculate the stress that has been applied to a C-ring specimen:

$$OD_f = OD - \Delta, \qquad \Delta = f\,\pi \frac{D2}{4EtZ}, \tag{1}$$

where OD is the outside diameter of C-ring before stressing, in. (or mm), OD_f is the outside diameter of stressed C-ring, in. (or mm), f is the desired stress, MPa (or psi) (within the proportional limit), Δ is the change of OD giving desired stress, mm (or in.), D is the mean diameter (OD $-$ t), mm (or in.), t is the wall thickness, mm (or in.), E is the modulus of elasticity, MPa (or psi), and Z is the a correction factor for curved beams.

After the specimens were fully prepared, they underwent immersion testing in sodium hydroxide (NaOH) 50% at 80°C for several different periods of time. It is worth noticing here that to reduce the effect of other corrosion modes, coating was applied to the samples except in the areas of interest.

5. Results and Discussion

The UTS values for the samples used were 421.3 MPa for ferrite pearlite sample, 949.4 MPa for martensite sample, 684.6 MPa for austenite sample, and 119 MPa for aluminum alloy 6XXX series sample, respectively. The representative results for the SCC test are shown in Figure 3. It is worth noting here that our intermediate observation revealed an indication of grain boundary thickening on the austenite sample and aluminum sample prior to crack formation. This could be a sign of hydrogen reacting with the material or a result of oxidation. This oxidation is preferable to occur at the grain boundary. Similar observations were also reported [20]. Chemically, the thickening indicates that there is segregation, concentrated near the grain, while physically it implies increasing of the dislocation density. At 95% of the UTS the ferrite pearlite sample started to develop small amount (1-2 grains) of SCC, while the martensite sample had more than 100 μm deep SCC and the austenite sample has almost twice that depth of SCC. For aluminum samples, both HT1 (precipitation hardened) and HT2 (annealed), they clearly show cracking taken place at the grain boundary. Reports of similar observation are also available although scarcely [21]. Further investigation also revealed that HT1 sample also shows some indication of general surface corrosion attack.

On all samples, the tendency of the crack to be intergranular was very obvious. Its appearance is similar to that of hydrogen damage, where the hydrogen atoms penetrate into the grain boundary.

The initiation type for the ferrite pearlite can be classified as local galvanic cells initiating dissolution, where pearlite has corroded more. Similarly, the martensite sample also has this tendency, with the carbide being corroded first. Austenite and aluminum samples, both the HT1 and HT2, on the other hand, show signs of pitting type crack initiation, where the grain boundary and the twin are the most extensively pitted.

Once the initiation was successful, the corrosion then moved to the second stage, propagation. The propagation type of cracks in ferrite pearlite occurred by film rupture mechanism, whereas corrosion tunnel model was visible in the austenite sample. The film rupture mechanism was supported by evidence of corrosion at the surface of the ferrite-pearlite in the form of general corrosion attack. On the other hand, the symptoms of general attack corrosion were not observed in the austenite nor aluminum HT2 sample. The austenite and the HT2 samples show clear evidence of corrosion tunnels, where the twin microstructure and the grain boundary in the figure are some of the entry points and subsequently become preferred locations for tunnel growth. This observation is solely based upon the morphology evaluation. The crack tip of HT1 sample similar to that of martensite sample. Although this does not necessarily mean that the mechanism is the same, this finding is important for computational bases, as the continuum mechanics and fracture mechanics differentiate the crack tip shapes.

The crack depth as a function of the applied stress is shown in Figure 4. Generally, aluminum shows significantly more susceptible to SCC than steel is. With respect to the absolute values of the applied stress, the rank of the crack depth is austenite, ferrite-pearlite, and martensite, for the steel and HT1 followed by HT2 for aluminum. However, when this applied stress is transferred to the percentage of the UTS, the ranking changes to austenite, martensite, and ferrite-pearlite (Figure 5) for the steel, and HT1 followed by HT2 for aluminum samples.

Another important finding is that when they are expressed in term of relative applied stress (to UTS), steel and aluminum have similar susceptibility. The laboratory results match the observation of the field failures. The morphology of the cracks on all different laboratory samples is very similar to that of samples from the field failures. The tendency to crack branching is clearly much stronger on a sample with austenite microstructure than on a sample with ferrite-pearlite. Similarly, for aluminum samples, the crack branching is visible on both the HT1 and HT2 samples. This information is very useful, especially to initiate continuum mechanics-based computational modeling for fracture mechanics [22, 23].

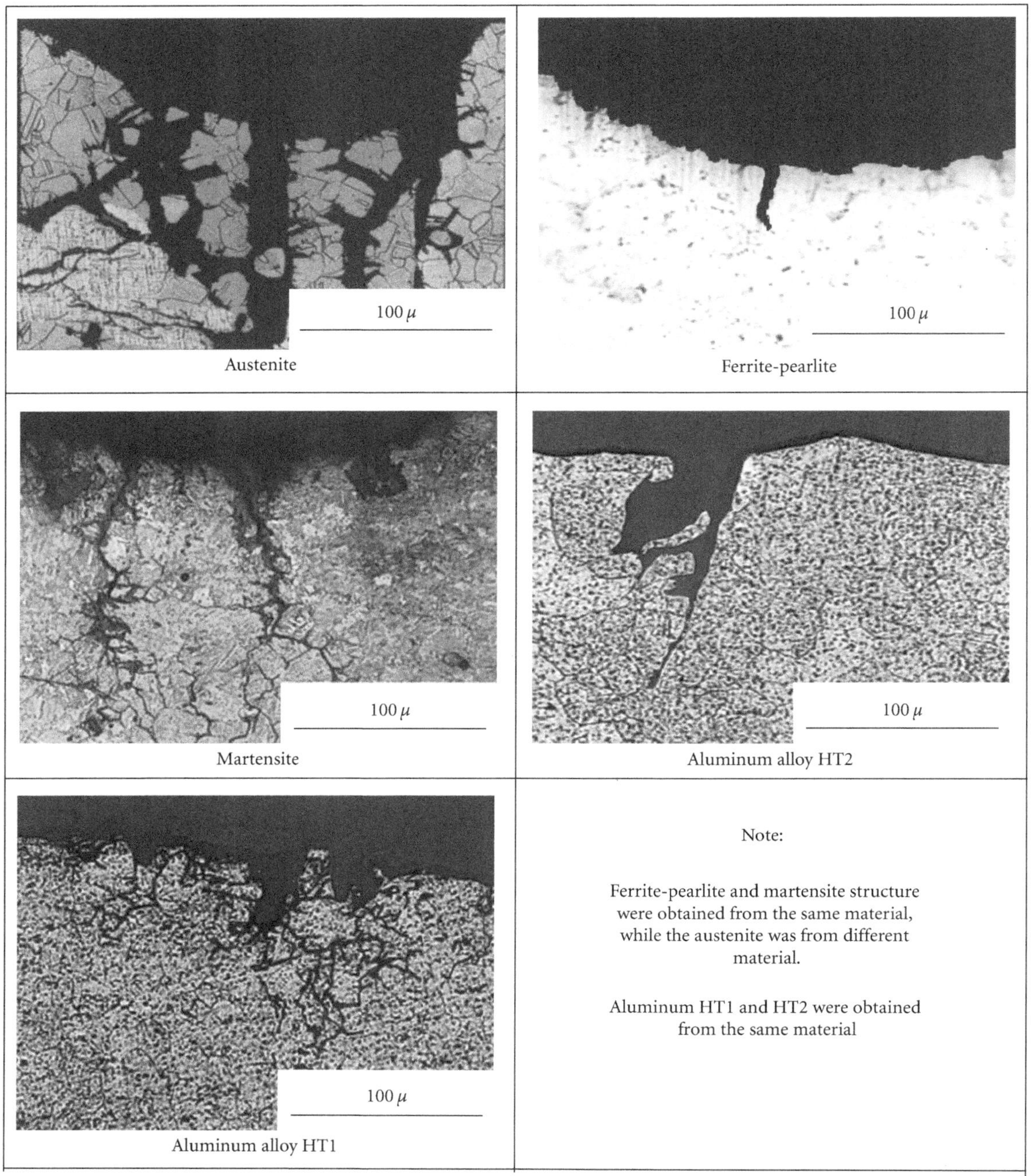

FIGURE 3: Appearance of the crack shapes for steels and aluminums.

6. Conclusion

Characterization of the SCC of steel and aluminum martensite in NaOH was performed. For steel samples, three different microstructures: austenite, ferrite-pearlite, while for the aluminum samples microstructural variation was obtained by HT1 and HT2. Expressed in absolute applied stress, aluminum has substantially higher susceptibility than steel. However, they become very similar when they are expressed in term of percentage of their UTS. For an applied load of constant percentage of ultimate tensile strength, among steel microstructures, austenite was found to be the most prone to SCC while for aluminum, HT1 is proven to be more susceptible to SCC than HT2 sample. Although there are limitations to this observation, such as neglecting the effect of the chemical composition of the material, the observation is useful for both computational purposes as well as for practicing engineers in the field. The result shows that each microstructure produces a unique crack shape that can be used as a basis for computational model construction.

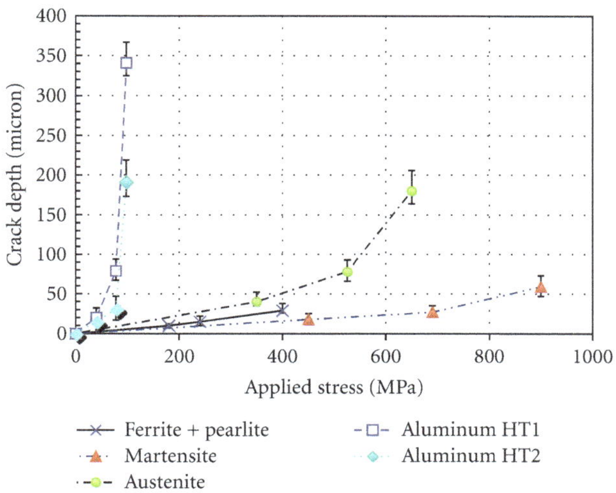

FIGURE 4: Crack depth as a function of applied stress.

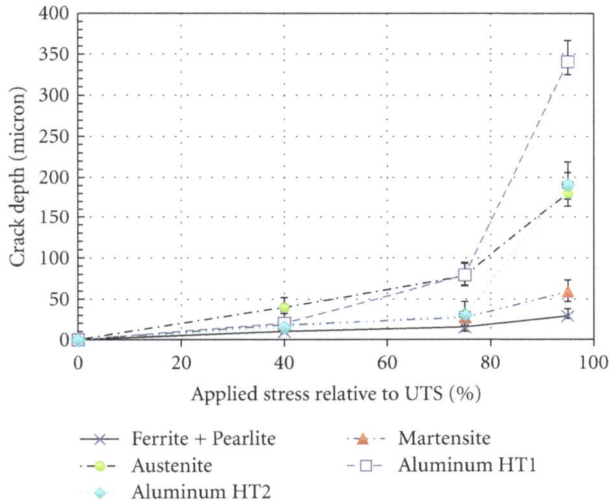

FIGURE 5: Crack depth as a function of applied stress.

Similarly, the results are also useful for practical engineers working in the field. The mechanism of SCC of the austenite and aluminum was found to be dominated by the corrosion tunnel model, while SCC of the other microstructures occurred by film rupture.

References

[1] T. P. Hoar and J. M. West, "Mechano-chemical anodic dissolution of austenitic stainless steel in hot chloride solution," *Proceeding of the Royal Society A*, vol. 268, pp. 304–315, 1962.

[2] T. P. Hoar and J. C. Scully, "Mechanochemical anodic dissolution of austenitic stainless steel in hot chloride solution at controlled electrode potential," *Journal of Electrochemical Society*, vol. 111, no. 3, pp. 348–352, 1964.

[3] J. L. Albarran, L. Martinez, and H. F. Lopez, "Effect of heat treatment on the stress corrosion resistance of a microalloyed pipeline steel," *Corrosion Science*, vol. 41, no. 6, pp. 1037–1049, 1999.

[4] N. Nakayama, "Inhibitory effects of nitrilotris(methylenephosphonic acid) on cathodic reactions of steels in saturated $Ca(OH)_2$ solutions," *Corrosion Science*, vol. 42, no. 11, pp. 1897–1920, 2000.

[5] A. El-Sayed, K. M. El-Sobki, and V. K. Gouda, "Stress corrosion behaviour of 410 stainless steel in boiling 70% NaOH solution," *Surface Technology*, vol. 14, no. 3, pp. 245–256, 1981.

[6] A. K. Jha, P. R. Narayanan, K. Sreekumar, and P. P. Sinha, "Cracking of Al–4.5Zn–1.5Mg aluminium alloy propellant tank—a metallurgical investigation," *Engineering Failure Analysis*, vol. 17, no. 2, pp. 562–570, 2010.

[7] A. Conde, B. J. Fernández, and J. J. de Damborenea, "Characterization of the SCC behaviour of 8090 Al–Li alloy by means of the slow-strain-rate technique," *Corrosion Science*, vol. 40, no. 1, pp. 91–102, 1998.

[8] A. F. Oliveira Jr., M. C. de Barros, K. R. Cardoso, and D. N. Travessa, "The effect of RRA on the strength and SCC resistance on AA7050 and AA7150 aluminium alloys," *Materials Science and Engineering A*, vol. 379, no. 1-2, pp. 321–326, 2004.

[9] R. P. Wei, M. Gao, and P. S. Pao, "The role of magnesium in CF and SCC of 7000 series aluminum alloys," *Scripta Metallurgica*, vol. 18, no. 11, pp. 1195–1198, 1984.

[10] Y. E. Wu and Y. T. Wang, "Enhanced SCC resistance of AA7005 welds with appropriate filler metal and post-welding heat treatment," *Theoretical and Applied Fracture Mechanics*, vol. 54, no. 1, pp. 19–26, 2010.

[11] J. C. Lin, H. L. Liao, W. D. Jehng, C. H. Chang, and S. L. Lee, "Effect of heat treatments on the tensile strength and SCC-resistance of AA7050 in an alkaline saline solution," *Corrosion Science*, vol. 48, no. 10, pp. 3139–3156, 2006.

[12] Y. Prawoto, A. Moin, M. Tadjuddin, and W. B. Wan Nik, "Effect of microstructures on SCC of steel: field failure analysis case study and laboratory test result," *Engineering Failure Analysis*, vol. 18, no. 7, pp. 1858–1866, 2011.

[13] Y. Prawoto, "Incorporating the morphological difference of corrosion cracks in computational fracture mechanics approach," *Computational Materials Science*, vol. 56, pp. 166–168, 2012.

[14] D. L. McDowell and F. P.E. Dunne, "Microstructure-sensitive computational modeling of fatigue crack formation," *International Journal of Fatigue*, vol. 32, no. 9, pp. 1521–1542, 2010.

[15] G. Rondelli, B. Vicentini, and E. Sivieri, "Stress corrosion cracking of stainless steels in high temperature caustic solutions," *Corrosion Science*, vol. 39, no. 6, pp. 1037–1049, 1997.

[16] A. S. Afolabi and J. O. Borode, "Stress corrosion cracking susceptibility of medium carbon steel in caustic and potash media," *Australian Journal of Technology*, vol. 10, pp. 165–170, 2007.

[17] R. O. Rihan and S. Nešić, "Erosion-corrosion of mild steel in hot caustic. Part I: NaOH solution," *Corrosion Science*, vol. 48, no. 9, pp. 2633–2659, 2006.

[18] J. Y. Zou and D. T. Chin, "Mechanism of steel corrosion in concentrated NaOH solutions," *Electrochimica Acta*, vol. 32, no. 12, pp. 1751–1756, 1987.

[19] ASTM Standard: G38-07, *Standard Practice for Making and Using C-Ring Stress-Corrosion Test Specimen*, ASTM Annual Book of Standards, 2007.

[20] F. Goutier, S. Valette, A. Vardelle, and P. Lefort, "Oxidation of stainless steel 304 L in carbon dioxide," *Corrosion Science*, vol. 52, no. 7, pp. 2403–2412, 2010.

[21] G. Svenningsen, M. H. Larsen, J. H. Nordlien, and K. Nisancioglu, "Effect of high temperature heat treatment on intergranular corrosion of AlMgSi(Cu) model alloy," *Corrosion Science*, vol. 48, no. 1, pp. 258–272, 2006.

[22] K. S. Chan, "Roles of microstructure in fatigue crack initiation," *International Journal of Fatigue*, vol. 32, no. 9, pp. 1428–1447, 2010.

[23] M. R. Wenman, K. R. Trethewey, S. E. Jarman, and P. R. Chard-Tuckey, "A finite-element computational model of chloride-induced transgranular stress-corrosion cracking of austenitic stainless steel," *Acta Materialia*, vol. 56, no. 16, pp. 4125–4136, 2008.

18

Dispersion Behavior of Gold-Nanocoated Dielectric Optical Fibers

P. K. Choudhury

Institute of Microengineering and Nanoelectronics (IMEN), Universiti Kebangsaan Malaysia, Bangi, 43600 Kuala Lumpur, Malaysia

Correspondence should be addressed to P. K. Choudhury, pankaj@ukm.my

Academic Editor: Mathrubutham Rajagopalan

Using a fairly rigorous approach, gold-nanocoated dielectric optical fibers are treated to evaluate the propagation characteristics corresponding to the transverse electric (TE) and hybrid EH modes, as supported by the guides with small and large dimensions. Situations with varying nanocoating thickness are considered for two different operating wavelengths. The results demonstrate a profound effect of the gold nanocoating on the propagation-related features—the number of existing modes decreases with increasing nanolayer thickness. Furthermore, the gold nanolayer also has effects on the modal propagation constants, which become more prominent with the increase in the order of modes. It has been found that, in fiber with smaller dimension, the EH_{11} mode hardly exhibits any effect due to the variation in existing nanolayer coating corresponding to 850 nm wavelength. Almost similar situations are observed for the TE_{01} mode in large core fiber operating at 1550 nm wavelength.

1. Introduction

Substantial amount of research on optical waveguides with varieties of new forms of cross-sectional geometries and/or composite materials appeared in the literature [1–10]. The electromagnetic (EM) behavior of various forms of optical fibers with different refractive index (RI) profiles and/or number of claddings have been paid much attention [11–16]. However, the reported investigations mostly implement oversimplified assumptions, because a strict analytical approach, owing to the presence of multiple fibre clads, becomes forbiddingly difficult to tackle with.

Fiber optic sensors are nowadays of much demand [17–20] owing to their ease of operation and environment-friendly nature. In this context, the gold-nanocoated fibers are of special mention as these can be used with enormous benefit in chemical sensing [21–23], because the optical sensitivity of the system can be enhanced on demand by controlling the nanolayer thickness. Further, such gold-nanocoated fibers remain much useful for the detection of weak chemical reactions without sacrificing the measurement sensitivity.

The knowledge of modal behavior of gold-nanocoated optical fibers has been the vital issue to secure their usefulness in sensing, particularly in medical diagnostics. This essentially needs a thorough investigation of these fibers in terms of the sustained modes and their propagation constants. Usually, in metal-nanocoated optical waveguides, the phenomenon of surface plasmon resonance (SPR) is commonly studied, and the features of modal propagation inside the guide are not given much attention. Keeping this in mind, the aim of the present communication is pivoted to the analytical treatment of gold-nanocoated optical fibers with the emphasis on the modal behavior of the guide, which forms the basics of any waveguiding medium. The structure of fiber is essentially a multilayered dielectric one coated with a metallic nanolayer. A considerable amount of refractive index (RI) difference between the different fiber regions is taken into account in order to validate the treatment with greater accuracy under the condition of minimum power *leak-off*. As such, the use of Maxwell's field equations becomes necessary, which makes the study rigorous, and the results are obtained with greater accuracy.

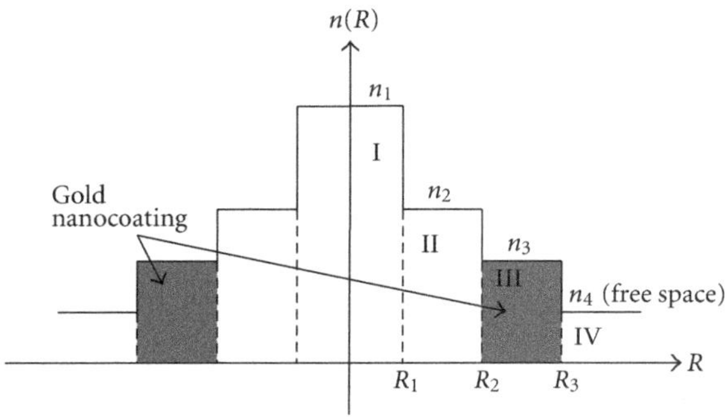

FIGURE 1: The RI profile of gold-nanocoated fiber.

The analytical treatment requires the estimation of fields in the different fiber regions, which essentially gets features due to the light-matter interaction in the guiding structure. Strictly speaking, there must be vanishing fields in the metallic region. However, as the thickness of the outermost gold-coated fiber region is of the order of nanoscale level, there must be some amount of field present in that particular region. This situation can be tackled by making use of Hankel function as one of the essential parts to represent fields in this region. Thus, a linear combination of the modified Bessel function of the second kind and Hankel function [24] is used to represent the field in the gold-nanocoated region. The present paper is aimed to investigate the effect of the gold-nanolayer thickness on the modal behavior of the guide, which ultimately provides the dispersion characteristics of the structure. In this stream, a rigorous analytical treatment is presented for gold-nanocoated fibers of small and large dimensions under the aforesaid assumptions in respect of fields, and the effect of the gold-nanolayer is visualized through the computations of the normalized propagation constants of the modes sustained in the nanocoated fiber. We consider two different operating wavelengths and investigate the effect on the propagation behavior for some of the low-order transverse electric (TE) and hybrid EH modes in the guide due to variation of the thickness of gold nanocoating.

The scope of the present paper is limited to the propagation behavior of the guide in terms of modal constraints. As such, the paper emphasizes on the dispersion characteristics of the guide in respect of the variations of the fiber dimensions and reports the observations in respect of the modal field behavior in terms of propagation cnstants. Apart from the dispersion behavior, further work in the direction is in progress and will be taken up in future communications.

2. Analytical Treatment

Figure 1 illustrates the RI profile of a four-layer dielectric optical fiber of which the third layer is assumed to be loaded with gold nanocoating, with z-axis being the optical axis. The outermost infinitely extended region is the free space (with the RI $n_4 = 1$). The other three layers have the RI values as n_1, n_2, and n_3 with $n_1 > n_2 > n_3$, and the interfaces at the radial distances $R = R_1$, $R = R_2$, and $R = R_3$ with $R_1 < R_2 < R_3$; the Au-nanolayer region is the one with the radial distance $R_3 - R_2 (= \delta)$. We treat the problem considering the wave equation obtained under cylindrical fiber geometry [25], and matching the solutions at the layer interfaces.

Now, corresponding to the case of a gold-nanocoated fiber, the fields in the different fiber regions may be considered as follows.

Bessel function $J_\nu(\bullet)$ of the first kind would represent the field in the fiber core region, whereas the decaying character of the field (as one moves away from the optical axis) can be represented by the modified Bessel function $K_\nu(\bullet)$ of the second kind. In the region next to the fiber core, the fields should be represented by a linear combination of $J_\nu(\bullet)$ and $Y_\nu(\bullet)$, where the latter one denotes Bessel function of the second kind. Finally, a linear combination of $K_\nu(\bullet)$ and $H_\nu(\bullet)$ will represent the field in the region just before the outermost clad; here $H_\nu^{(1)}(\bullet)$ stands for Hankel function. Strictly speaking, the fields must be vanishing in the metallic region. However, the existence of some fields may always be expected in nanocoated domains, which stimulates one to implement Hankel function to partially represent the field in the gold-nanocoated region of the fiber. In all the aforesaid functions, ν is the mode designating parameter representing the azimuthal periodicity, and it can take only discrete values.

Following the field representations as described above, after a few mathematical steps, the use of the longitudinal components of the electric/magnetic fields will yield the tangential field components in the different regions of the guide, as expressed here.

Region I: Core ($0 \le R \le R_1$):

$$\begin{aligned}
E_{R1} &= \left\{-A_1 \frac{j\beta}{q^2} Q_1 J_\nu'(Q_1 R) + A_2 \frac{\nu\omega\mu}{q^2 R} J_\nu(Q_1 R)\right\} e^{j\nu\phi},\\
H_{R1} &= -\left\{A_1 \frac{\nu\omega\varepsilon}{q^2 R} J_\nu(Q_1 R) + A_2 \frac{j\beta}{q^2} Q_1 J_\nu'(Q_1 R)\right\} e^{j\nu\phi},\\
E_{\phi 1} &= \left\{A_1 \frac{\beta\nu}{q^2 R} J_\nu(Q_1 R) + A_2 \frac{j\omega\mu}{q^2} Q_1 J_\nu'(Q_1 R)\right\} e^{j\nu\phi},\\
H_{\phi 1} &= \left\{-A_1 \frac{j\omega\varepsilon}{q^2} Q_1 J_\nu'(Q_1 R) + A_2 \frac{\beta\nu}{q^2 R} J_\nu(Q_1 R)\right\} e^{j\nu\phi}.
\end{aligned} \tag{1}$$

Region II: Inner cladding ($R_1 \le R \le R_2$)

$$\begin{aligned}
E_{R2} = &\left\{-A_3 \frac{j\beta}{q^2} Q_2 J_\nu'(Q_2 R) - A_4 \frac{j\beta}{q^2} Q_2 Y_\nu'(Q_2 R)\right.\\
&\left. + A_5 \frac{\nu\omega\mu}{q^2 R} J_\nu(Q_2 R) + A_6 \frac{\nu\omega\mu}{q^2 R} Y_\nu(Q_2 R)\right\} e^{j\nu\phi},\\
H_{R2} = -&\left\{A_3 \frac{\omega\varepsilon\nu}{q^2 R} J_\nu(Q_2 R) + A_4 \frac{\omega\varepsilon\nu}{q^2 R} Y_\nu(Q_2 R)\right.\\
&\left. + A_5 \frac{j\beta}{q^2} Q_2 J_\nu'(Q_2 R) + A_6 \frac{j\beta}{q^2} Q_2 Y_\nu'(Q_2 R)\right\} e^{j\nu\phi},
\end{aligned}$$

$$E_{\phi 2} = \left\{ A_3 \frac{\beta\nu}{q^2 R} J_\nu(Q_2 R) + A_4 \frac{\beta\nu}{q^2 R} Y_\nu(Q_2 R) + A_5 \frac{j\omega\mu}{q^2} Q_2 J_\nu'(Q_2 R) + A_6 \frac{j\omega\mu}{q^2} Q_2 Y_\nu'(Q_2 R) \right\} e^{j\nu\phi},$$

$$H_{\phi 2} = \left\{ -A_3 \frac{j\omega\varepsilon}{q^2} Q_2 J_\nu'(Q_2 R) - A_4 \frac{j\omega\varepsilon}{q^2} Q_2 Y_\nu'(Q_2 R) + A_5 \frac{\beta\nu}{q^2 R} J_\nu(Q_2 R) + A_6 \frac{\beta\nu}{q^2 R} Y_\nu(Q_2 R) \right\} e^{j\nu\phi}. \quad (2)$$

Region III: Gold-nanocoated layer ($R_2 \le R \le R_3$)

$$E_{R3} = \left\{ -A_7 \frac{j\beta}{q^2} Q_3 K_\nu'(Q_3 R) - A_8 \frac{j\beta}{q^2} Q_3 H_\nu^{(1)\prime}(Q_3 R) + A_9 \frac{\nu\omega\mu}{q^2 R} K_\nu(Q_3 R) + A_{10} \frac{\nu\omega\mu}{q^2 R} H_\nu^{(1)}(Q_3 R) \right\} e^{j\nu\phi},$$

$$H_{R3} = -\left\{ A_7 \frac{\nu\omega\varepsilon}{q^2 R} K_\nu(Q_3 R) + A_8 \frac{\nu\omega\varepsilon}{q^2 R} H_\nu^{(1)}(Q_3 R) + A_9 \frac{j\beta}{q^2} Q_3 K_\nu'(Q_3 R) + A_{10} \frac{j\beta}{q^2} Q_3 H_\nu^{(1)\prime}(Q_3 R) \right\} e^{j\nu\phi},$$

$$E_{\phi 3} = \left\{ A_7 \frac{\beta\nu}{q^2 R} K_\nu(Q_3 R) + A_8 \frac{\beta\nu}{q^2 R} H_\nu^{(1)}(Q_3 R) + A_9 \frac{j\omega\mu}{q^2} Q_3 K_\nu'(Q_3 R) + A_{10} \frac{j\omega\mu}{q^2} Q_3 H_\nu^{(1)\prime}(Q_3 R) \right\} e^{j\nu\phi},$$

$$H_{\phi 3} = \left\{ -A_7 \frac{j\omega\varepsilon}{q^2} Q_3 K_\nu'(Q_3 R) - A_8 \frac{j\omega\varepsilon}{q^2} Q_3 H_\nu^{(1)\prime}(Q_3 R) + A_9 \frac{\beta\nu}{q^2 R} K_\nu(Q_3 R) + A_{10} \frac{\beta\nu}{q^2 R} H_\nu^{(1)}(Q_3 R) \right\} e^{j\nu\phi}. \quad (3)$$

Region IV: outer cladding ($R \ge R_3$)

$$E_{R4} = \left\{ -A_{11} \frac{j\beta}{q^2} Q_4 K_\nu'(Q_4 R) + A_{12} \frac{\nu\omega\mu}{q^2 R} K_\nu(Q_4 R) \right\} e^{j\nu\phi},$$

$$H_{R4} = -\left\{ A_{11} \frac{\nu\omega\varepsilon}{q^2 R} K_\nu(Q_4 R) + A_{12} \frac{j\beta}{q^2} Q_4 K_\nu'(Q_4 R) \right\} e^{j\nu\phi},$$

$$E_{\phi 4} = \left\{ A_{11} \frac{\beta\nu}{q^2 R} K_\nu(Q_4 R) + A_{12} \frac{j\omega\mu}{q^2} Q_4 K_\nu'(Q_4 R) \right\} e^{j\nu\phi},$$

$$H_{\phi 4} = \left\{ -A_{11} \frac{j\omega\varepsilon}{q^2} Q_4 K_\nu'(Q_4 R) + A_{12} \frac{\beta\nu}{q^2 R} K_\nu(Q_4 R) \right\} e^{j\nu\phi}, \quad (4)$$

In (1)–(4), A_1–A_{12} are the unknown constants to be determined by the use of electromagnetic boundary conditions, and the suffices 1, 2, 3, and 4 illustrate the corresponding fiber regions. Also, the prime represents the differentiation with respect to the argument, and the parameter q is determined as

$$q^2 = k^2 - \beta^2 = \omega^2 \mu\varepsilon - \beta^2. \quad (5)$$

Furthermore, the core/clad parameters Q_1, Q_2, Q_3 and Q_4 are defined as

$$\begin{aligned} Q_1^2 &= k_1^2 - \beta^2 = \omega^2 \mu \varepsilon_1 - \beta^2, \\ Q_2^2 &= \beta^2 - k_2^2 = \beta^2 - \omega^2 \mu \varepsilon_2, \\ Q_3^2 &= \beta^2 - k_3^2 = \beta^2 - \omega^2 \mu \varepsilon_3, \\ Q_4^2 &= \beta^2 - k_4^2 = \beta^2 - \omega^2 \mu \varepsilon_4. \end{aligned} \quad (6)$$

In Equations (6), ε_1, ε_2, ε_3, and ε_4 are the dielectric constants, μ is the permeability of the medium, and β represents the longitudinal component of the propagation constant.

Implementing the continuity condition of fields, the tangential components of the electric field E and the magnetic field H will exhibit smooth match at the layer boundaries represented by the parametric coordinates $R = R_1$, $R = R_2$, and $R = R_3$. This yields twelve equations altogether, the consistency of which will be determined by the vanishing (12×12) determinant (Δ) formed by the coefficients of those equations. This indicates

$$\Delta = 0, \quad (7)$$

the solutions to which will provide the modal propagation constants of the guide under consideration. Since Δ is complex, situations of the simultaneously vanishing real and imaginary parts of Δ will determine the allowed values of the propagation constants sustained by the gold-nanocoated fiber.

3. Results and Discussion

Equation (7) represents the dispersion relation corresponding to the gold-nanocoated fiber, as deduced by using Maxwell's field equations. The analysis becomes much rigorous to identify the sustained TE and EH modes. The numerical computations implement two different values of the operating wavelength λ as 850 nm and 1550 nm—the two wavelength regions where optical waveguides are usually operated. We consider the RI values n_1 and n_2 as 1.4488 and 1.444, respectively. Also, the gold-nanocoated layer has the RI as $n_3 = 1.42$. In order to attain a comparative study, fibers with small and large dimensions are taken into consideration. In this context, for smaller dimensioned fiber, we assume parametric boundaries R_1 and R_2 as 4.1 μm and 62.5 μm, respectively, and the outermost layer is considered to be infinitely extended. As to the fiber with large dimension, we chose the parameters as $R_1 = 50\,\mu$m, $R_2 = 12\,\mu$m along with the infinite extension of the outermost layer. In both the cases, we perform investigations for varying R_3—the thickness of gold-nanolayer. We further consider that the dielectric regions of the fiber are homogeneous and isotropic along with their nature to be nonmagnetic, that is, $\mu_1 = \mu_2 = \mu_3 = \mu_4 \approx \mu_0$, the free-space permeability.

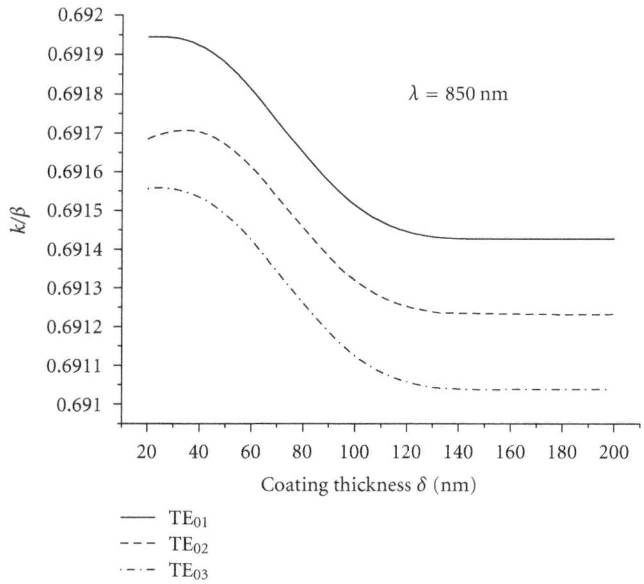

FIGURE 2: Variation of the normalized propagation constant k/β of TE modes with the nanocoating thickness (the case of fiber with smaller dimension operating at 850 nm).

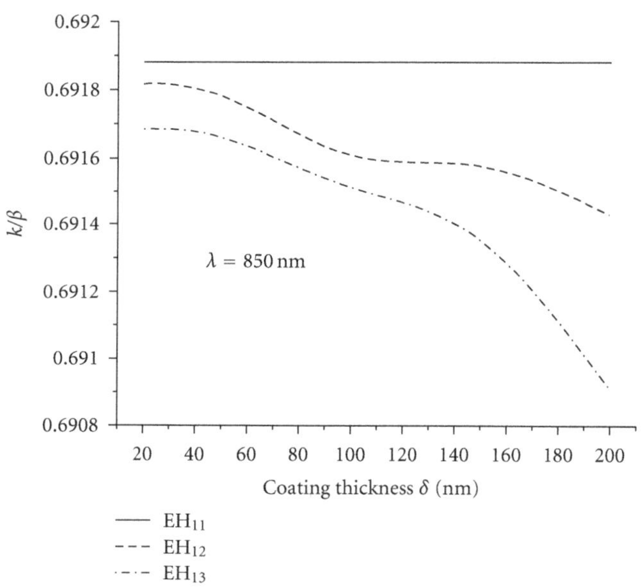

FIGURE 3: Variation of the normalized propagation constant k/β of EH modes with the nanocoating thickness (the case of fiber with smaller dimension operating at 850 nm).

Figures 2 and 3, respectively, illustrate the plots (corresponding to the case of small core gold-nanocoated fiber operating at 850 nm wavelength) of the normalized propagation constant k/β against the nanocoating thickness δ. Within the context, it is to be noted that the values of the propagation constant β are limited by the condition of sustained guidance; that is, $n_1 k \geq \beta \geq n_2 k$, with k being the free-space propagation constant.

In order to observe the effect of the presence of gold nanolayer on the number of sustained modes, we performed computations for several values of δ, and from the study of the eigenvalue equation (i.e., (7)), we find that the number of sustained modes decreases with the increase in nanocoating thickness. This essentially reflects the feature that the increase in gold-nanolayer thickness makes the fiber to attain the property of mode suppression as the fiber gets loaded with a metallic layer in this case.

In Figure 2, we demonstrate the variation of the normalized propagation constants corresponding to the first three TE modes, namely, TE_{01}, TE_{02}, and TE_{03} excited at 850 nm wavelength. We notice that the value of k/β initially decreases with the increase in δ. This essentially indicates that, for all the three types of modes, the values of k/β are relatively large corresponding to the situation when the gold-nanolayer thickness is small. In other words, the β values are larger for smaller coating thickness, and the TE_{01} mode attains the lowest value of β (or the highest value of k/β). This is very much obvious looking at the conceptual understanding of wave propagation in bounded mediums. Figure 2 also depicts that the propagation constants of TE modes saturate after the coating thickness is increased to 120 nm onwards. The thickness of gold nanocoating can be conceptualized through correlating it with the size of gold nanoparticles. The nanocoating thickness must have the impact on the size of grown nanoparticles—the thicker the coating is, the larger the nanoparticle size will be, affecting thereby the confinement of modes in the gold nanolayer. Figure 2 illustrates that the propagation of TE modes becomes saturated with the increase in nanocoating thickness.

Figure 3 corresponds to the situation when the EH modes are excited into the fiber. We observe that the propagation constants of the EH_{11} mode remain almost unaffected due to the variation in nanocoating thickness, which might be attributed to the specific field patterns the mode attains in the guide. However, the next higher order EH_{12} and EH_{13} modes exhibit marginal decrease in normalized propagation constant (or the increase in propagation constant) with increasing nanocoating thickness. Furthermore, the excitation of EH modes does not indicate any kind of saturation in the values of propagation constants with increasing δ, which is in contrast to the situation of TE mode excitation, wherein TE_{01}, TE_{02}, and TE_{03} modes show a kind of saturation in the values of k/β with the increase in coating thickness.

Increasing the operating wavelength (to 1550 nm) results into changes in the propagation constants of the sustained TE and EH modes, as depicted in Figures 4 and 5, respectively. In both the cases the first-order modes exhibit the maximum value of k/β, which is as usual. We observe that the TE_{01} mode initially shows almost unchanged propagation constant values with increasing coating thickness until it reaches a value close to 120 nm (Figure 4). Further increase in δ results into a decrease in the values of k/β. This demonstrates partial effect of the gold-nanoparticle size on the propagation of modes for which both the field pattern as well as the nanoparticle thickness are responsible. EH modes also exhibit trends of initial decrease in the values of

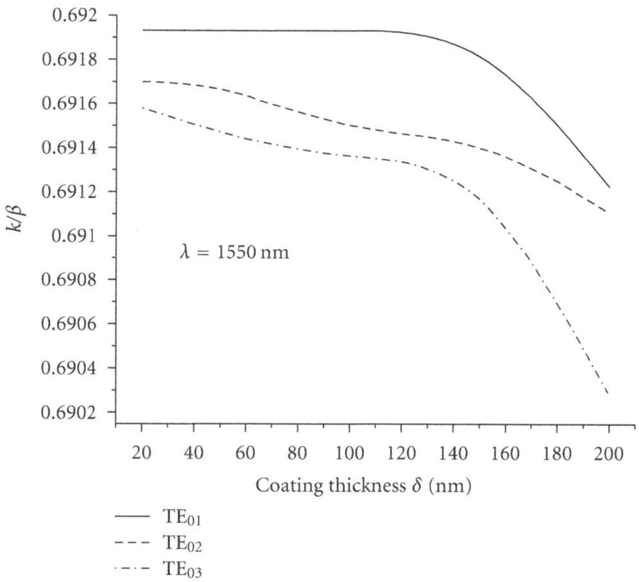

FIGURE 4: Variation of the normalized propagation constant k/β of TE modes with the nanocoating thickness (the case of fiber with smaller dimension operating at 1550 nm).

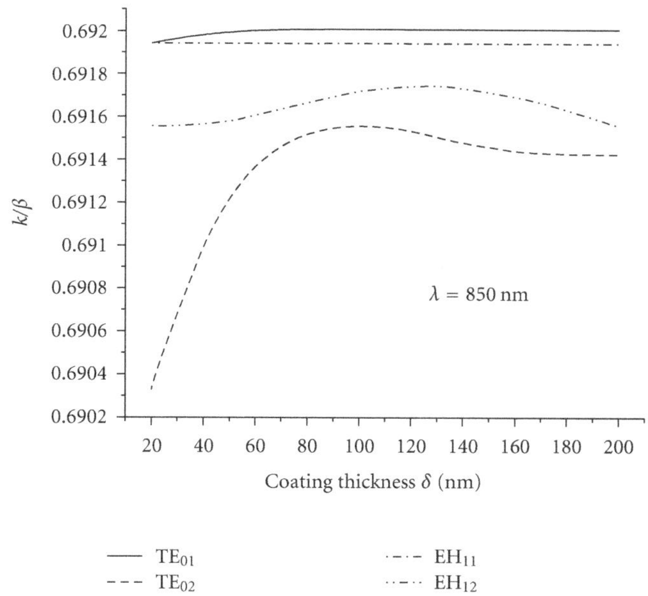

FIGURE 6: Variation of the normalized propagation constant k/β of TE and EH modes with the nanocoating thickness (the case of fiber with larger dimension operating at 850 nm).

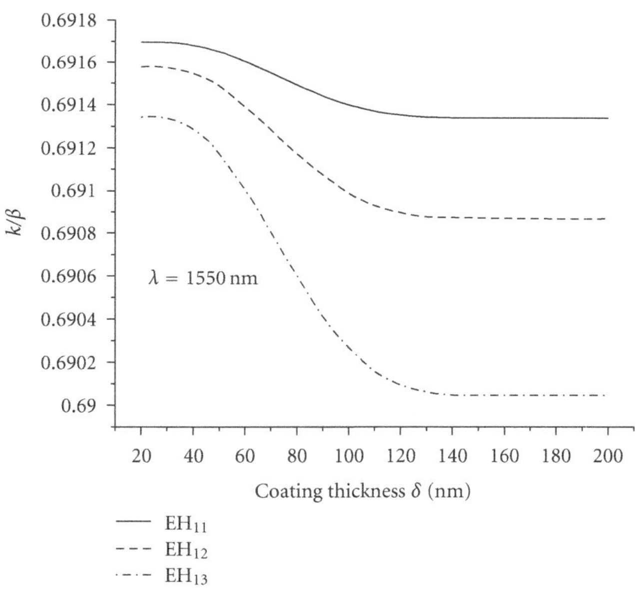

FIGURE 5: Variation of the normalized propagation constant k/β of EH modes with the nanocoating thickness (the case of fiber with smaller dimension operating at 1550 nm).

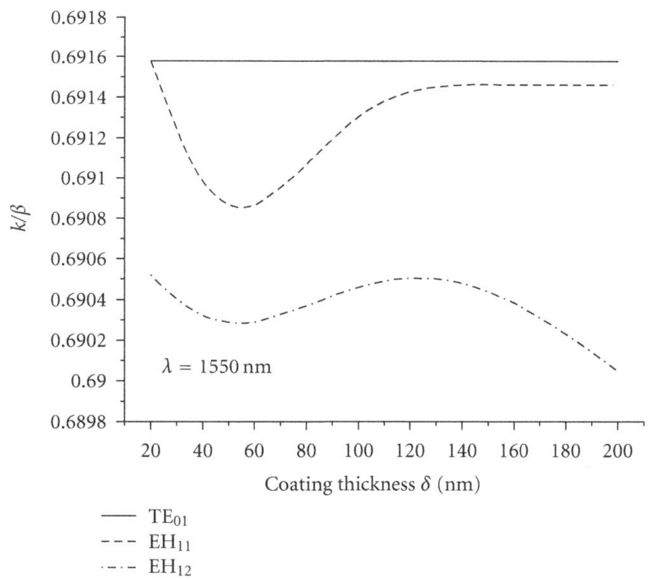

FIGURE 7: Variation of the normalized propagation constant k/β of TE and EH modes with the nanocoating thickness (the case of fiber with larger dimension operating at 1550 nm).

k/β with increasing coating thickness and become ultimately saturated. In all the cases, we observe that the normalized propagation constant decreases with the increase in the order of mode.

The aforesaid results correspond to the situation when the gold-nanocoated fiber core size is small. However, studies can also be performed corresponding to the situation when the fiber core attains a relatively large core so that the field functions can be expressed in terms of their asymptotic forms under the limit of large arguments. We made computations considering such a case for operating wavelengths as 850 nm and 1550 nm, and the results are illustrated in Figures 6 and 7, respectively, in the form of $k/\beta - \delta$ plots.

Interestingly, under the asymptotic limit of fields, we notice that there exist only the TE_{01} and TE_{02} modes in the guide (Figure 6) with the TE_{01} mode having marginal variation in the value of k/β with increasing gold nanocoating thickness. The TE_{02} mode, however, shows larger initial increase in its normalized propagation constant, which becomes almost saturated after the coating thickness is 100 nm and above. We also observe from Figure 6 that the EH_{11} mode has hardly any dependence on the thickness value in terms of its propagation constants, and the EH_{12} mode exhibits a minor dependence only in similar terms. The numerical results do not reflect the existence of further higher-order modes.

Operation of large core gold-nanocoated fiber at 1550 nm yields further interesting results in term of the existence of modes. We find that the TE_{01} is the only mode that falls under the TE category, and no other mode has been observed to belong in this category. Moreover, the propagation constants of the TE_{01} mode remain unaffected due to the variation in the nanocoating thickness. In the EH mode group, as in the case of 850 nm operating wavelength, only the EH_{11} and EH_{12} modes are confined into the guide, and higher order modes are non existing. The trend of variation of the values of k/β in this case has some sort of similarity with the situation when the fiber is excited at 850 nm operating wavelength.

The propagation of TE and EH modes in the fiber under consideration emphasizes the fact that the gold-nanolayer essentially has its effect on the modal propagation constants, which become more effective when the nanolayer thickness is relatively small. For larger values of coating thickness, the propagation constants exhibit marginal dependence on the coating thickness as the larger nanoparticle size physically has less impact on the propagating field.

4. Conclusion

The foregoing analytical study reports on the observation of the dispersion behavior of the guide. Investigation reveals that the gold-nanocoating on dielectric optical fibers has the effect in the form of reducing the number of modes propagating in the fiber. We studied the cases of TE and EH modes. Relating the thickness of the nanolayer with the nanoparticle size, the effect of nanocoating has been found to be more pronounced under its small values, whereas larger coating thickness drives the system towards the situation when the larger nanoparticles exhibit less impact on the propagating fields making thereby the modal propagation constants almost saturated. Studying the case under the asymptotic limit of large core, making thereby the related field functions to be assumed to have large arguments, it has been observed that the number of modes becomes greatly reduced. Furthermore, when the fiber is operated at 850 nm wavelength, the low-order TE and TM modes exhibit almost unchanged values of propagation constants with the variation in nanocoating thickness. Increasing the operating wavelength to 1550 nm, however, makes the TE mode only to remain unaffected in terms of propagation constant values due to the changes in coating thickness.

The study essentially throws information on the dispersion characteristics of gold-nanocoated optical fiber under variations of the fiber dimensions. Further studies in line with the above discussion concepts will be taken up in future communications.

Acknowledgment

The author is thankful to an anonymous reviewer for making some comments on the text, which caused to enhance the descriptions.

References

[1] P. K. Choudhury and O. N. Singh, "Some multilayered and other unconventional lightguides," in *Electromagnetic Fields in Unconventional Materials and Structures*, O. N. Singh and A. Lakhtakia, Eds., chapter 8, John Wiley & Sons, New York, NY, USA, 2000.

[2] D. Kumar, P. K. Choudhury, and F. A. Rahman, "Towards the characteristic dispersion relation for step-index hyperbolic waveguide with conducting helical winding," *Progress in Electromagnetics Research*, vol. 71, pp. 251–275, 2007.

[3] C. Yeh and I. F. Shimabukuro, *The Essence of Dielectric Waveguides*, Springer, New York, NY, USA, 2008.

[4] D. Kumar, P. K. Choudhury, and O. N. Singh, "Towards the dispersion relations for dielectric optical fibers with helical windings under slow- and fast-wave considerations—a comparative analysis," *Progress in Electromagnetics Research*, vol. 80, pp. 409–420, 2008.

[5] A. S. N. Amin, M. Mirhosseini, and M. Shahabadi, "Modal analysis of multilayer conical dielectric waveguides for azimuthal invariant modes," *Progress in Electromagnetics Research*, vol. 105, pp. 213–229, 2010.

[6] V. Kesari and J. P. Keshari, "Analysis of a circular waveguide loaded with dielectric and metal discs," *Progress in Electromagnetics Research*, vol. 111, pp. 253–269, 2011.

[7] P. K. Choudhury and W. K. Soon, "On the tapered optical fibers with radially anisotropic liquid crystal clad," *Progress in Electromagnetics Research*, vol. 115, pp. 461–475, 2011.

[8] P. K. Choudhury and P. T. S. Ping, "On the dispersion relations of tapered coreoptical fibers with liquid crystal clad," *Progress in Electromagnetics Research*, vol. 118, pp. 117–133, 2011.

[9] D. Zarifi, A. Abdolali, M. Soleimani, and V. Nayyeri, "Inhomogeneous planar layered chiral media: analysis of wave propagation and scattering using Taylor's series expansion," *Progress in Electromagnetics Research*, vol. 125, pp. 119–135, 2012.

[10] J. Dong and J. Li, "Characteristics of guided modes in uniaxial chiral circular waveguides," *Progress in Electromagnetics Research*, vol. 124, pp. 331–345, 2012.

[11] V. K. Chaubey, K. K. Dey, S. P. Ojha, and P. Khastgir, "Modal characteristics of a doubly clad step-index optical fibre: a general analytical approach," *Canadian Journal of Physics*, vol. 66, no. 9, pp. 796–802, 1988.

[12] P. K. Choudhury, P. Khastgir, S. P. Ojha, and L. K. Singh, "Weak guidance in bent and unbent four-layer planar waveguides. A comparative study of dispersion curves," *Japanese Journal of Applied Physics, Part 2*, vol. 31, no. 1, pp. L39–L42, 1992.

[13] P. K. Choudhury and R. A. Lessard, "An estimation of power transmission through a doubly clad optical fiber with an annular core," *Microwave and Optical Technology Letters*, vol. 29, no. 6, pp. 402–405, 2001.

[14] Z. Duan, Y. Wang, X. Mao, W. X. Wang, and M. Chen, "Experimental demonstration of double-negative metamaterials partially filled in a circular waveguide," *Progress in Electromagnetics Research*, vol. 121, pp. 215–224, 2011.

[15] D. Chen and B. Sun, "Multi-wavelength fiber optical parametric oscillator based on a highly nonlinear fiber and a sagnac loop filter," *Progress in Electromagnetics Research*, vol. 106, pp. 163–176, 2010.

[16] B. Zhu, J. Stiens, V. Matvejev, and R. Vounckx, "Inexpensive and easy fabrication of multi-mode tapered dielectric circular probes at millimeter wave frequencies," *Progress in Electromagnetics Research*, vol. 126, pp. 237–254, 2012.

[17] P. H. Paul and G. Kychakoff, "Fiber-optic evanescent field absorption sensor," *Applied Physics Letters*, vol. 51, no. 1, pp. 12–14, 1987.

[18] A. Messica, A. Greenstein, and A. Katzir, "Theory of fiber-optic, evanescent-wave spectroscopy and sensors," *Applied Optics*, vol. 35, no. 13, pp. 2274–2284, 1996.

[19] P. K. Choudhury and T. Yoshino, "On the fiber-optic chlorine sensor with enhanced sensitivity based on the study of evanescent field absorption spectroscopy," *Optik*, vol. 115, no. 7, pp. 329–333, 2004.

[20] P. K. Choudhury and O. N. Singh, "An overview of optical sensors and their applications," in *Frontiers in Optical Technology: Materials and Devices*, chapter 9, Nova Science, New York, NY, USA, 2007.

[21] S. F. Cheng and L. K. Chau, "Colloidal gold-modified optical fiber for chemical and biochemical sensing," *Analytical Chemistry*, vol. 75, no. 1, pp. 16–21, 2003.

[22] A. K. Sharma, R. Jha, and B. D. Gupta, "Fiber-optic sensors based on surface plasmon resonance: a comprehensive review," in *IEEE Sensors Journal*, vol. 7, pp. 1118–1129, 2007.

[23] F. A. Rahman, P. K. Choudhury, D. Kumar, and Z. Yusoff, "An analytical investigation of four-layer dielectric optical fibers with Au nano-coating—a comparison with three-layer optical fibers," *Progress in Electromagnetics Research*, vol. 90, pp. 269–286, 2009.

[24] M. Abramowitz and I. A. Stegun, *Handbook of Mathematical Functions*, chapter 9, Dover Publications, New York, NY, USA, 1965.

[25] G. Keiser, *Optical Fiber Communications*, chapter 2, McGraw-Hill, Singapore, 1986.

Challenges of the Modeling Methods for Investigating the Interaction between the CNT and the Surrounding Polymer

Roham Rafiee,[1] Timon Rabczuk,[2] Reza Pourazizi,[1] Junhua Zhao,[2] and Yancheng Zhang[2]

[1] *Composites Research Laboratory, Faculty of New Sciences and Technologies, University of Tehran, Tehran 1439955941, Iran*
[2] *Institute of Structural Mechanics, Bauhaus University Weimar, Marienstraße 15, 99423 Weimar, Germany*

Correspondence should be addressed to Roham Rafiee; roham.rafiee@ut.ac.ir

Academic Editor: Philip Harrison

The interaction between the carbon nanotubes (CNT) and the polymer is a key factor for determining the mechanical, thermal, and electrical properties of the CNT/polymer nanocomposite. However, it is difficult to measure experimentally the interfacial bonding properties between the CNT and the surrounding polymer. Therefore, computational modeling is used to predict the interaction properties. Different scale models, from atomistic to continuum, are critically reviewed addressing the advantages, the disadvantages, and the future challenges. Various methods of improvement for measuring the interaction properties are described. Finally, it is concluded that the semicontinuum modeling may be the best candidate for modeling the interaction between the CNT and the polymer.

1. Introduction

Predicting the mechanical properties and modeling the mechanical behavior of the CNT Reinforced Polymer (CNTRPs) have attracted the great attention of many researchers [1–5], especially considering their rapid development in various industries such as automotive industry, aerospace, packaging, and wind turbines. Therefore, over the past two decades, extensive studies have been executed to predict the mechanical properties of CNTRPs.

Different factors affect the performance of CNTRPs. The main governing issue which characterizes the efficiency of CNT reinforcement at nano-/microscale can be attributed to the load transfer mechanism from the matrix to the CNTs. As a result, prior to understanding the mechanical behavior of CNTRP at macrolevel, the corresponding physical performance has to be addressed. For this purpose, a proper simulation technique, which captures the required details sufficiently and reasonably, needs to be employed in accordance with the investigated scale.

Interfacial bonding in the interphase region between the embedded CNT and its surrounding polymer is a crucial issue for the load transferring and reinforcement phenomena. The atomic structure of carbon nanotubes consisting of sp^2 hybridized carbons hinders the formation of strong covalent bonds with the surrounding polymer matrix. Although functionalization can improve the load transfer of the interphase by providing covalent cross-links between the carbon atoms of CNTs and the molecules of polymer, this procedure has a main drawback in providing defects in the structure of carbon nanotubes due to formation of sp^3 hybridized sites. It can significantly reduce the beneficial properties of carbon nanotubes. Therefore, CNTs naturally interact with the polymer chains of the matrix just through weakly nonbonded van der Waals (vdW) and electrostatic forces. Mechanical interlocking, which is useful to improve the adhesion between the fiber and the matrix in fibrous composites, is not pertinent to CNTRPs due to the smooth surface of the CNTs.

Up to now, various methods have been employed to investigate the interaction behavior between the CNT and the surrounding polymer, including experimental studies, atomistic modeling, and continuum modeling. Through systematically studying the shortcomings and advantages of all the available methods from the literature, the objective of this review is to find the best solution for evaluating the interaction behavior between the CNT and the polymer matrix. On one side, the accuracy must be guaranteed; on the other side, the method should be simple and have minimal computational costs.

This paper aims to review recently developed modeling techniques to study the interaction between CNT and surrounding polymer. The organization of the paper is as follows: Section 2 is a brief explanation of challenges imposed to the experimental studies at the scale of Nano. Section 3 discusses performed theoretical studies for investigating interaction between CNT and polymer. This section includes two main groups of available modeling techniques as atomistic modeling Section 3.1 and continuum modeling Section 3.2. Molecular dynamics, coarse grain simulation, and density functional theory are reviewed in Sections 3.1.1, 3.1.2, and 3.1.3, respectively. Then, associated obstacles and shortcomings of atomistic modeling are presented in Section 3.1.4 which are the main reasons encouraging researchers to employ continuum modeling approaches. Categorized under continuum modeling group, Sections 3.2.1 and 3.2.2 present analytical and numerical modeling techniques. Finally, Section 4 concludes the review by highlighting pros and cons of available methods.

2. Experimental Studies

The performed experimental studies for characterizing the interaction between CNT and surrounding polymer can be divided into three main categories.

Some researchers have performed experiments on CNT-polymer bulk composites at the macroscale and observed the enhancements in mechanical properties (like elastic modulus and tensile strength) and tried to correlate the experimental results and phenomena with continuum theories like micromechanics of composites or Kelly Tyson shear lag model [1–5, 32, 33].

Other researchers have used the spectroscopy techniques. Spectroscopy is a powerful technique for characterizing the properties of the CNT. This technique includes Raman spectroscopy, Fourier Transform infrared spectroscopy (FTIR), and X-ray scattering. The main principle of this method is that by straining the CNT/polymer, the C–C bonds will vibrate, and this will affect the peaks in the Raman spectrum. Therefore, Raman spectroscopy can determine the characteristics of CNTs such as the diameters and the level of dispersion in the matrix [34–38].

In situ TEM (Transmission Electron Microscopy) straining has also been used to understand the mechanics, fracture, and failure processes of the interface. In these techniques, the CNT-polymer composite (an electron-transparent thin specimen) is strained inside a TEM and simultaneously imaged to get real-time and spatially resolved (1 nm) information [39, 40].

Rahmat and Hubert [41] and also Desai and Haque [42] have conducted a comprehensive review on the experimental methods.

Generally, an experimental study at the scale of nano is a challenging task due to the following reasons:

(i) the lack of proper direct measuring techniques at the nanometer scale;

(ii) tremendous limitations on the specimen size;

(iii) an uncertainty in the data obtained from indirect measurements;

(iv) inadequate preparation techniques of test specimens and the lack of control over the alignment and distribution of nanotubes.

Therefore researchers have a tendency to employ theoretical methods for predicting properties of CNT-reinforced composites. Moreover, it is not possible to measure the interfacial bonding properties between CNT and surrounding polymer through direct measurement, and this would be another source of error.

3. Theoretical Studies

In this category of studies, the interaction between the CNT and the surrounding polymer which plays a key factor in transferring the load from the matrix to the CNT is examined using theoretical and computer models. Up to now, the employed theoretical studies to simulate the interaction and measure the mechanical properties of the interaction are mainly divided into two major groups: atomistic modeling and continuum modeling.

3.1. Atomistic Modeling. Atomistic modeling is accomplished using molecular dynamic (MD) simulation, density functional theory (DFT), or coarse grain (CG) simulation. These techniques are widely used to resolve the relevant fracture problems on different scales.

3.1.1. Molecular Dynamics. The MD method solves Newton's equations of motion for a molecular system [43], which provides the methodology for detailed microscopic modeling on the molecular scale [44]. For the nanocomposite interaction studies, MD has shown strong potential, due to its accuracy, flexibility, and various force fields [10, 13, 45, 46]. Xu and Buehler [10] adopted the AIREBO potential to simulate the mechanical behavior of the CNT-based nanocomposite. In the work of Zhang et al. [13], the UA (united atom) potential is adopted to address the similar problem of the nanocomposites. Moreover, the coarse-grained potential is also employed to study the interaction among particle-particle and particle-polymer [45, 46].

Additionally, the comparable sizes of nanotubes and polymer chains make MD a great candidate to simulate nanocomposites without sacrificing the accuracy of the results, especially with the rapid development of high-performance supercomputers.

The MD simulations can also investigate the effects of various parameters and types of interaction on mechanical properties like elastic moduli, interfacial shear strength, and bonding energy.

The interface properties of CNT/polymer were widely investigated based on MD simulation from the available literatures [6, 8–10, 12, 13]. Referring to the nonbonded CNT/polymer interaction, Chen et al. [9] performed MD simulation of single polymer molecular with CNT, and they found that armchair single-walled carbon nanotubes

(SWNTs) present the strongest adhesion. Liao and Li [6] and Rahmat and Hubert [12] studied the noncovalent interaction between the CNTs and the matrix; the shear strength and the binding energy of the composite interfaces were given in their research. Wong et al. [8] found that the locally nonuniformed distribution of CNT-induced mechanical interlocking leads to high interfacial shear stress. The MD simulation also showed that the mismatch of the coefficients of thermal expansions between CNT and polymer matrix also promotes the stress transfer ability. Xu and Buehler [10] have studied the thermal resistance of the two interacting CNTs, with and without polymer, which shows that the thermal resistance of the CNT junctions can be significantly improved through altering the number of polymer molecules. Recently, Zhang et al. [13] have studied the effects of the dispersion of polymer-wrapped two neighbouring SWNTs on nanoengineering load transfer. They found that dispersion angle of the polymer wrapped SWNTs dominates the load transfer. Besides, the self-repairing function of the system is found due to the nonbond interaction between the polymer and the SWNTs.

Frankland et al. [7] modeled a single armchair CNT by MD and found that the shear strength increases 10 times just by forming less than 1% cross-links between the CNT and the polymer. Through the MD simulation, Chowdhury and Okabe [47] also investigated the influence of the chemical cross-links at the interface and found that forming a few cross-links between the CNT and the polymer, the shear strength increases from 310 MPa to 1630 MPa. Liu et al. [11] obtained the shear strength by pullout test of SWNT from covalent bonded polymer with an MD simulation. They found that the increased interface shear strength can either enhance or reduce the damping stability of the composite. Other researchers like Kuang, Tsai, and Odegard adopted MD methods to study the properties of the nanocomposites and interface between the CNT and the polymer [48–50].

However, the large number of particles and the short time steps involved in MD simulations call for powerful computational facilities, which are typically provided through parallel computing. The simulation time may be shortened by employing typical modeling strategies, such as adopting periodic boundary conditions [51] and reducing the cutoff distance assumption [52]. A more detailed discussion can be found in the Table 1.

3.1.2. Coarse Grain Simulation. A coarse grain (CG) approach is used to develop the tools for investigating the material properties. It is also used to study the underlying mechanical behavior typically required for material design that systematically integrates characteristic chemical responses. The CG models provide efficient means to simulate and investigate systems in which the desired behavior, property, or response is inherently at the mesoscale—those that are both inaccessible to full atomistic representations and inapplicable to continuum theory. The CG techniques are developed to map atomistic systems onto systems with coarser resolution. Therefore, a developed CG model can only reflect the behavior included in their governing potentials and associated parameters, and consequently, the source of such parameters typically determines the accuracy and utility of the CG model [53, 54]. It is emphasized that not all systems will benefit from a coarse-grain representation, and prudent consideration must be given regarding the system characterization and the intent of the simulations. Typical motivations for a CG approach include the following:

(1) inaccessible time scale and length scale for phenomenon or behavior via full atomistic representation;

(2) focus on bulk properties and/or mechanical behavior rather than on molecular structure and/or chemical interactions;

(3) desire for a direct simplified analysis of simulation results and system behavior.

In the past decade, various CG molecular dynamics methods were extensively used to study the thermomechanical properties of polymers, proteins, lipids, and biomaterials [53–56].

Müller-Plathe's group reported many interesting properties of different polymers (e.g., thermal conductivity, chain stiffness) by the CG MD simulations [57–59], in which the CG potentials were developed and their parameters were provided based on the reverse-mapping procedure by validating from full-atom MD simulations. Kremer's group developed two kinds of CG potentials for polystyrene, and the thermodynamical and the chemical properties were studied by CG MD simulations [60, 61]. Theodoru and Fermeglia investigated the multiscale modeling for polymer systems using CG MD and full-atom MD methods as well as Monte Carlo (MC) method [62, 63]. Nielsen et al. [64] calculated surface tension in bulk PE by CG MD simulations, which considerably reduces the amount of numerical computing required in the MD simulations. Recently, Zhao et al. [65] have developed a more accurate CG potential to study the thermomechanical properties (such as bulk density, glass transition temperature, expansion coefficient, Young's moduli, and yield stress) of bulk polyethylene (PE). The size-dependent mechanical properties of compressive stress-strain curves for PE particles have also been obtained by the CG MD method [66].

As it can be seen, all aforementioned studies are performed on pure polymer. The main reason for this is that the first step in studying interaction between CNT and surrounding polymer using CG MD is constructing a reasonable model for the chain of polymer. Therefore, numerous studies have been conducted on pure polymer.

Buehler's group developed the various CG MD potentials to characterize the nanoindentation of single- and double-carbon nanotubes (CNTs) buckypaper [67]. Based on the CG model, the mechanical properties of CNT networks were obtained, and the microstructural evolution and failure mechanism were discussed [68]. Afterwards, the viscoelasticity of the CNTs networks was reported by the CG MD simulations [69].

However, all the previous CG models are used in polymers and CNTs, respectively, while the CNT reinforced polymer composites are scarcely studied by CG MD simulations since the Lennard-Jones (LJ) potentials between CG beads of polymers and CNTs are not obtained yet. Nevertheless,

TABLE 1: Conducted molecular dynamics modeling of interaction between the CNT and the polymer.

Researcher(s)	Year	Reinforcement	Matrix	Simulation size	Simulation setup	Conclusion	Shortcomings/ advantages
Liao and Li [6]	2001	SWNT and DWNT	One chain of 2 to 80 mer polystyrene	20 Å long CNT	Pullout without cross-link between CNTs and matrix	Full-atom and nonbond interaction. Pullout shear strength of 160 MPa	More accurate while limited to small size
Frankland et al. [7]	2002	53 Å long (10, 10) SWNT	178 chains of 42 mer polyethylene	53 Å long SWNT	Pullout for both crystalline and amorphous polymer	An order of magnitude increase in shear strength with less than 1% cross-link between SWNT polymer	Close to the actual situation, limited to small scale
Wong et al. [8]	2003	SWNT	Epoxy, polystyrene	31 Å long CNT	Ununiformed SWNT with two different diameters.	Local nonuniformity of CNT and mismatch of the coefficients of thermal expansions lead to high interfacial shear stress	New aspect for analyzing the enhanced effect
Chen et al. [9]	2008	49.19 Å long (10, 10) SWNT	One chain of 10 mer polyphenylacetylene	49.19 Å CNT with Single polymer chain	SWNTs with different chirality	Strongest adhesion for armchair-type SWNT	Clear analysis but the size is too small
Xu and Buehler [10]	2009	(10, 10) SWNT	Polyethylene, various chain lengths	25 to 75 nm long CNT	Overlapped SWNTs with or without polymers	The thermal resistance of CNT junctions can be significantly improved through modifying the molecular structure at the interface	New idea with full-atom analysis, heavy calculations
Liu et al. [11]	2011	(10, 10) SWNT	EPON 862 and EPI-CURE curing agent	4.6 nm × 4.6 nm × 6.3 nm	SWNT pullout test is preformed by MD method, then the shear strength is applied to a micromechanical damping model	The increased interface shear strength can either enhance or reduce the damping stability of the composite	Useful method to determine the parameters for multiscale hierarchical model
Rahmat and Hubert [12]	2012	52.69 Å (10, 0) SWNT	9 Chains of 32 mer PMMA	Varied	Introduction and pullout test with optimized simulation phases	Interfacial binding energy of 0.39 kcal/mol $Å^2$ for no covalent interaction	Common MD simulation.
Zhang et al. [13]	2013	59.03 Å long armchair (5, 5)	Chain length of 20, 40, 60	59.03 Å long SWNT	Varying the dispersion angle of two neighbouring SWNTs	Dispersion angle of polymer wrapped SWNTs dominates the load transfer	United atom (UA) method, can apply to large scale

this issue will likely be addressed soon, since studies in this direction show high potential and important applications.

Because the CG models normally lose more detailed information among the beads (such as dihedral and improper information), accurate quantitative nonlinear mechanical properties of materials cannot be accurately given in the CG MD simulations.

3.1.3. Density Functional Theory. Density functional theory (DFT) studies the electronic structure of atomic scale systems based on quantum mechanics. The DFT models commonly use approximations, such as local density approximation (LDA), in which only the value of the electronic density at each point in the space is considered compared to the MD simulation.

Ab initio simulations use DFT and are able to converge to the accurate solution. However, these simulations are computationally expensive and grow in size with the power 3–7 of the number of basic functions. Therefore, the ab initio simulations are fairly small in size and use various techniques to reduce the computational costs. For example, during some stages of the simulation, only some part of the system may be optimized, or atomic relaxation may not be allowed and molecules may be considered to be rigid.

Ab initio models have studied various types of nanotube-polymer interaction including van der Waals and chemical functionalization. The results of these models are typically presented in the form of local density of state for the nanotube and the nanocomposite, band structures, charge density contours, and binding energy as a function of distance [70, 71]. In the work of Kim et al. [70], the encapsulation energy of a single (C_2H_2) molecule into the nanotube and the formation energy of (C_2H_2) @ CNT are calculated. In the work of Mylvaganam and Zhang [71], the possible ways of functionalizing CNTs are investigated with the aid of the density functional theory. Using the methoxy radical and the secondary butyl anion as initiators and their reactions with ethylene and epoxide, this leads to polyethylene- and polyepoxide-grafted nanotubes.

Most of DFT models are based on static or zero Kelvin structures and are still limited to small systems. Hence, when using DFT models, the strength of these models (accurate prediction of atomic structure) along with the inevitably expensive computations should be considered.

3.1.4. The Challenges of Atomistic Modeling. As it has been stated, atomic modeling has the accuracy and high flexibility to observe effects of various parameters on the mechanical properties of the nanocomposites and the interphase between the CNT and the polymer. But a lot of limiting factors are affecting these methods. For example, most of the published papers that have used these methods are using SWCNTs with a length less than 10 nm. In addition, the number of polymer chains considered is less than 100.

This is because atomistic modeling methods are not able to handle a large number of atoms due to increasing computational costs. Also the formulas used in these methods often have high complexity. Due to these limitations, continuum models have been used to overcome the aforementioned difficulties.

3.2. Continuum Modeling. Continuum modeling of CNT and surrounding polymer interactions is accomplished either analytically or numerically.

3.2.1. Analytical Modeling. In studies by analytical methods to model interphase between the carbon nanotube and the polymer, most of the researchers have tried to use the existing microscale rules in the field composites. These rules known as micromechanics rules cannot be directly applied to the nanoscale due to assumptions which are no longer valid. These involve the uniform distribution of fibers in polymer and perfect bonding between CNT and polymer, as well as considering reinforcement phase as a continuum medium which is applicable to fibers. In other words, direct use of micromechanics equations for CNTRP will simply neglect the lattice structure of CNT and consider CNT as a solid fiber. Moreover, micromechanics equations assume perfect bonding between the reinforcing agents and the surrounding polymer. This is not pertinent to the interphase between the CNT and the surrounding polymer, which interact through non-bonded van der Waals (vdW) forces.

Some researchers have proposed that a material region between CNT and surrounding polymer should be considered, but the way to treat the properties of this specific material region remains to be determined. The associated properties are selected between matrix and CNT Young's modulus without any experimental basis. Actually, at the scale of nano, there is no material region for interphase between CNT and polymer, but instead, the region is a molecular interaction space. Some others suggested some corrective factors in order to reflect the imperfection bonding between CNT and polymer, but these factors are also obtained on the basis of curve-fitting data.

Actually, CNTs interact with the surrounding polymer through vdW forces. This will cause the CNT to slide in the polymer matrix when subjected to axial loading. Therefore, the assumption of perfect bonding between CNT and polymer is not valid. Thus, some researchers try to incorporate vdW interactions between CNT and polymer in their modeling methods. Cohesive zone models have been widely used in the continuum study of interface debonding and sliding in composites. A cohesive zone model assumes a relation between the normal (and shear) traction(s) and the opening (and sliding) displacement(s). When implemented in the finite element method, the cohesive zone model is capable of simulating interface debonding and sliding. The existing cohesive models are all phenomenological, because it is difficult to obtain directly the cohesive laws for interfaces. Jiang et al. [14] have developed the cohesive zone model for the interface between the CNT and the polymer. They have calculated the total energy of all the atoms by integrating over the volume and by differentiation from the movement of the plates in the vertical and in the tangential direction. Finally, they report the relation between the normal (and shear) traction(s) and the opening (and sliding) displacement(s) across the interface in terms of the area density of CNT and the volume density of polymer, as well as the parameters in the vdW force.

Assuming an interface which contains vdW interactions, Tan et al. [15] have attempted to calculate the relationship between the stress and the strain in nanocomposites. In this regard, the rules set by Jiang have helped to describe the relationship between the stress and the strain at the interface.

Seidel and Lagoudas [16] tried to evaluate the effect of an interphase on the mechanical properties of nanocomposites. They have considered interphase as a physical environment and assumed elastic modulus of this area 0.1 to 10 times the elastic modulus of the matrix. Then, the effects of interphase properties on nanocomposites were examined using a concentric cylinder model.

The Shear-Lag model was also employed by some other researchers to study the interaction between CNT and

TABLE 2: Conducted analytical modeling of interaction between CNT and polymer.

Researcher(s)	Year	Interface or interphase	Type of interaction	Thickness	Interphase modulus	Method of modeling
Jiang et al. [14]	2006	Interface	vdW	N/A	N/A	Cohesive energy
Tan et al. [15]	2007	Interface	vdW	N/A	Imperfection coefficient	Elasticity
Seidel and Lagoudas [16]	2006	Interphase	Perfect	$0.5 \times R_{NT}$ to $4 \times R_{NT}$	$0.1 \times E_m$ to $10 \times E_m$	Composite cylinders micromechanics
Tsai and Lu [17]	2009	Interphase	Perfect	0.01 nm	G_{int} = 0.13, 0.286, 4.2, 370 GPa	Elasticity (shear lag)
Nairn [18]	2011	Interface or interphase	Imperfect	N/A	Imperfection coefficient	Shear lag (elasticity)
Shokrieh and Mahdavi [19]	2011	Interphase	vdW	0.17 nm	Function of l_{NT} and R_{NT}	Equivalent fiber (elasticity)
Barai and Weng [20]	2011	Interface	Perfect	N/A	Imperfection coefficient	Elasticity

polymer [17, 18]. This model is usually used for long fiber composites at microscale, and it is not suitable for CNT reinforced composites wherein the effective scale is nano. This model has been basically developed for continuum reinforcing agent, and it could not be applied to the lattice structure of CNT. Tsai and Lu [17] took into account the interphase as a physical environment. They used different values for the shear modulus of interphase and investigated the effect of interphase on load transfer efficiency in the CNTs reinforced nanocomposites using the conventional shear lag model. Nairn [18] considered the effect of the interphase by a coefficient varying between zero and infinity in the formula for calculating the modulus of the nanocomposite.

Shokrieh and Mahdavi [19] considered a virtual material region for the interphase. They have obtained the properties of an equivalent fiber using the elasticity approach when the interactions are assumed to be vdW. The obtained results were in a very good agreement with the developed multiscale FE modeling by Shokrieh and Rafiee [29].

Barai and Weng [20] studied the mechanical properties of carbon nanotube with elasticity equations. They fit an imperfect factor to their equations and investigated the effect of the interphase by changing this ratio. It was reported that when the imperfection at the interface increases, the effective Young's modulus of the composite decreases and those CNTs with very weak interfaces will behave like voids. It should be pointed out that since Barai and Weng [20] have considered perfect bonding between CNT and polymer, they are addressing higher stress transferability from matrix to CNT in contrast to what has been reported by Tan et al. [15] through weak vdW interactions.

An overview of conducted analytical modeling to study the interaction between the CNT and the surrounding polymer is summarized in Table 2.

As it can be understood from the literature survey, different values have been considered for the modulus and also for the thickness of interphase. In some investigations, the influence of the interphase region has just been simulated using some correction factor to reflect the imperfection.

3.2.2. Numerical Modeling. Several researchers have used numerical modeling employing the finite element analysis to study the interaction between the CNT and the surrounding polymer. Some investigators considered an intermediate region between the CNT and the polymer as a separate phase and used various elements like link, truss, rod, beam, and spring with different properties describing the molecular interaction between the CNT atoms and the surrounding polymer chains. Some others have ignored this region in the modeling and simply assumed a perfect bonding between nanotube and the polymer [21, 22]. However, since it is well known that perfect bonding between the nanotubes and the resin is not a valid assumption, these studies will not be reviewed in this report.

Haque and Ramasetty [21] obtained shear and axial stresses at the interface without considering interphase using the equilibrium and the elastic equations.

Wan and his coworkers [27] calculated the properties of the RVE of CNTRP considering interphase with the thickness of 1.357 nm. The modulus of interphase has been varied between 0.3 and 10 times the elastic modulus of the matrix. They considered the interphase as an isotropic material with constant properties.

Li and Chou [23] consider the vdW interactions between the nanotube and the inner surface of the surrounding polymeric matrix and used nonlinear truss elements to simulate vdW interactions. They considered the thickness of this region as 0.17 nm and achieved the properties of the elements using Lennard-Jones potential.

Pourakbar and his coworkers [24] have studied the effect of functionalized nanotube on the properties of a RVE. They modeled covalent bonds between the nanotubes and the resin with beam elements and used properties that were derived from molecular mechanics approach. They found out that by increasing the population of covalent bonds at the interphase, the stress transferability from matrix to CNT is increased and it will converge to the results achieved from the rule of mixture, even though the rule of mixture is not valid at nanoscale.

Table 3: Finite element modeling of CNT and polymer interaction.

Researcher(s)	Year	Interface or interphase	Type of interaction	Thickness	E_{int} and G_{int}	Element
Haque and Ramasetty [21]	2005	Interface	Perfect	N/A	N/A	N/A
Wan et al. [27]	2005	Interphase	Perfect	1.357 nm	$10 \times E_m$ and $0.3 \times E_m$	N/A
Li and Chou [23]	2006	Interphase	vdW	0.17 nm	Derived from LJ potential	Truss rod
Pourakbar et al. [24]	2008	Interphase	Functionalized	1.522 nm	From MD	Beam
Giannopoulos et al. [25]	2009	Interphase	vdW	0.17 nm	Parametric	Joint element
Montazeri and Naghdabadi [26]	2009	Interphase	vdW	0.34 nm	Derived from LJ potential	Spring
Kulkarni et al. [22]	2010	Interface	Perfect	0	Different value	Interface element
Needleman et al. [28]	2010	Interphase and interface	Perfect	Variable	Linearly between E_m and E_n	N/A
Shokrieh and Rafiee [29]	2010	Interphase	vdW	0.17 nm	Derived from LJ potential	Nonlinear spring
Ayatollahi et al. [30]	2011	Interphase	Perfect	0.15 nm	1, 5, 10 and 15 GPa	Solid
Wernik and Meguid [31]	2011	Interphase	vdW	0.38 nm	Derived from LJ potential	Truss rod

Giannopoulos et al. [25] considered the modulus of interphase as a function of the matrix and CNT's modulus. In their work, joint elements were used to simulate the interphase between the CNT and the matrix in radial direction along the nanotube. The results reveal a high dependence of the longitudinal modulus of the nanocomposites on the stiffness of the interfacial region. The high stiffness values of the joint elements (perfect bonding between the CNT and matrix) lead to the similar results obtained by the rule of mixture. In contrast, when the nanotube is very softly bonded to the matrix, no substantial enhancement in mechanical properties is observed.

Montazeri and Naghdabadi [26] have used one-dimensional spring elements to model the interphase. They have achieved the associated properties of the spring elements from Lennard-Jones potential.

Kulkarni et al. [22] have studied the properties of the RVE by considering the interface and the force region between the nanotubes and the resin. They have used interface elements in the modeling and used varied strength and investigated the effect of that on the properties of RVE. They reported that the modulus of the nanocomposite increases with increasing interface strength. The rate of increase in the modulus of the nanocomposite becomes smaller, as the interface strength increases.

Needleman and his coworkers [28] calculated the properties of RVEs. They have simulated the interphase with a variable thickness and nonlinear properties between nanotubes and the resin. They considered a perfect bonding between the nanotubes and the resin and investigated the influence of interphase strength on the properties of investigated RVE.

Shokrieh and Rafiee [29] proposed a model for interphase where bonds between nanotubes and the resin would be updated when the applied strain evolves. Consequently, the model is called Adaptive vdW Interaction (AVI). They have used nonlinear spring elements capturing vdW interactions. In other words, when the applied strain is increased, the instantaneous status of each vdW bond is updated in accordance with the governing equation of the vdW interaction. They have observed that the rule of mixture will overestimate the results. This stems from this fact that the rule of mixture considers perfect bonding between CNT and polymer. The results demonstrate the importance of considering vdW based interphase.

Wernik and Meguid [31] offered the same model as Shokrieh and Rafiee's model for interphase. The only difference was that they used truss elements instead of spring elements for interphase. They reported the same trend reported by Shokrieh and Rafiee in comparison with the rule of mixture.

Ayatollahi et al. [30] obtained properties of RVE considering interphase with 3 times the thickness of the carbon nanotube and different elastic moduli as 0.2, 2, and 20 GPa. They investigated the effect of the interphase on the properties of the RVE with different values of elasticity modulus of the interphase. They found out that generally, by increasing the modulus of interphase, the load transfer between CNT and polymer will be improved and the modulus of nanocomposites will be increased.

An overview of the finite element modeling studies on the interphase region between the CNT and the polymer is presented in Table 3.

It can be seen from Table 3 that those who have assumed the interactions between CNT and surrounding polymer take place through force environment also used interface to simulate the interaction between CNT and polymer. They have observed that increasing the interface strength leads to convergence of the modulus of nanocomposite to that obtained by the rule of mixtures, since they are addressing perfect bonding between CNT and polymer.

On the other side, when the interaction between CNT and surrounding polymer is treated as a physical environment, an interphase between CNT and surrounding polymer is modeled. The modeling of interphase appears either in the form of intermediate material region between CNT and polymer or in the form of connection links between CNT and polymer capturing by spring, truss, or beam elements. The former is called continuum modeling of interphase, while the latter is called semicontinuum modeling of the interphase region.

4. Conclusion

The modeling of interface/interphase region between the CNT and the polymer is reviewed in this scientific review. The theoretical studies are divided into two main groups: atomistic modeling and continuum modeling.

Atomistic modeling can well capture the shear strength and bonding energy of the interface between the CNT and the surrounding polymer matrix in nanoscale. Moreover, the obtained properties can be upscaled to the macroscale model. The united atom (UA) and coarse grain (CG) methods are promising solutions to extend the simulation to microscale. However, the beads-incorporated simulation still requires a huge amount of calculation and is limited to small length scale compared to the continuum model. Moreover, obtaining the accurate nonlinear mechanical properties is one of the limitations of using the CG method. As a consequence, continuum modeling is used by different researchers to overcome mentioned difficulties. But the degree to which the continuum modeling can truly predict the actual behavior of CNT at nanoscale has to be carefully studied.

Continuum modeling can be done either analytically or numerically. The analytical modeling of the continuum model usually considers simply perfect bonding between the CNT and the surrounding polymer and tries to capture the influence of the imperfect bonding between the CNT and the polymer by introducing some correction factor in the micromechanics rules.

The numerical continuum modeling which takes place using FEM is divided into three main groups. The first group takes into account the force interaction between CNT and polymer through interface without considering any distinct material region. In this group, the researchers usually replace CNT with a continuum medium for sake of simplicity, and thus their results approach the values reported by rule of mixture. It is worth mentioning that the basic assumptions of the rule of mixture are not valid for the case of CNT. Micromechanics consider continuum medium for the reinforcing agent, which is not pertinent to the case of CNT, and also it assumes perfect bonding between CNT and surrounding polymer when instead CNT naturally interacts with the polymer through weak vdW interactions.

The second group considers an intermediate material region between the CNT and the polymer as a continuum medium and uses different values for the properties of this distinct region. Both the thickness and the mechanical properties of this region are under question in this category of study, as the selection has no experimental basis. Almost all studies selected Young's modulus of the interphase region as a varying parameter between Young's modulus of resin and CNT.

The first two groups can be also found in the field of analytical modeling as well.

The third group uses different elements consisting of joint, bar, truss, spring, and beam to construct the interaction between the CNT and the surrounding polymer. The properties of these elements are obtained from molecular space, and therefore equivalent structural members are employed. Due to the specific fact that a continuum region has not been considered for the interphase region and actually continuum elements are used for the purpose of molecular interaction, this group of study is also called as semicontinuum modeling. Instead of considering a continuum material region, just vdW interactions or covalent bonds between CNT and surrounding polymer are simulated using continuum elements. The former is the natural interaction between CNT and polymer in the absence of chemical functionalization, while the latter captures functionalized CNT. Semicontinuum modeling has received more interest among researchers in the recent years.

It can be concluded that despite the fact that atomistic modeling can provide an accurate result on local properties and interactions between CNT and polymer, it is computationally intensive, and it is limited to very short time and small length scales. However, continuum modeling can be considered as a reasonable compromise in modeling, while it is focusing on global behavior rather than local one. Among developed continuum modeling, the more preferred approach is semicontinuum modeling, wherein CNT is treated as a lattice structure, and the interactions between CNT and surrounding polymer are replaced by continuum elements.

References

[1] J. Gou, B. Minaie, B. Wang, Z. Liang, and C. Zhang, "Computational and experimental study of interfacial bonding of single-walled nanotube reinforced composites," *Computational Materials Science*, vol. 31, no. 3-4, pp. 225–236, 2004.

[2] E. T. Thostenson and T. Chou, "Aligned multi-walled carbon nanotube-reinforced composites: processing and mechanical characterization," *Journal of Physics D*, vol. 35, no. 16, pp. L77–L80, 2002.

[3] C. Bower, R. Rosen, L. Jin, J. Han, and O. Zhou, "Deformation of carbon nanotubes in nanotube-polymer composites," *Applied Physics Letters*, vol. 74, no. 22, pp. 3317–3319, 1999.

[4] C. A. Cooper, S. R. Cohen, A. H. Barber, and H. D. Wagner, "Detachment of nanotubes from a polymer matrix," *Applied Physics Letters*, vol. 81, no. 20, pp. 3873–3875, 2002.

[5] D. Qian and E. C. Dickey, "In-situ transmission electron microscopy studies of polymer-carbon nanotube composite deformation," *Journal of Microscopy*, vol. 204, no. 1, pp. 39–45, 2001.

[6] K. Liao and S. Li, "Interfacial characteristics of a carbon nanotube-polystyrene composite system," *Applied Physics Letters*, vol. 79, no. 25, pp. 4225–4227, 2001.

[7] S. J. V. Frankland, A. Caglar, D. W. Brenner, and M. Griebel, "Molecular simulation of the influence of chemical cross-links on the shear strength of carbon nanotube-polymer interfaces," *Journal of Physical Chemistry B*, vol. 106, no. 12, pp. 3046–3048, 2002.

[8] M. Wong, M. Paramsothy, X. J. Xu, Y. Ren, S. Li, and K. Liao, "Physical interactions at carbon nanotube-polymer interface," *Polymer*, vol. 44, no. 25, pp. 7757–7764, 2003.

[9] H. Chen, Q. Xue, Q. Zheng, J. Xie, and K. Yan, "Influence of nanotube chirality, temperature, and chemical modification on the interfacial bonding between carbon nanotubes and polyphenylacetylene," *Journal of Physical Chemistry C*, vol. 112, no. 42, pp. 16514–16520, 2008.

[10] Z. Xu and M. J. Buehler, "Nanoengineering heat transfer performance at carbon nanotube interfaces," *ACS Nano*, vol. 3, no. 9, pp. 2767–2775, 2009.

[11] A. Liu, K. W. Wang, and C. E. Bakis, "Effect of functionalization of single-wall carbon nanotubes (SWNTs) on the damping characteristics of SWNT-based epoxy composites via multiscale analysis," *Composites A*, vol. 42, no. 11, pp. 1748–1755, 2011.

[12] M. Rahmat and P. Hubert, "Molecular dynamics simulation of single-walled carbon nanotube-PMMA Interaction," *Journal of Nano Research*, vol. 18-19, pp. 117–128, 2012.

[13] Y. Zhang, J. Zhao, N. Wei, J. Jiang, Y. Gong, and T. Rabczuk, "Effects of the dispersion of polymer wrapped two neighbouring single walled carbon nanotubes (SWNTs) on nanoengineering load transfer," *Composites B*, vol. 45, no. 1, pp. 1714–1721, 2013.

[14] L. Y. Jiang, Y. Huang, H. Jiang et al., "A cohesive law for carbon nanotube/polymer interfaces based on the van der Waals force," *Journal of the Mechanics and Physics of Solids*, vol. 54, no. 11, pp. 2436–2452, 2006.

[15] H. Tan, L. Y. Jiang, Y. Huang, B. Liu, and K. C. Hwang, "The effect of van der Waals-based interface cohesive law on carbon nanotube-reinforced composite materials," *Composites Science and Technology*, vol. 67, no. 14, pp. 2941–2946, 2007.

[16] G. D. Seidel and D. C. Lagoudas, "Micromechanical analysis of the effective elastic properties of carbon nanotube reinforced composites," *Mechanics of Materials*, vol. 38, no. 8–10, pp. 884–907, 2006.

[17] J.-L. Tsai and T.-C. Lu, "Investigating the load transfer efficiency in carbon nanotubes reinforced nanocomposites," *Composite Structures*, vol. 90, no. 2, pp. 172–179, 2009.

[18] J. A. Nairn, "Aspect ratio requirements for nanotube-reinforced, polymer-matrix composites," *Composites A*, vol. 42, no. 11, pp. 1850–1855, 2011.

[19] M. M. Shokrieh and S. M. Mahdavi, "Micromechanical model to evaluate the effects of dimensions and interphase region on the elastic modulus of CNT/polymer composites," *Modarres Mechanical Engineering*, vol. 3, pp. 13–25, 2011.

[20] P. Barai and G. J. Weng, "A theory of plasticity for carbon nanotube reinforced composites," *International Journal of Plasticity*, vol. 27, no. 4, pp. 539–559, 2011.

[21] A. Haque and A. Ramasetty, "Theoretical study of stress transfer in carbon nanotube reinforced polymer matrix composites," *Composite Structures*, vol. 71, no. 1, pp. 68–77, 2005.

[22] M. Kulkarni, D. Carnahan, K. Kulkarni, D. Qian, and J. L. Abot, "Elastic response of a carbon nanotube fiber reinforced polymeric composite: a numerical and experimental study," *Composites B*, vol. 41, no. 5, pp. 414–421, 2010.

[23] C. Li and T. Chou, "Multiscale modeling of compressive behavior of carbon nanotube/polymer composites," *Composites Science and Technology*, vol. 66, no. 14, pp. 2409–2414, 2006.

[24] K. Pourakbar, N. JamilPour, A. R. Najafi, G. Rouhi, A. Arshi, and A. Fereidoon, "A finite element model for estimating Young's modulus of carbon nanotube reinforced composites incorporating elastic cross-links," *International Journal of Mechanical, Industrial and Aerospace Engineering*, pp. 172–175, 2008.

[25] G. I. Giannopoulos, S. K. Georgantzinos, and N. K. Anifantis, "A semi-continuum finite element approach to evaluate the Young's modulus of single-walled carbon nanotube reinforced composites," *Composites B*, vol. 41, no. 8, pp. 594–601, 2010.

[26] A. Montazeri and R. Naghdabadi, "Study the effect of viscoelastic matrix model on the stability of CNT/polymer composites by multiscale modeling," *Polymer Composites*, vol. 30, no. 11, pp. 1545–1551, 2009.

[27] H. Wan, F. Delale, and L. Shen, "Effect of CNT length and CNT-matrix interphase in carbon nanotube (CNT) reinforced composites," *Mechanics Research Communications*, vol. 32, no. 5, pp. 481–489, 2005.

[28] A. Needleman, T. L. Borders, L. C. Brinson, V. M. Flores, and L. S. Schadler, "Effect of an interphase region on debonding of a CNT reinforced polymer composite," *Composites Science and Technology*, vol. 70, no. 15, pp. 2207–2215, 2010.

[29] M. M. Shokrieh and R. Rafiee, "Prediction of mechanical properties of an embedded carbon nanotube in polymer matrix based on developing an equivalent long fiber," *Mechanics Research Communications*, vol. 37, no. 2, pp. 235–240, 2010.

[30] M. Ayatollahi, S. Shadlou, and M. Shokrieh, "Multiscale modeling for mechanical properties of carbon nanotube reinforced nanocomposites subjected to different types of loading," *Composite Structures*, vol. 93, no. 9, pp. 2250–2259, 2011.

[31] J. M. Wernik and S. A. Meguid, "Multiscale modeling of the nonlinear response of nano-reinforced polymers," *Acta Mechanica*, vol. 217, no. 1-2, pp. 1–16, 2011.

[32] L. S. Schadler, S. C. Giannaris, and P. M. Ajayan, "Load transfer in carbon nanotube epoxy composites," *Applied Physics Letters*, vol. 73, no. 26, pp. 3842–3844, 1998.

[33] H. Wagner, "Nanotube-polymer adhesion: a mechanics approach," *Chemical Physics Letters*, vol. 361, no. 1-2, pp. 57–61, 2002.

[34] P. Ajayan, L. Schadler, C. Giannaris, and A. Rubio, "Single-walled carbon nanotube-polymer composites: strength and weakness," *Advanced Materials*, vol. 12, no. 10, pp. 750–753, 2000.

[35] C. A. Cooper, R. J. Young, and M. Halsall, "Investigation into the deformation of carbon nanotubes and their composites through the use of Raman spectroscopy," *Composites A*, vol. 32, no. 3-4, pp. 401–411, 2001.

[36] V. Hadjiev, M. Iliev, S. Arepalli, P. Nikolaev, and B. S. Files, "Raman scattering test of single-wall carbon nanotube composites," *Applied Physics Letters*, vol. 78, no. 21, pp. 3193–3195, 2001.

[37] A. Paipetis, C. Galiotis, Y. C. Liu, and J. A. Nairn, "Stress transfer from the matrix to the fibre in a fragmentation test: Raman experiments and analytical modeling," *Journal of Composite Materials*, vol. 33, no. 4, pp. 377–399, 1999.

[38] L. Valentini, J. Biagiotti, J. M. Kenny, and M. A. López Manchado, "Physical and mechanical behavior of single-walled carbon nanotube/polypropylene/ethylene-propylene-diene rubber nanocomposites," *Journal of Applied Polymer Science*, vol. 89, no. 10, pp. 2657–2663, 2003.

[39] D. Qian, G. J. Wagner, W. K. Liu, M. Yu, and R. S. Ruoff, "Mechanics of carbon nanotubes," *Applied Mechanics Reviews*, vol. 55, no. 6, pp. 495–532, 2002.

[40] D. Qian, W. K. Liu, and R. S. Ruoff, "Load transfer mechanism in carbon nanotube ropes," *Composites Science and Technology*, vol. 63, no. 11, pp. 1561–1569, 2003.

[41] M. Rahmat and P. Hubert, "Carbon nanotube-polymer interactions in nanocomposites: a review," *Composites Science and Technology*, vol. 72, no. 1, pp. 72–84, 2011.

[42] A. V. Desai and M. A. Haque, "Mechanics of the interface for carbon nanotube-polymer composites," *Thin-Walled Structures*, vol. 43, no. 11, pp. 1787–1803, 2005.

[43] W. F. van Gunsteren and H. J. C. Berendsen, "Computer simulation of molecular dynamics: methodology, applications, and

perspectives in chemistry," *Angewandte Chemie*, vol. 29, no. 9, pp. 992–1023, 1990.

[44] D. C. Rapaport, *The Art of Molecular Dynamics Simulation*, Cambridge University Press, Cambridge, UK, 2004.

[45] V. Kalra and Y. L. Joo, "Coarse-grained molecular dynamics study of block copolymer/nanoparticle composites under elongational flow," *Journal of Chemical Physics*, vol. 131, no. 21, Article ID 214904, 2009.

[46] M. Rahmat, H. Ghiasi, and P. Hubert, "Interaction energy and polymer density profile in nanocomposites: a coarse grain simulation based on interaction stress," *Polymer Chemistry*, vol. 3, no. 5, pp. 1158–1167, 2012.

[47] S. C. Chowdhury and T. Okabe, "Computer simulation of carbon nanotube pull-out from polymer by the molecular dynamics method," *Composites A*, vol. 38, no. 3, pp. 747–754, 2007.

[48] Y. Kuang and X. He, "Young's moduli of functionalized single-wall carbon nanotubes under tensile loading," *Composites Science and Technology*, vol. 69, no. 2, pp. 169–175, 2009.

[49] J.-L. Tsai, S.-H. Tzeng, and Y.-T. Chiu, "Characterizing elastic properties of carbon nanotubes/polyimide nanocomposites using multi-scale simulation," *Composites B*, vol. 41, no. 1, pp. 106–115, 2010.

[50] G. Odegard, T. Gates, K. Wise, C. Park, and E. J. Siochi, "Constitutive modeling of nanotube-reinforced polymer composites," *Composites Science and Technology*, vol. 63, no. 11, pp. 1671–1687, 2003.

[51] S. J. V. Frankland and V. M. Harik, "Analysis of carbon nanotube pull-out from a polymer matrix," *Surface Science*, vol. 525, no. 1–3, pp. L103–L108, 2003.

[52] M. Al-Haik, M. Y. Hussaini, and H. Garmestani, "Adhesion energy in carbon nanotube-polyethylene composite: effect of chirality," *Journal of Applied Physics*, vol. 97, no. 7, Article ID 074306, 2005.

[53] R. E. Rudd, "Coarse-grained molecular dynamics for computer modeling of nanomechanical systems," *International Journal for Multiscale Computational Engineering*, vol. 2, no. 2, p. 157, 2004.

[54] S. Cranford and M. J. Buehler, "Coarse-graining parameterization and multiscale simulation of hierarchical systems: part I: theory and model formulation," in *Multiscale Modeling: From Atoms to Devices*, P. Derosa and T. Cagin, Eds., Taylor & Francis, London, UK, 2010.

[55] P. Derreumaux and N. Mousseau, "Coarse-grained protein molecular dynamics simulations," *Journal of Chemical Physics*, vol. 126, no. 2, Article ID 025101, 2007.

[56] N. K. Vaidya, H. Huang, and S. Takagi, "Coarse grained molecular dynamics simulation of interaction between hemagglutinin fusion peptides and lipid bilayer membranes," *Advances in Applied Mathematics and Mechanics*, vol. 2, pp. 430–450, 2010.

[57] M. Zhang and F. Müller-Plathe, "The Soret effect in dilute polymer solutions: influence of chain length, chain stiffness, and solvent quality," *Journal of Chemical Physics*, vol. 125, no. 12, Article ID 124903, 2006.

[58] A. Di Matteo, F. Müller-Plathe, and G. Milano, "From mesoscale back to atomistic models: a fast reverse-mapping procedure for vinyl polymer chains," *Journal of Physical Chemistry B*, vol. 111, no. 11, pp. 2765–2773, 2007.

[59] T. Terao, E. Lussetti, and F. M. Müller-Plathe, "Non-equilibrium molecular dynamics methods for computing the thermal conductivity: application to amorphous polymers," *Physical Review E*, vol. 75, no. 5, Article ID 057701, 2007.

[60] M. Fermeglia and S. Pricl, "Multiscale modeling for polymer systems of industrial interest," *Progress in Organic Coatings*, vol. 58, no. 2-3, pp. 187–199, 2007.

[61] C. D. Wick and D. N. Theodorou, "Connectivity-altering Monte Carlo simulations of the end group effects on volumetric properties for poly(ethylene oxide)," *Macromolecules*, vol. 37, no. 18, pp. 7026–7033, 2004.

[62] V. A. Harmandaris, N. P. Adhikari, N. F. A. Van Der Vegt, and K. Kremer, "Hierarchical modeling of polystyrene: from atomistic to coarse-grained simulations," *Macromolecules*, vol. 39, no. 19, pp. 6708–6719, 2006.

[63] V. A. Harmandaris, D. Reith, N. F. A. Van Der Vegt, and K. Kremer, "Comparison between coarse-graining models for polymer systems: two mapping schemes for polystyrene," *Macromolecular Chemistry and Physics*, vol. 208, no. 19-20, pp. 2109–2120, 2007.

[64] S. Nielsen, C. F. Lopez, G. Srinivas, and M. L. Klein, "A coarse grain model for n-alkanes parameterized from surface tension data," *Journal of Chemical Physics*, vol. 119, no. 14, pp. 7043–7049, 2003.

[65] J. Zhao, S. Nagao, and Z. Zhang, "Thermomechanical properties dependence on chain length In bulk polyethylene: coarse-grained molecular dynamics simulations," *Journal of Materials Research*, vol. 25, no. 3, pp. 537–544, 2010.

[66] J. Zhao, S. Nagao, Z. L. Zhang, and H. Kristiansen, "Coarse-grained molecular dynamics simulations on size effect of glassy polyethylene particles," *Journal of Nanoscience and Nanotechnology*, vol. 10, no. 11, pp. 7340–7342, 2010.

[67] S. W. Cranford and M. J. Buehler, "In silico assembly and nanomechanical characterization of carbon nanotube buckypaper," *Nanotechnology*, vol. 21, no. 26, Article ID 265706, 2010.

[68] B. Xie, Y. Liu, Y. Ding, Q. Zheng, and Z. Xu, "Mechanics of carbon nanotube networks: microstructural evolution and optimal design," *Soft Matter*, vol. 7, no. 21, pp. 10039–10047, 2011.

[69] Y. Li and M. Kröger, "Viscoelasticity of carbon nanotube buckypaper: zipping-unzipping mechanism and entanglement effects," *Soft Matter*, vol. 8, pp. 7822–7830, 2012.

[70] G. Kim, Y. Kim, and J. Ihm, "Encapsulation and polymerization of acetylene molecules inside a carbon nanotube," *Chemical Physics Letters*, vol. 415, no. 4–6, pp. 279–282, 2005.

[71] K. Mylvaganam and L. C. Zhang, "Nanotube functionalization and polymer grafting: an ab initio study," *Journal of Physical Chemistry B*, vol. 108, no. 39, pp. 15009–15012, 2004.

High Spatial Resolution Time-of-Flight Secondary Ion Mass Spectrometry for the Masses: A Novel Orthogonal ToF FIB-SIMS Instrument with *In Situ* AFM

James A. Whitby,[1] Fredrik Östlund,[1,2] Peter Horvath,[1] Mihai Gabureac,[1] Jessica L. Riesterer,[1,3] Ivo Utke,[1] Markus Hohl,[2] Libor Sedláček,[4] Jaroslav Jiruše,[4] Vinzenz Friedli,[1,5] Mikhael Bechelany,[1,6] and Johann Michler[1]

[1] *EMPA, Swiss Federal Laboratories for Materials Science and Technology, Laboratory for Mechanics of Materials and Nanostructures, Feuerwerkerstrasse 39, 3602 Thun, Switzerland*
[2] *TOFWERK AG, Uttigenstrasse 22, 3600 Thun, Switzerland*
[3] *Idaho National Laboratory, Idaho Falls, ID 83415, USA*
[4] *Tescan a. s., Libušina třída 21, 623 00 Brno, Czech Republic*
[5] *Specs Zürich GmbH, Technoparkstrasse 1, 8005 Zürich, Switzerland*
[6] *Institut Européen des Membranes (UR CNRS 5635), Université Montpellier 2, Place Eugène Bataillon, 34095 Montpellier, France*

Correspondence should be addressed to James A. Whitby, james.whitby@empa.ch

Academic Editor: Adam Georg Balogh

We describe the design and performance of an orthogonal time-of-flight (TOF) secondary ion mass spectrometer that can be retrofitted to existing focused ion beam (FIB) instruments. In particular, a simple interface has been developed for FIB/SEM instruments from the manufacturer Tescan. Orthogonal extraction to the mass analyser obviates the need to pulse the primary ion beam and does not require the use of monoisotopic gallium to preserve mass resolution. The high-duty cycle and reasonable collection efficiency of the new instrument combined with the high spatial resolution of a gallium liquid metal ion source allow chemical observation of features smaller than 50 nm. We have also demonstrated the integration of a scanning probe microscope (SPM) operated as an atomic force microscope (AFM) within the FIB/SEM-SIMS chamber. This provides roughness information, and will also allow true three dimensional chemical images to be reconstructed from SIMS measurements.

1. Introduction and Design Philosophy

The ion microprobe or secondary ion mass spectrometer (SIMS) has existed for almost fifty years [1, 2] but although great improvements have been made in sensitivity and ease of use, the lateral resolution has not improved quite as much. Early papers reported measurements from fifty micron areas for example, [2] but submicron image resolution using gallium liquid metal ion sources (LMIS) was reported even thirty years ago [3] and the separation of features as small as 20 nm wide has been reported [4, 5], close to the limits set by sensitivity [4, 6] and lateral ion straggling [7]. However, liquid metal gallium ion beams fell out of favour for SIMS because the limited secondary ion yields did not in general allow the full lateral resolution of the small ion beam spot-size to be used, and because many applications simply demanded more sensitivity—most easily achieved by using primary ions such as Cs^+ or O_2^+ to improve the secondary ion yield by up to two orders of magnitude. That the effective spatial resolution in SIMS is often signal limited, especially so for liquid metal gallium ion sources, is well documented, for example, [4, 6, 8]. Nonetheless, the enormous increase in interest in nanomaterials over the last decade, and the limited spatial resolution of some common chemical analysis techniques such as electron-beam-induced X-ray emission (typically restricted to a information volume of about 1 μm diameter by electron scattering) have prompted a new look at high spatial resolution SIMS [9, 10]. A further motivation to re-examine the use of a gallium LMIS for SIMS is the increasingly widespread adoption of Focused Ion Beam (FIB) instruments for sample preparation (both milling and deposition). These almost universally use gallium LMIS beams to achieve excellent spatial resolutions with a high beam brightness (corresponding to acceptable milling speeds). The coupling of a secondary ion mass spectrometer to a dedicated FIB instrument has been previously described by others for example, [5, 11, 12].

Advantages of the combination of a FIB/SEM with a SIMS include excellent sample imaging (secondary electrons from either electron or ion beams, the latter giving better channelling contrast) and positioning, and the use of the FIB to precisely expose buried regions of interest for further analysis by SIMS (and other techniques). It is far more common for FIB/SEM instruments to be designed from the start to accommodate extra analytical techniques such as EDX, EBSD, and cathodoluminescence. and these additional techniques which are complementary to SIMS can be decisive in answering analytical questions posed by complex samples. Having several types of analytical measurement possible with a single vacuum chamber is not only a great time and money saver, but also avoids uncertainty due to atmospheric contamination. The advantages of SIMS measurements over the EDX measurements more commonly found on FIB/SEM instruments are (i) better limits of detection (including fewer interferences for light elements), (ii) better spatial resolution, (iii) ability to make isotopic and molecular measurements. It is, however, more difficult in general to quantify SIMS measurements than is the case for EDX measurements on sufficiently large and thick materials.

In this paper, we describe an implementation of an FIB-SIMS instrument and show examples of measurements made with the prototype. Our focus has not been upon the best possible SIMS performance, but rather upon a pragmatic, cost-effective approach to sub-μm resolution chemical imaging that take as its starting points a standard commercial FIB instrument and a tried and tested mass analyser design. The FIB-SIMS functionality was developed as part of the European Union funded FIBLYS (see http://www.fiblys.eu/) project which aims to create a modular suite of analytical instruments which can be fitted to an optimised FIB/SEM chamber. A consequence of the decision to use a commercial FIB/SEM instrument is that the vacuum chamber is only designed to work at high vacuum (above 10^{-6} mbar) rather than the ultrahigh vacuum usual for secondary ion mass spectrometers. The implications of this are discussed below in the performance section.

2. The Instrument

The FIB/SEM chamber, columns, and FIB/SEM software are commercially available parts from Tescan a.s. (a Vela SEM fitted with a FIB column using a liquid metal gallium ion source). The mass spectrometer is a commercial orthogonal extraction time-of-flight instrument from Tofwerk AG (model C-TOF) with a 0.5 m nominal drift length, including

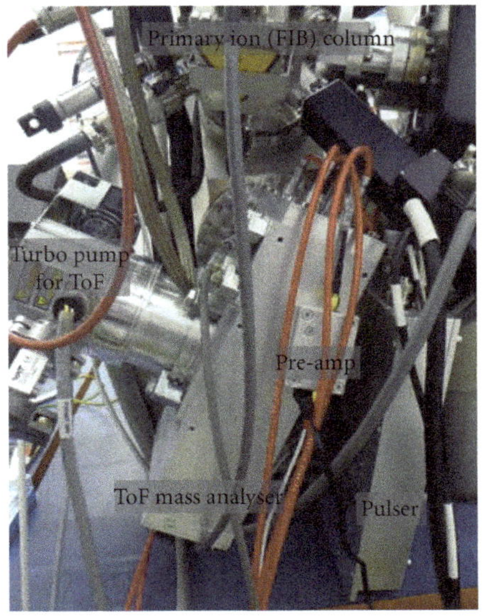

Figure 1: A closer look at the mass analyser in position on the back of the chamber; there is very little space available. The turbo pump on the mass analyser (to protect the detector) will be replaced by an ion pump. The pulser electronics for the orthogonal extraction are placed close to the mass analyser in order to preserve the pulse shape.

the reflectron. This mass spectrometer was chosen because of its compact dimensions (see Figures 1 and 2) and because of the performance advantages and price/performance ratio of time-of-flight mass analysis compared to quadrupole mass analysis (as the samples expected to be studied by FIB-SIMS instruments are often very small and unique, the simultaneous detection of all masses in a time-of-flight instrument offers efficiency gains compared to scanning a quadrupole filter). Because of the small size of the instrument, the mass resolving power $M/\Delta M$ (FWHM) is no more than 800 at 69 Th, just sufficient to distinguish many molecular interferences from isobaric monatomic ions (Th denotes the unit Thomson for charge to mass ratio [13]). Orthogonal extraction [14], typically using a 2 μs pulse, provides a high effective duty cycle and more importantly avoids the need to pulse the primary beam (for TOF operation) or to use monoisotopic gallium (necessary to preserve mass resolution when pulsing the primary beam to define the TOF start signal), thus no changes to the FIB hardware are necessary. Secondary ions are detected in the mass spectrometer using a pair of chevron mounted microchannel plates (Photonis) and the resulting analogue signal passed through a preamplifier, an amplifier/constant fraction discriminator (RoentDek ATR19-2) and time-to-digital converter (Cronologic GmbH HPTDC8-PCI) before being passed to computer memory. Essentially the same model of mass spectrometer has been previously described in more detail in [15] for a similar application, although in the current implementation all hardware is from Tofwerk AG. Either positive or negative secondary ions can be detected, but not simultaneously. The postacceleration voltage towards

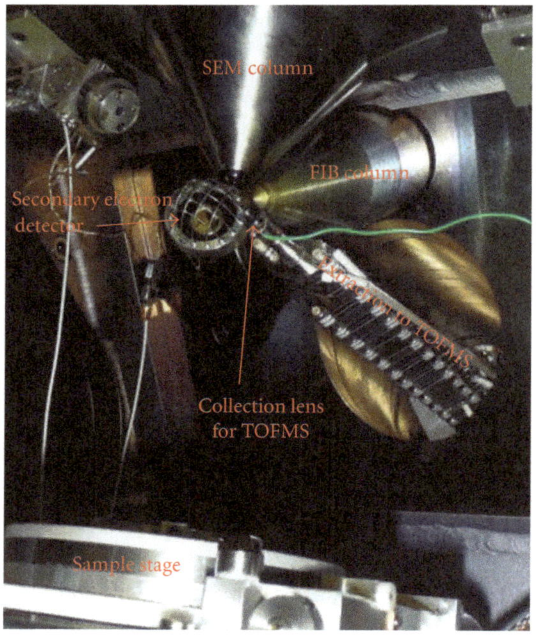

Figure 2: Interior of FIB/SEM-SIMS prototype. The sample stage has been lowered to make the arrangement of the collection lens and charged particle columns clearer. For SIMS work, the sample is usually positioned at a working distance of 9 mm below the SEM column, and tilted to 55° to present the sample surface normal to the incident gallium ions. Additional instrumentation such as nanomanipulators, gas injection systems, and EDX or BSE detectors can be added as required. For SPM measurements, the SPM assembly was mounted on the sample stage.

the microchannel plate detector is typically 5 kV, high enough to still give detection efficiencies of about 20% for ions with masses up to 2000 Da [16].

In order to make secondary ion measurements possible, collection and transfer ion topics were designed and built to take ions from the sample to the mass analyser (mounted on a flange of the FIB/SEM chamber). A set of Einzel lenses acting as relay lenses is used to transmit ions from the collection optics to the mass analyzer; all potentials are below 200 V. A grounded shroud around the collection lens is used to improve collection efficiency and to minimize electrostatic disturbance to other instruments in the chamber. When SIMS measurements are not required, the front part of the extraction ion optics nearest the sample and pole pieces simply clips off (labelled "collection lens" in Figure 2), thus allowing the full range of sample sizes to be used by the FIB. For SIMS operation, it is advantageous to extend the ion optics so that the entrance aperture is nearer to the sample and ion beam: in the current design, the collection optics extend to within 8 mm of the sample/primary ion beam intersection. This leaves enough space for simultaneous operation of the SIMS functionality with a scanning probe microscope or gas injection system (e.g., for focused ion beam induced processing).

To collect mass spectra, software was written to (i) control the functions of the mass spectrometer, (ii) to initiate an appropriate scan series by the FIB, and (iii) to collect secondary electron images in parallel with the secondary ion mass spectra. The same software is used to view and analyse the mass spectra and to export the results in other formats (as ion images, mass spectra from selected regions, or raw data). All data are stored using the publicly available HDF5 file standard [17] which is supported by many data analysis packages such as Matlab, IDL, and Igor Pro. The TOF-SIMS software is, for performance reasons, hosted on a separate computer to that which controls the FIB/SEM. The two computers communicate via Tescan proprietary protocols using an Ethernet connection. Custom electronics and a timer card from National Instruments (model NI6602) are used to adapt hardware signals from the FIB/SEM (e.g., level shifting, logic operations, resynchronisation of clock cycles) in order to control the time-to-digital converter card and mass spectrometer extraction pulser.

During operation, the user uses the FIB/SEM to locate the point of interest and to adjust focus and magnification. Then, the desired image resolution, dwell time (mass to charge range) and number of frames are selected on the computer hosting the mass spectrometer software and the measurement is started from the latter. All triggers from the subsequent acquisition series are generated by the FIB/SEM instrument until the predetermined number of frames has been reached or the measurement is stopped manually. The FIB beam remains at each point in a frame for the defined dwell time before rapidly moving to the next point. The beam is not blanked during the move between neighbouring points, but between lines and frames it is. Although the SIMS capability has only been tested with FIB/SEM instruments from Tescan, it should be possible to fit the mass analysis hardware into instruments from other manufacturers, but the software and signalling interfaces might have to be modified. Example results and different ways of displaying the data are shown in Figures 4, 5, and 6 for a nickel finder grid.

A note on nomenclature: we refer to one complete x-y raster of the primary beam when collecting SIMS data as a "frame". Each frame is composed of data from an evenly spaced Cartesian grid of points projected onto the sample surface. From each of these points, one mass spectrum is obtained from one extraction cycle of the TOFMS. Because dynamic SIMS implies sputtering, we may consider each measurement point as having some depth, and thus being a volume element or voxel. These are referred to as "raw voxels". In practice, because the number of ions detected per raw voxel is rather low, these raw voxels are binned into "effective voxels" for visualization although the raw data is kept. It should be noted that SIMS measurements alone do not provide depth information, thus the use of "voxel" might be argued with. The meaning of "depth profile" and "chemical image" is illustrated in Figure 3.

3. Performance

Experience with this design of mass spectrometer suggests that the dynamic range of the instrument is limited by internal ion scattering to about eight decades (implying 10 ppb detection limits if all secondary ions were produced with the

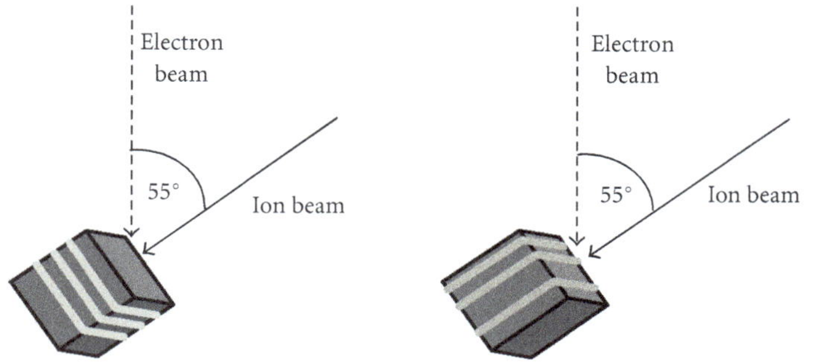

FIGURE 3: Clarification of the primary ion beam and sample orientation for the BAM L200, VCSEL, and ALD samples described in the text, all of which have planar structures. Depth profiles are reconstructed using the middle quarter of each image slice ("frame") in order to minimise edge effects. The resolution in the ion beam direction (depth resolution) is limited only by ion-induced mixing and roughening and redeposition/edge effects and is in general better than the resolution in the directions normal to the ion beam (lateral resolution) which depends heavily on the spot size and shape.

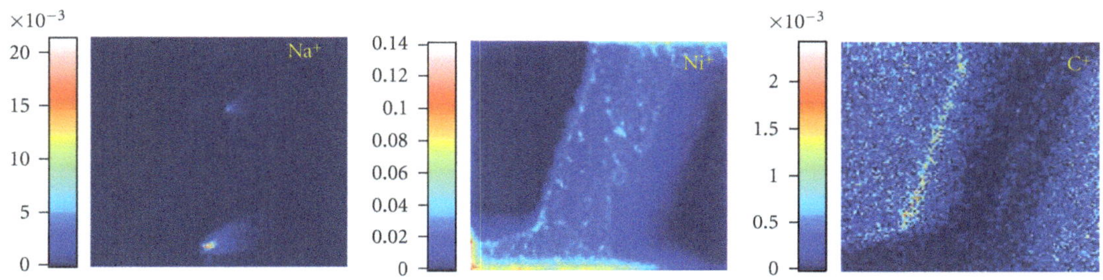

FIGURE 4: Secondary ion images for sodium, nickel, and carbon from part of a nickel finder grid on an adhesive carbon patch. Field of view is 20 × 20 microns. Oxygen flooding was used to enhance the nickel signal. Because the full dataset is three dimensional, it could be shown that the sodium signal was only present at the surface of the sample. The colour scale intensity units are ions per raw voxel (1024 × 1024 × 178); for display purpose the data has been binned to 128 × 128 × 178).

same useful yield). In practice, for most samples integration times sufficient to make 10 ppb peaks clear above the scattered ion background are impractical using primary ion currents suited to submicron resolution imaging (<200 pA). Figure 6 shows in excess of six decades of dynamic range, and Figure 7 shows a calibration curve for boron in silicon from which a limit of detection of 5.5 ppm is inferred. The limit of detection is a strong function of the secondary ion yield, so it will vary for different secondary ions and matrices.

The useful ion yield was calculated for aluminium secondary ions from a distributed Bragg reflector (DBR) structure consisting of alternating AlGaAs and GaAs layers of known thickness (see Figure 12), and a figure of 4×10^{-6} detected ions per sputtered atom was obtained using a primary ion energy of 20 keV. We are not aware of a comparable useful yield measurement in the literature for the same material and primary ion beam. It is well known that secondary ion yields vary enormously with the ion and matrix (as well as primary ion element, energy, and incident angle) and surface quality. Useful ion yields are further dependent upon the tuning of the mass analyser and extraction optics (it is for example common practice to use an energy filter to suppress polyatomic ions despite the resulting reduction in useful ion yield for the analyte ions); a comparison to the (sparse) literature is, therefore, somewhat difficult. Reported values for useful ion yields for gallium ion SIMS include 1.4×10^{-2} for Si^+ from silicon oxide, 6×10^{-5} for Si^+ from silicon, 1.93×10^{-3} for In^+ from indium phosphide and 7.8×10^{-6} for P^+ from indium phosphide [18, 19]. Bearing in mind that aluminium ions seem to be detected more sensitively than most on our instrument and that the chamber pressure is always above 4×10^{-6} mbar (and so some signal enhancement due to oxygen possible) and there is a suggestion that the observed useful ion yield could be improved. We did not use silicon or silicon dioxide to calculate a useful ion yield, because it was not possible to obtain a stable signal under normal conditions at the pressure used in the instrument described here. The implications of high vacuum operation of a SIMS instrument (instead of ultrahigh vacuum) are discussed below.

Chemical images in which lines spaced by less than 100 nm are separated have been obtained (see Figures 9 and 13), and we are certain that this figure can be improved upon in future by better focussing of the FIB beam. The depth resolution can be, as expected, much better than 20 nm; see Figure 15.

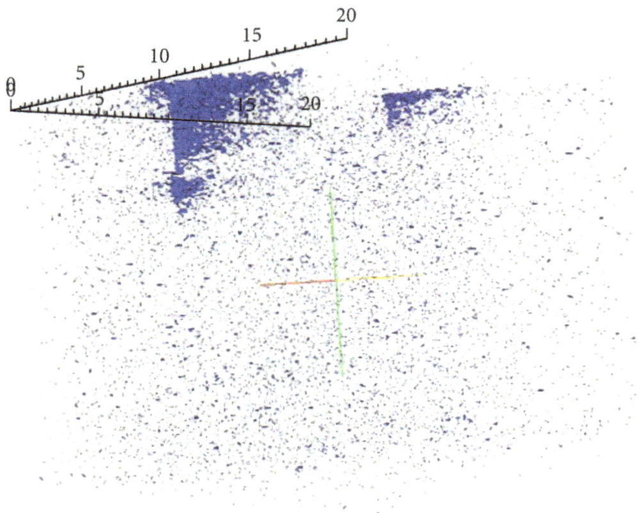

FIGURE 5: Three-dimensional display of sodium secondary ions from the nickel finder grid shown in Figure 4. It can be seen that the sodium is concentrated in two locations, both at the surface of the sample. The tailing in the z direction may be due to the shape of the sodium deposit on the nickel grid or may also be in part due to ion-beam-induced roughening and mixing. The x and y dimensions are 20 μm as for Figure 4; the z dimension is unknown, but is orders of magnitude less than this. The view is of isointensity surfaces after thresholding to remove background counts.

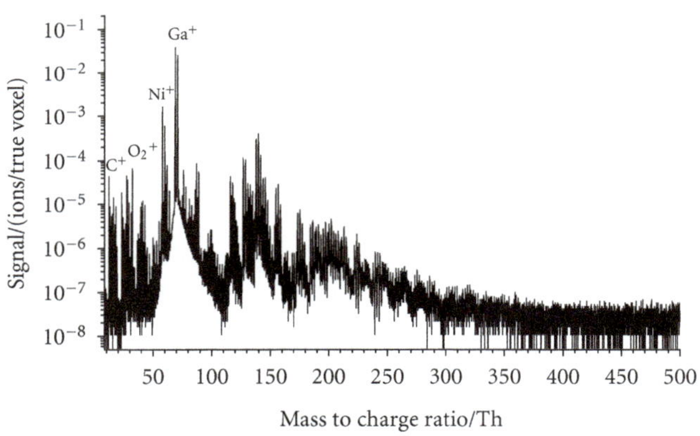

FIGURE 6: Mass spectrum from nickel finder grid on adhesive carbon patch (see also Figure 4). About six decades of dynamic range are exhibited by this spectrum. Note the relatively large molecular oxygen peak due to the use of oxygen flooding to enhance the nickel ion signal.

Mass Discrimination. The gallium isotope ratio from a GaAs target (so the gallium signal consists of secondary ions from both the sample and from implanted primary ions) was measured as 1.46 ± 0.02, the literature value is 1.507 ± 0.0004 [20]. Mass discrimination is therefore of the order of 15‰ Da for this example. Similar magnitudes for the mass discrimination have been observed for iron and nickel from metal alloys.

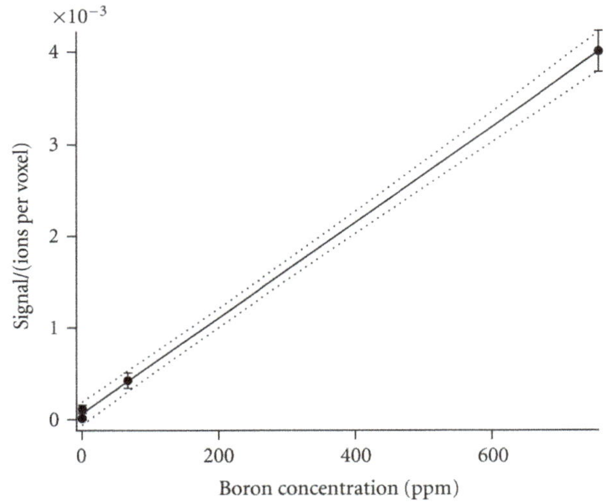

FIGURE 7: Calibration plot for boron in doped silicon wafers showing a linear fit to the measured points (three standards and one blank), and 95% confidence limits. An oxygen partial pressure of 6.0×10^{-5} mbar was used to enhance the signal; the primary beam was 30 keV, >14.7 nA with a scan area of 10 by 10 μm (the current cannot be measured above 14.7 nA). The limit of detection under these conditions for this material was 5.5 ppm.

Mass Resolution and Abundance Sensitivity. The measurements reported here were obtained using a mass resolving power of $M/\Delta M$ (FWHM) of 400 Th/Th at mass to charge ratio 69 Th. The abundance sensitivity was 2.5×10^{-5}.

3.1. Effect of Chamber Pressure. Commercial FIB/SEM instruments are mostly designed to work at high vacuum rather than ultrahigh vacuum, and this has several consequences when performing secondary ion measurements in such an instrument.

(i) Because secondary ion yields depend upon the partial pressure of for example, oxygen (e.g., [21]), there will be a change in sensitivity (drift) as the instrument pumps down after opening the chamber, for example, to change samples. The presence of oxygen in conjunction with ion beam sputtering can complicate the appearance of oxide interfaces [22].

(ii) Even at the lowest pressure the chamber can attain (>1×10^{-6} mbar), there will be a sufficient partial pressure of chemically active gases (e.g., water and oxygen) to affect the secondary ion yield for some materials [23, 24]. This may improve sensitivity, but complicates comparison of useful yields to published values. See Figure 8 for an example of this effect.

(iii) The absorption of chamber gases onto the sample surface contributes a background to the secondary ion mass spectrum, and causes a nonlinear dependence of the SIMS signal upon the primary beam current and scan area [25, 26].

(iv) The presence of active gases changes the extent to which surfaces roughen under ion bombardment.

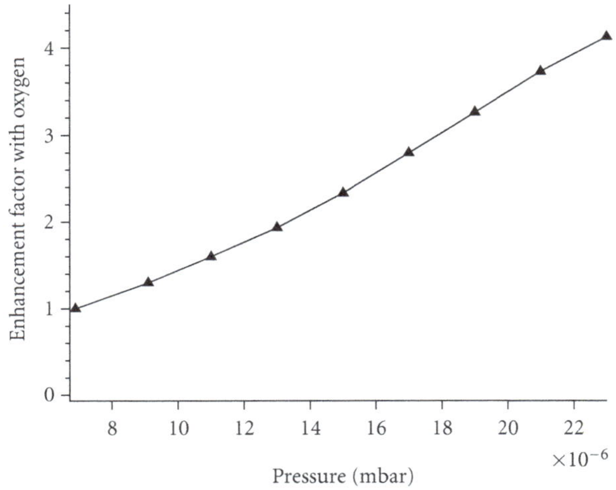

Figure 8: Effect of oxygen flow rate on SIMS signal for a cobalt alloy; the graph shows the Co^+ secondary ion signal as oxygen was bled into the chamber through a needle valve. The plot has been normalised to the signal when the needle valve was closed (i.e., no oxygen flow, chamber pressure 6.9×10^{-6} mbar). 5 keV, 1270 pA Ga^+ primary beam, $10 \times 10\,\mu m$ scan area. It can be seen from the gradient that there is likely a significant oxygen enhancement effect for cobalt secondary ions even at the lowest pressure (with no oxygen flow).

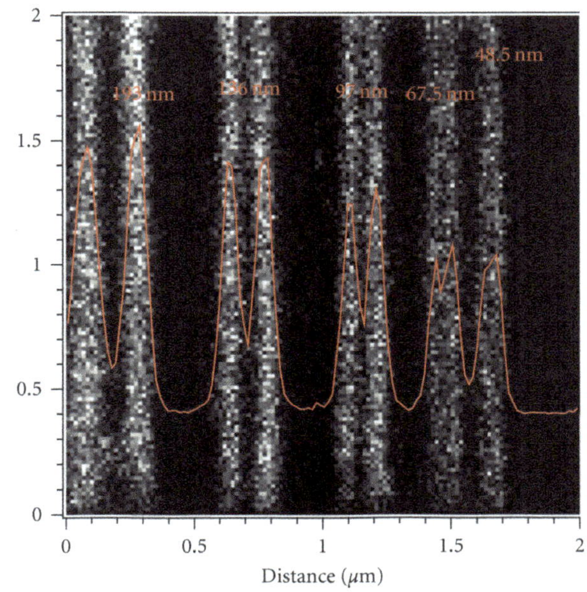

Figure 9: Aluminium secondary ion image from the BAM L200 aluminium/indium/gallium arsenide standard. Field of view $2 \times 2\,\mu m$, displayed pixel size 15.6 nm, 30 keV, <1 pA primary gallium ion beam. Superimposed in red are the integrated profile, and labels denoting the period of each line pair (see bam_l200repe.pdf from http://www.rm-certificates.bam.de/ for more information). The 67.5 nm period line pair (i.e., lines of width 34 nm, separated by 34 nm) is barely split, but the 97 nm period pair (49 nm wide lines) is well resolved.

All of these effects may be minimised (but not avoided completely) by venting with an inert gas or better by using a load-lock chamber. Normalisation to the gallium ion signal can sometimes be used to correct for oxidation effects on SIMS yields [27]. In general, however, for good reproducibility it is necessary to record (and to control) the pressure in the chamber when performing SIMS measurements at high vacuum. For samples which are sensitive to the residual gases in the chamber (e.g., silicon), we suggest deliberately adding a known partial pressure of for example, oxygen in order to improve stability. Oxygen flooding can however have side-effects such as distortion of depth profiles [24].

4. Applications

To illustrate the performance of our first FIB-SIMS prototype, we present here some example applications. When an image is captioned as a depth profile, it means that the data has been integrated from a region of each x-y plane (frame) and is plotted as a function of frame number (equivalent to depth or z-direction). When an image is captioned as an "image", it means that the x-y (frame) data is shown, and that the data has been integrated along the z/depth direction. See also Figure 3. For all of the data shown here, the primary ion beam was incident normal to the mean sample surface. Depth profiles use only the centre part of each frame in order to minimise crater edge effects (this conservative approach throws away 75% of the signal).

4.1. Spatial Resolution Standard BAM L200. Figure 9 shows an aluminium secondary ion image from an unmounted, unpolished chip of the BAM L200 resolution test reference material. BAM define the spatial resolution to be that of the smallest period grating that can be distinguished, in this case P7 with period 67.5 nm (the widths and separation of the aluminium rich stripes in P7 are 34 nm). The ion beam FWHM is calculated to be about 50 nm (see below). Noteworthy is that the chemical image was achieved from a rather rough surface—the existence of surface topography (scratches from sawing the sample) did not distort the chemical image. It was also possible to detect (but not resolve) the single aluminium rich stripes W8 (38 nm), W5 (19. nm) W10 (14.2 nm) and W9 (3.6 nm). In a round-robin study of SIMS resolution [10] only ten out of sixteen laboratories could detect a 5 nm wide aluminium rich stripe in the similar BAM L002 standard, and the smallest calculated beam FWHM was 69 nm.

Figure 10 shows the corresponding indium ion image. The signal here is nosier both because indium is less sensitively detected than aluminium from this material and because the indium-rich stripes are very thin (<5 nm, much smaller than a single pixel in this image).

The lateral resolution of an image can be defined in various ways, as discussed by [28] and will, in general, depend also upon the signal to noise ratio of the image. The ISO definition for surface chemical analysis makes reference to the distance between the 12% and 88% intensity points in a line scan across a well-defined step function [29].

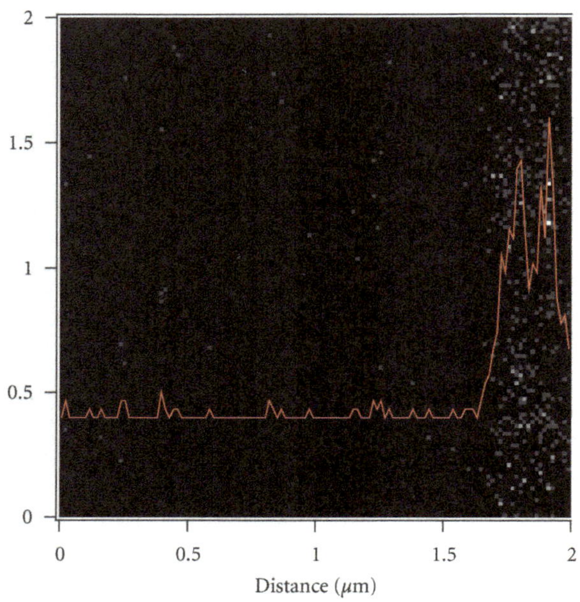

FIGURE 10: Indium positive secondary ion image from BAM L200 (same data set as Figure 9). The unresolved indium features actually consist of two line pairs (each with 5 nm wide indium rich lines)—the distance between the line pairs is not documented.

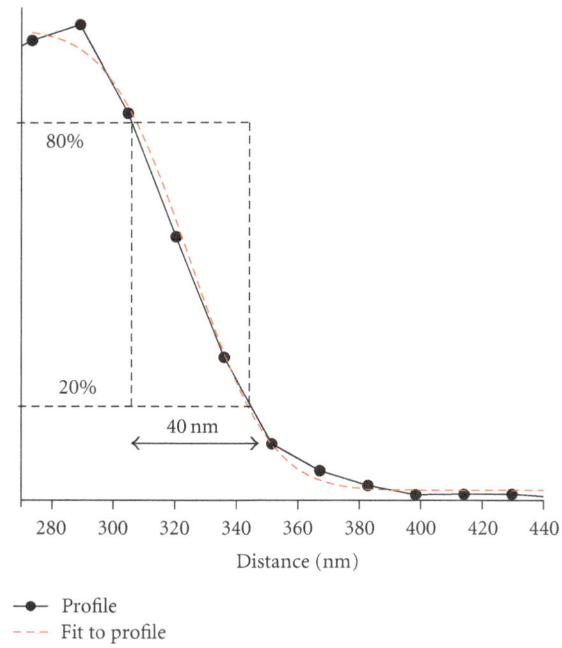

FIGURE 11: Enlargement of part of the profile from Figure 9 showing a superimposed fit of a complementary error function and the 20% and 80% points. The inferred spot size for the FIB beam (and so the spatial resolution of the SIMS signal) is between 40 and 50 nm for this dataset.

For a Gaussian beam, a good approximation to gallium LMIS beam spot distributions [10], this distance corresponds to the full width at half-maximum (FWHM). Note that some FIB manufacturers prefer to use the 20% and 80% intensity points. Figure 11 shows the result of fitting the complementary error function (the convolution of a Gaussian with a step function) to part of the aluminium profile shown in Figure 9. The aluminium rich stripe in P4 has a width of 97 nm; just large enough compared to the spot size to regard the profile as a measured edge spread function. A reasonable fit is obtained suggesting that the beam profile is well approximated by a Gaussian distribution. Using either the ISO definition or a deconvolution of the data gives a FWHM for the primary beam of about 50 nm (the 20%–80% definition gives a spot size of about 40 nm). With experience, we expect further increases in performance, as according to the manufacturer it should be possible to obtain this beam size at much higher currents than we used (the specification is for 50 pA into a 20 nm spot, or 500 pA into a 50 nm spot).

The above analysis assumes that the resolution is determined by the primary beam size and not by any other physical processes. We can neglect lateral straggling of primary ions as this is of the order of a few nanometers [7], possibly up to 20 nm [9]. Although differential sputtering and the resulting increase in rugosity of the surface might contribute to a larger effective spot size, because secondary ion yields depend on surface roughness and the surface-primary beam angle, we do not see any change of lateral resolution with depth.

4.2. Semiconductor Laser. The performance achieved with the BAM L200 standard is particularly relevant to applications involving semiconductor or photonic devices fabricated in gallium arsenide. Figures 12 to 14 show depth profiles and chemical images of a Vertical Cavity Surface Emitting Laser constructed from an AlGaAs/GaAs Distributed Bragg Reflector (DBR) fused to an AlGaInAs/InP gain wafer. The device has been previously described in more detail in [30] and was the first example of a wafer-fused semiconductor disk laser operating at the 1.3 μm wavelength. The specimen described here is denoted VECSEL 3/1. Figure 12 shows a positive secondary ion depth profile through the device. On the right hand side is part of the AlGaAs/GaAs DBR structure (35 repeats of a 100 nm layer of GaAs and a 100 nm layer of $Al_{0.9}Ga_{0.1}As$). On the left-hand side (left of frame 70 on the x-axis) are the structures in the AlGaInAs/InP wafer. The expected structure is shown schematically in Figure 13. Although much of the compositional structure of the device is visible, the structure within the five quantum well regions was not resolved. (Each of these 44 nm thick units consists of three 10 nm thick $Al_{0.28}Ga_{0.26}In_{0.46}As$ layers separated by two 7 nm thick layers of $Al_{0.14}Ga_{0.18}In_{0.68}As$; each unit is then spaced apart by InP layers). The depth resolution in this profile is apparently of the order of 20 nm. As reported previously for a SIMS depth profile of a similar material [31] using different experimental conditions, we see a decrease in the modulation depth of the repeated layers with depth (i.e., the peak to valley difference of the aluminium signal) which was shown to be due to roughening of the surface (ripple formation [31]). The effect is less severe for the conditions reported here, consistent with the lower extent of

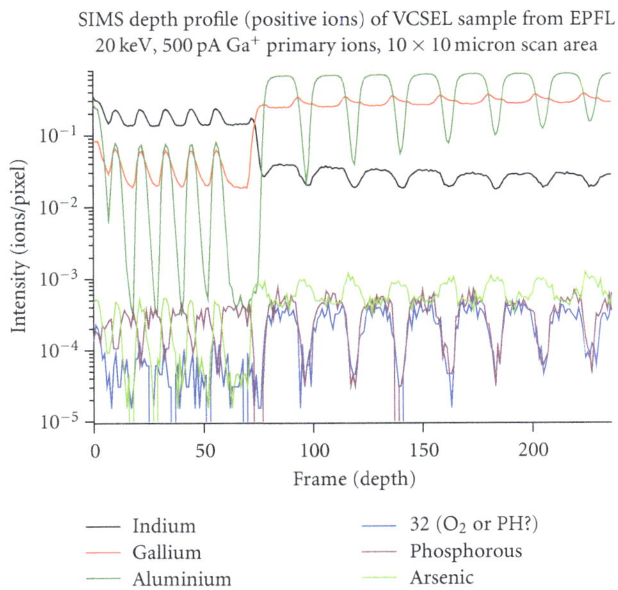

FIGURE 12: Depth profile of VECSEL 3/1 (see text). The well resolved aluminium features on the right hand side of the figure are each 100 nm in width and 100 nm apart. Similar results were achieved using a range of primary ion energies and currents. The modulation depth of the gallium signal is restricted because of the use of a gallium ion primary beam, which necessarily results in a strong gallium secondary ion background signal (due to secondary ions from implanted gallium).

roughening as measured by scanning probe microscopy (see section below).

Figure 13 shows the same structure, but in a different experiment viewed from the side. The 100 nm thick aluminium stripes in the DBR are clearly separated from each other, but do not have flat tops, and the modulation depth is only about 80%, so the resolution in this image is a little better than 100 nm.

That the FIB-SIMS cannot only visualise expected chemical structures, but also reveal new information, is shown in Figure 14. This overlay of depth profiles obtained from both positive and negative secondary ions shows an unexpected small peak in the oxygen ion signal at frame 60—this is very close to the wafer interface and suggests that a surface oxide layer may have been incorporated during the wafer fusion process.

4.3. Atomic Layer Deposition Fabricated Multilayers. Atomic layer deposition (ALD) is a coating technique using self-limiting surface reactions of alternately supplied volatile precursor molecules to make highly conformal films with excellent control of thickness [32], almost to single atomic layers. Conformal ALD coatings have been used for example to coat nanopillars for novel applications in photovoltaic materials and nanomechanics, as well as to make transparent conductive oxide films. Zinc oxide has potential as a semiconductor material for optoelectronics, and aluminium-doped zinc oxide can be used as a thin transparent conducting oxide (TCO) film. The FIB-SIMS was used to investigate ALD multilayers of zinc oxide and aluminium oxide that were prepared as part of a study into their photoelectric properties, and in particular to study some fundamental characteristics of the ALD produced materials.

The films were produced at EMPA in a home-made reactor using dimethyl zinc (95% purity from STREM Chemicals, Inc.), trimethyl aluminium (97% purity from Sigma-Aldrich Co.) and water vapour as reagents, and argon as a carrier gas. The substrate was (111) silicon carefully cleaned to remove organic contaminants from the surface. The deposition was carried out at 65°C with the following cycle times: 0.1 s pulse, 20 s exposure, and 45 s purge. A 20 nm thickness of Al_2O_3 required 154 cycles and a 20 nm thickness of ZnO required 100 cycles. Figure 15 shows a FIB-SIMS depth profile through five pairs of nominally 20 nm zinc oxide and 20 nm aluminium oxide layers on a silicon substrate. All of the layers are clearly seen in the depth profile and the aluminium signal shows a "top-hat" profile suggesting that the depth resolution in aluminium oxide is much better than 20 nm. However, the modulation depth of the aluminium signal is poor and the zinc signal does not have a "top-hat" profile. Figure 16 shows that the zinc oxide layers are actually thinner than the aluminium oxide layers, furthermore we expect the zinc oxide layers to be sputtered faster. As a result, the zinc oxide layers although clearly visible are not thick enough to give a "top-hat" profile. It is not clear why the zinc signal from the uppermost layer (and to a lesser extent the aluminium signal too) is so much stronger than for subsequent layers. We suspect that this is a result of adsorbed atmospheric constituents (molecular oxygen, water) in the ALD material. The signature is probably not simply due to a build up to the equilibrium concentration of implanted gallium, because it does not repeat at each subsequent interface. (For 20 keV primary ion energy, we do observe repeated structure in the SIMS depth profile within each aluminium oxide layer; this structure is believed to be a measurement artefact. SIMS measurements at interfaces can be complicated because sputtering can cause chemical changes (e.g., reduction) in the sample material at the same time that the implanted ion dose and damage approach a steady state [22]). The high signal for the mass to charge ratio 28 Th between the last aluminium oxide layer and the silicon substrate is due to chemistry at the substrate interface resulting in (i) either a stronger AlH^+ ion signal (which interferes strongly with $^{28}Si^+$, and in practice dominates in this sample) or (ii) a stronger $^{28}Si^+$ signal due, for example, to the presence of a thin layer of silicon oxide. Isotope ratios suggest that the latter explanation (enhancement of silicon secondary ion signal at the interface) is more likely. An unexpected observation was that of a weak boron signal (see Figure 17), correlating with the zinc oxide layers. The presence of boron in ALD fabricated zinc oxide layers was subsequently confirmed by glow discharge optical emission spectroscopy (GD-OES), and was found to be due to an impurity in the zinc precursor.

Limiting Depth Resolution of SIMS. Secondary ions come from very near the surface (<0.5 nm) and the depth

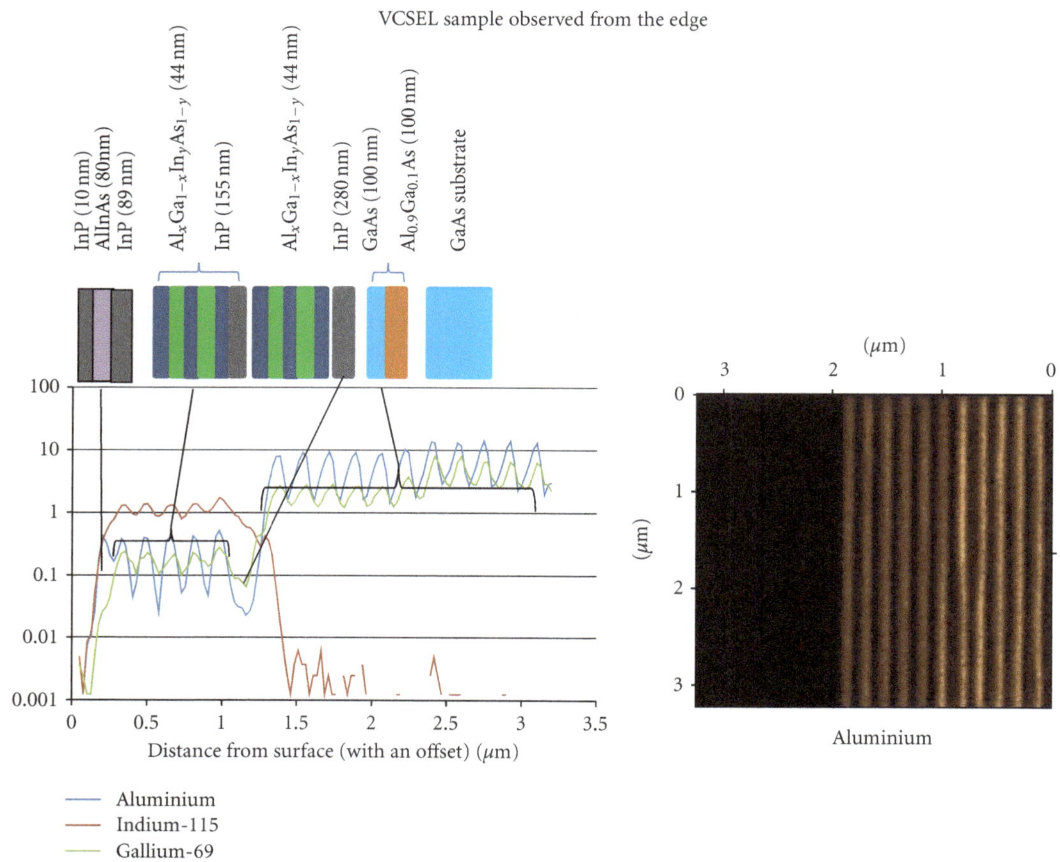

Figure 13: Side-on view of VECSEL 3/1 (see text). As for the depth profile, all features can be discerned, expect for the structure within each quantum well gain region (each 44 nm thick containing three 10 nm bands richer in aluminium). The 100 nm thick aluminium rich regions in the distributed Bragg reflector (right-hand side of figure and image) are well separated but do not have a flat-top. The lateral resolution is thus of the order of 100 nm in this example.

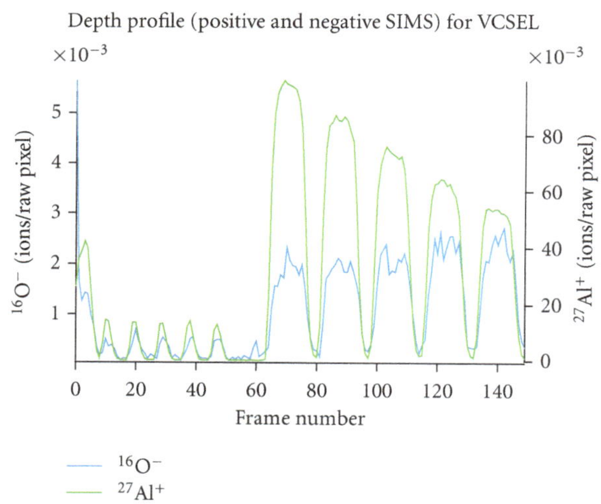

Figure 14: Overlay of depth profiles of aluminium and oxygen for VECSEL 3/1. Note the oxygen peak at frame 60, close to the interface between the two wafers. The smooth drop in the aluminium signals (not evident in Figure 12) was later discovered to be due to a temporary contamination of the ion optics.

resolution of a secondary ion depth profile, given a sufficient signal to noise ratio, is in principle only limited by ion-induced mixing (usually of the order of a few nanometers but can be more depending on primary ion energy and angle and sample composition) and ion-induced roughening, in common with other techniques which use an ion beam to sputter away material in order to obtain a depth profile (e.g., XPS, Auger). Changes in the composition of the sample usually affect secondary ion yields, and differential sputtering (in the chemical sense rather than the topographical sense) will have an effect on depth profiles at interfaces [33]. The processes that limit SIMS depth resolution are discussed in detail in [34]. It is worth noting that we have no evidence for any background due to sputtered material being resputtered and ionised from the SIMS ion optics (i.e., no memory effect is detectable). Redeposition of material from the crater walls onto the crater floor can affect depth profiles [34, 35], the extent to which this matters will depend upon the chemical structure of the sample, the size of the crater and the size and shape of the primary ion beam. With appropriate mathematical models, monolayer thicknesses can be reconstructed from SIMS depth profiles [36]. Note

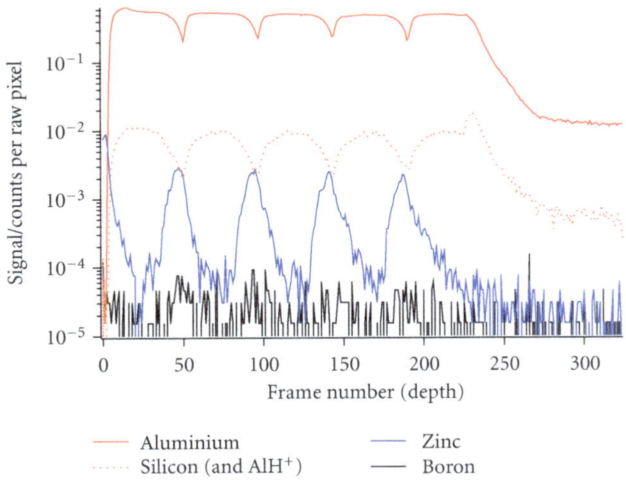

FIGURE 15: FIB-SIMS depth profile of ALD multilayer (five repeats of nominally 20 nm zinc oxide/20 nm aluminium oxide on a silicon substrate). 240 pA, 5 keV normally incident Ga+ bombardment. For clarity, each trace is on a differently scaled linear vertical axis. The surface of the sample is to the left of the figure, the substrate to the right.

that the mixing-roughness-information depth (MRI) model proposed by Hofmann [36] ideally requires knowledge of the original surface topography, any interface topography, and of the sputtering induced roughness. A scanning probe microscope can provide at least two of these.

4.4. SPM-SIMS Integration. Because dynamic secondary ion measurements imply significant sputtering, in general ions originating from different depths will be measured at different times and thus the recorded dataset is in some sense three dimensional [37, 38]. However, for heterogeneous samples there is no simple relationship between sputtering time or ion dose and the depth. This is because different materials and different crystal orientations will in general have different sputtering rates [39]. Furthermore, if the sample does not initially have a flat polished surface, then there will always be ambiguity about the spatial relationship between observed features as sputtering proceeds. Even for polished samples where the matrix is homogenous, surface roughening and swelling [40] due to the ion beam can distort the relationship between sputtering time or dose and the depth being sampled. Artefacts in the SIMS image may also arise due to imperfectly normal crater walls, because of edge effects when topography develops on the surface being measured and due to redeposition of sputtered atoms [35], and because local topography affects useful secondary ion yields [41]. For all of these reasons it is highly desirable to have available during sputtering a measurement of the surface profile relative to the original surface before sputtering commenced. The best way of doing this is to use a scanning probe microscope to measure the surface (other options such as stereo-pair electron beam imaging do not have sufficient depth resolution or accuracy; laser interferometry has been used to obtain sputtering rates [42] but gave no information on roughening or differential sputtering within a crater and can be difficult to interpret [43]). The use of an *ex situ* SPM to improve SIMS 3D images has been previously described for example, by [31, 39, 44]. *In situ* measurements are highly desirable to increase the speed of measurements and to avoid concerns about contamination or reactions at the surface. This is however not commonplace because (i) few scanning probe microscopes are intended to work under vacuum, (ii) fewer still can easily be integrated with an electron microscope and focused ion beam (limited space, possible interactions), (iii) scanning probe microscopes are expensive, and (iv) combining SPM measurements with SIMS measurements can enormously increase the required measurement time. However, If an SPM is integrated with a SIMS measurement, then it opens up the possibility of not only sensing topography but also other properties of the surface (e.g., contact potential, magnetisation, and friction) complementary to the composition measured by SIMS. Because we had prior experience with SPM/SEM integration, it was natural to do some "proof of concept" experiments to demonstrate SPM/FIB/SEM-SIMS integration within a single vacuum chamber. We are unaware of any prior publications that show *in situ* SPM measurements of SIMS craters.

The *in situ* scanning probe microscope was developed together with SPECS Zurich GmbH. It utilised SmarAct-GmbH positioners for coarse control of the tip and closed-loop Nanodrive piezo stages from Mad City Labs Inc. for z tip control and x-y fine positioning (x-y scanning was done by moving the sample rather than the probe tip). Frequency modulation atomic force microscopy (AFM) was performed with an Akiyama probe from Nanoworld AG [45]. A Nanonis SPM control system (base package and oscillation controller) was used for the probe amplitude and phase feedback loops, and to monitor the frequency shift of the probe. The geometry and self-sensing properties of the Akiyama probe are particularly well suited to a compact integration within the vacuum chamber close to the intersection of the charged particle beams with the sample. When not in use, the SPM probe tip was retracted from the measurement region using the coarse positioner, allowing free access of the charged particle beams to the sample, and unhindered collection of secondary ions and electrons. The SPM was mounted directly onto the sample stage of the FIB/SEM as shown in Figure 18. Although the Akiyama probe signal can be transduced in different ways, we illustrate here only the calculated Z displacement of the tip due to the surface being measured. The free software WSXM [46] was used to visualise and analyse the SPM datasets. An SPM scan in both directions over a 5 × 5 micron area typically took about 20 minutes (this obviously depends on the roughness, whether the crater edge is part of the scan, and the desired resolutions). The time to automatically withdraw the SPM tip so that FIB scanning can continue is about 15 seconds; more critical to combining the use of FIB-SIMS and SPM is the successful automatic coarse repositioning of the SPM tip.

The semiconductor laser sample described above was used as a test specimen. We were particularly interested to

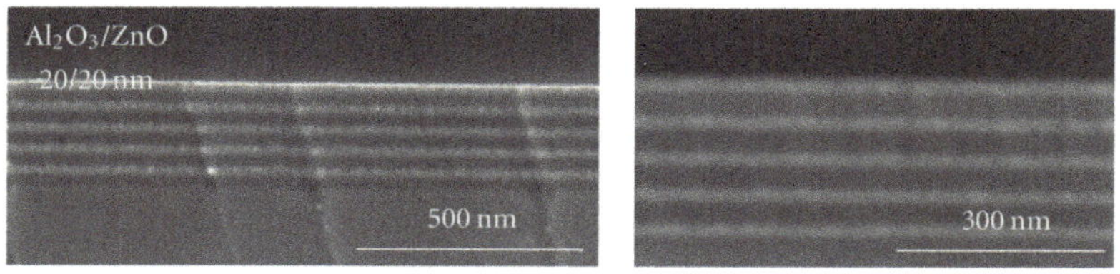

Figure 16: SEM image of the nominal 20 nm/20 nm zinc oxide/aluminium oxide multilayer sample for which SIMS results are shown in Figure 15. The bright stripes are the zinc oxide layers, the dark stripes are the aluminium oxide layers, and the substrate is silicon. It can be seen that the zinc oxide layers are thinner than the aluminium oxide layers.

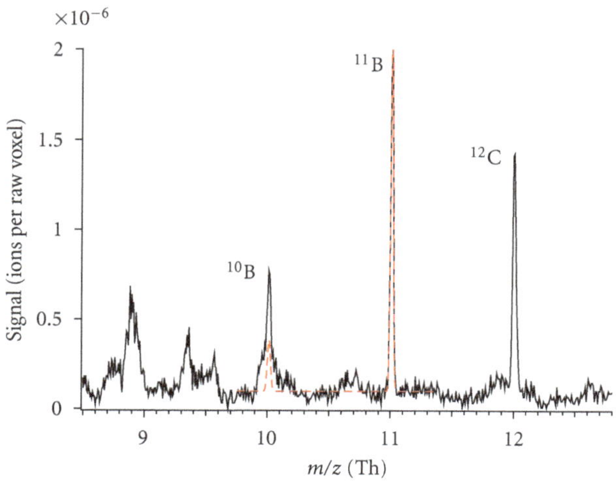

Figure 17: Part of the secondary ion mass spectrum from the multilayer zinc oxide/aluminium oxide ALD dataset shown in Figure 15. The red trace is a fit to boron with natural isotopic abundance, scaled and offset to the $^{11}B^+$ peak. The roughness in the baseline, and the apparent broad peaks below 10 Th are due to electrical interference from the extraction pulse (boron rich samples show the expected isotope ratio more clearly). There are no possible isobaric interferences for boron, so the identification is unambiguous. The fitted ^{10}B signal corresponds to about 350 detected ions in total.

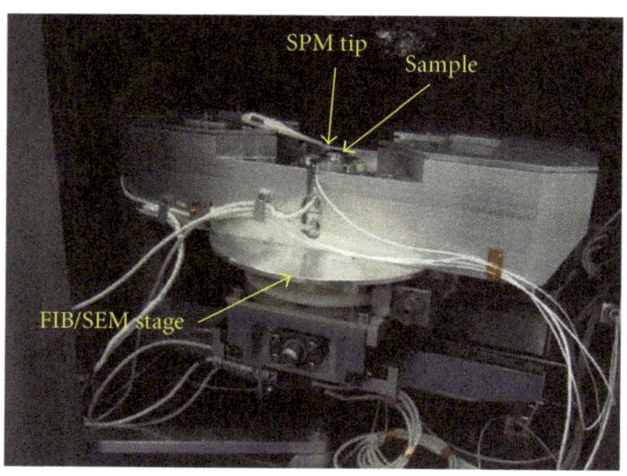

Figure 18: SPM tip and assembly in position on top of the FIB/SEM stage, with the stage in its "home" position. The SPM assembly is simply bolted to the stage.

find out if surface roughening could have been affecting the depth resolution of the SIMS measurements. AFM scans were performed before, during, and after the SIMS measurement. ("during" meaning that the FIB beam was briefly turned off, and the AFM tip moved into position, before reverting to SIMS mode and continuing to sputter).

Figures 19 and 20 illustrate that the crater floor was fairly smooth (RMS roughness 4.9 nm at a depth of 750 nm after 200 frames), so ion-induced roughening cannot have been significantly limiting the extent to which the aluminium-rich stripes in the DBR were resolved although this degree of roughening will certainly have been detrimental to the resolution of the 7 and 10 nm deep features in the laser gain medium (quantum well structures). The crater walls, as imaged by the AFM, are not perfectly perpendicular to the sample surface, but there is little evidence for significant redeposition of material around the crater. It seems likely that the main reason for the unexpectedly poor modulation and resolution of the 100 nm thick aluminium structures is a combination of sputtering of redeposited material from within the crater and primary beam halo effects.

5. Conclusions

We have shown that it is possible to adapt a commercial FIB/SEM instrument for SIMS operation by adding a time-of-flight mass analyser without any detrimental effect upon normal operation. A depth resolution of <20 nm and a lateral resolution of <100 nm (see Figure 9) have been demonstrated (both for normally incident ions), and we expect to improve upon both of these figures in the near future. We have also demonstrated that the additional integration of a scanning probe microscope is possible, allowing true depth and roughness information to be obtained in series with SIMS measurements.

Acknowledgments

The authors gratefully acknowledge the financial support of the European Union for part of this work (FP7 Project FIBLYS, Grant Agreement no. CP-TP 214042-2) and funding from the Swiss Commission for Technology and Innovation with SPECS Zürich supporting the SEM/SPM development, (CTI-Project no. 9208.1 PFNM-NM). The authors would also like to thank Gerhard Bürki of EMPA for technical

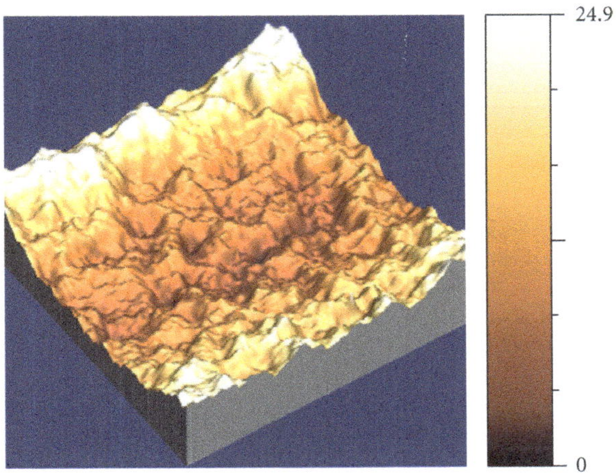

FIGURE 19: $5 \times 5\,\mu$m SPM scan from within an FIB/SIMS crater after 200 10×10 micron frames at 104 pA and 20 kV Ga^+. The RMS roughness is 4.9 nm. The dataset has been corrected for global tilt. For a higher primary beam current of 401 pA (similar to that used for the depth profiles and images shown above) RMS roughnesses of between 1 and 6 nm were measured.

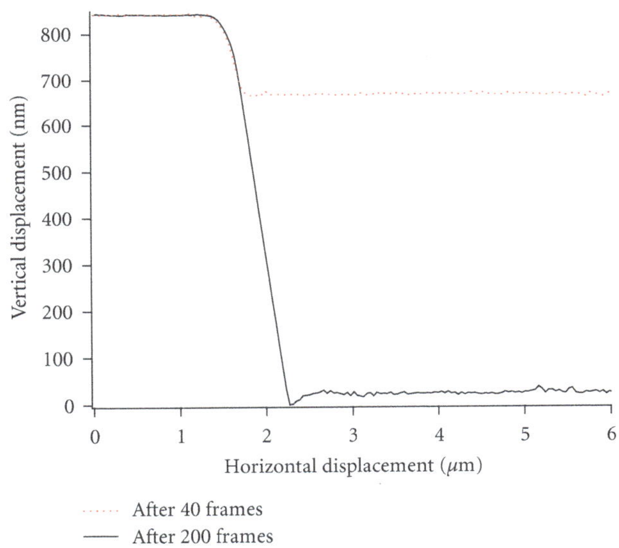

FIGURE 20: SPM scans of the crater edge in semiconductor laser after 40 and 200 frames using 100 pA 20 keV Ga^+. The crater wall is not perfectly normal to the crater floor, but the slope remains constant as the crater becomes deeper.

assistance with the FIB/SEM and Dr V Iakovlev and Dr. A. Sirbu of EPFL for providing the VCSEL sample. For "Advances in Materials Science and Engineering" Dr. A. G. Balogh and an anonymous reviewer helped to improve the quality of this paper.

References

[1] C. A. Andersen and J. R. Hinthorne, "Ion microprobe mass analyzer," *Science*, vol. 175, no. 4024, pp. 853–860, 1972.

[2] R. Castain and G. Slodzian, "Optique corpusculaire—premiers essais de microanalyse par emission ionique secondaire," *Comptes Rendus Hebdomadaires des Séances de l'Academie des Sciences*, vol. 255, pp. 1893–1895, 1962.

[3] A. R. Bayly, A. R. Waugh, and K. Anderson, "SIMS microanalysis with a gallium ion microprobe," *Nuclear Instruments and Methods In Physics Research*, vol. 218, no. 1-3, pp. 375–382, 1983.

[4] J. M. Chabala, R. Levi-Setti, and Y. L. Wang, "Practical resolution limits of imaging microanalysis with a scanning ion microprobe," *Applied Surface Science*, vol. 32, no. 1-2, pp. 10–32, 1988.

[5] D. N. Dunn and R. Hull, "Reconstruction of three-dimensional chemistry and geometry using focused ion beam microscopy," *Applied Physics Letters*, vol. 75, no. 21, pp. 3414–3416, 1999.

[6] L. B. Li, D. S. McPhail, N. Yakovlev, and H. Seng, "Strategies for improving the sensitivity of FIB-SIMS," *Surface and Interface Analysis*, vol. 43, pp. 495–497, 2011.

[7] G. Betz and F. Rüdenauer, "Collision cascade limit to spatial resolution in focused ion beam processes," *Applied Surface Science*, vol. 51, no. 1-2, pp. 103–112, 1991.

[8] D. S. McPhail, L. B. Li, R. J. Chater, N. Yakovlev, and H. Seng, "From FIB-SIMS to SIMS-FIB. The prospects for a 10 nm lateral resolution SIMS instrument with full FIB functionality," *Surface and Interface Analysis*, vol. 43, pp. 479–483, 2011.

[9] M. Nojima, M. Toi, A. Maekawa et al., "Evaluation of the nano-beam SIMS apparatus," *Applied Surface Science*, vol. 231-232, pp. 930–935, 2004.

[10] M. Senoner and W. E. S. Unger, "Lateral resolution of secondary ion mass spectrometry-results of an inter-laboratory comparison," *Surface and Interface Analysis*, vol. 39, no. 1, pp. 16–25, 2007.

[11] F. A. Stevie, "Focused ion beam secondary ion mass spectrometry (FIB-SIMS)," in *Introduction to Focused Ion Beams*, L. A. Gianuzzi and F. A. Stevie, Eds., chapter 13, Springer, New York, NY, USA, 2005.

[12] L. A. Giannuzzi and M. Utlaut, "A review of Ga^+ FIB/SIMS," *Surface and Interface Analysis*, vol. 43, pp. 475–478, 2010.

[13] R. G. Cooks and A. L. Rockwood, "The Thomson–a suggested unit for mass spectroscopists," *Rapid Communications in Mass Spectrometry*, vol. 5, p. 93, 1991.

[14] M. Guilhaus, "Principles and instrumentation in time-of-flight mass spectrometry: physical and instrumental concepts," *Journal of Mass Spectrometry*, vol. 30, no. 11, pp. 1519–1532, 1995.

[15] A. Tempez, J. A. Schultz, S. Della-Negra et al., "Orthogonal time-of-flight secondary ion mass spectrometric analysis of peptides using large gold clusters as primary ions," *Rapid Communications in Mass Spectrometry*, vol. 18, no. 4, pp. 371–376, 2004.

[16] I. S. Gilmore and M. P. Seah, "Ion detection efficiency in SIMS: dependencies on energy, mass and composition for microchannel plates used in mass spectrometry," *International Journal of Mass Spectrometry*, vol. 202, no. 1-3, pp. 217–229, 2000.

[17] HDF5 Group, http://www.hdfgroup.org.

[18] H. N. Migeon, F. Saldi, Y. Gao, and M. Schuhmacher, "Ion microscope and ion microprobe analysis under oxygen, cesium and gallium bombardment," *International Journal of Mass Spectrometry and Ion Processes*, vol. 143, no. C, pp. 51–63, 1995.

[19] G. Frache, B. El Adib, J.-N. Audinot, and H.-N. Migeon, "Evaluation of ionization yields under gallium bombardment," *Surface and Interface Analysis*, vol. 43, pp. 639–642, 2010.

[20] L. A. Machlan, J. W. Gramlich, L. J. Powell, and G. M. Lambert, "Absolute isotopic abundance ratio and atomic weight of a reference sample of gallium," *Journal of Research of the National Bureau of Standards*, vol. 91, no. 6, pp. 323–331, 1986.

[21] J. Maul and K. Wittmaack, "Secondary ion emission from silicon and silicon oxide," *Surface Science*, vol. 47, no. 1, pp. 358–369, 1975.

[22] A. Licciardello, L. Renna, and S. Pignataro, "Interfacial phenomena during depth profiling with gallium ions: a ToF-SIMS approach," *Surface and Interface Analysis*, vol. 30, no. 1, pp. 243–246, 2000.

[23] A. E. Morgan, H. A. M. de Grefte, N. Warmoltz, H. W. Werner, and H. J. Tolle, "The influence of bombardment conditions upon the sputtering and secondary ion yields of silicon," *Applications of Surface Science*, vol. 7, no. 4, pp. 372–392, 1981.

[24] P. C. Zalm and C. J. Vriezema, "The influence of bombardment conditions upon the sputtering and secondary ion yields of silicon," *Nuclear Instruments and Methods in Physics Research*, vol. 64, pp. 626–631, 1992.

[25] Y. Kudriavtsev, A. Villegas, A. Godines, and R. Asomoza, "SIMS analysis of residual gas elements with a Cameca IMS-6f ion microprobe," *Applied Surface Science*, vol. 252, no. 10, pp. 3406–3412, 2006.

[26] T. Sakamoto, B. Tomiyasu, M. Owari, and Y. Nihei, "Ambient oxygen effect in Ga^+ FIB-SIMS," *Surface and Interface Analysis*, vol. 22, no. 1, pp. 106–110, 1994.

[27] T. Sakamoto, M. Owari, and Y. Nihei, "Behavior of gallium secondary ion intensity in gallium focused ion beam secondary ion mass spectrometry," *Japanese Journal of Applied Physics Part 1*, vol. 36, no. 3, pp. 1287–1291, 1997.

[28] M. Senoner, T. Wirth, and W. E. S. Unger, "Imaging surface analysis: lateral resolution and its relation to contrast and noise," *Journal of Analytical Atomic Spectrometry*, vol. 25, no. 9, pp. 1440–1452, 2010.

[29] ISO 181115: 2001, Surface Chemical Analysis–Vocabulary, ISO, Geneva, Switzerland.

[30] J. Lyytikäinen, J. Rautiainen, L. Toikkanen et al., "1.3-μm optically-pumped semiconductor disk laser by wafer fusion," *Optics Express*, vol. 17, no. 11, pp. 9047–9052, 2009.

[31] G. Friedbacher, D. Schwarzbach, P. K. Hansma, H. Nickel, M. Grasserbauer, and G. Stingeder, "A combination of atomic force microscopy and secondary ion mass spectrometry for investigation of AlxGa1-xAs/GaAs superlattices," *Fresenius' Journal of Analytical Chemistry*, vol. 345, no. 8-9, pp. 615–617, 1993.

[32] S. M. George, A. W. Ott, and J. W. Klaus, "Surface chemistry for atomic layer growth," *Journal of Physical Chemistry*, vol. 100, no. 31, pp. 13121–13131, 1996.

[33] H. H. Andersen, "The depth resolution of sputter profiling," *Applied Physics*, vol. 18, no. 2, pp. 131–140, 1979.

[34] C. W. Magee and R. E. Honig, "Depth profiling by SIMS–depth resolution, dynamic range and sensitivty," *Surface and Interface Analysis*, vol. 4, no. 2, pp. 35–41, 1982.

[35] F. G. Rüdenauer and W. Steiger, "Sputter redeposition as a limit to spatially three-dimensional SIMS microanalysis," *Ultramicroscopy*, vol. 24, no. 2-3, pp. 115–123, 1988.

[36] S. Hofmann, "Profile reconstruction in sputter depth profiling," *Thin Solid Films*, vol. 398-399, pp. 336–342, 2001.

[37] A. J. Patkin and G. H. Morrison, "Secondary ion mass spectrometric image depth profiling for three-dimensional elemental analysis," *Analytical Chemistry*, vol. 54, no. 1, pp. 2–5, 1982.

[38] R. G. Rudenauer, "Spatially multidimensional secondary ion mass spectrometry analysis," *Analytica Chimica Acta*, vol. 297, no. 1-2, pp. 197–230, 1994.

[39] M. L. Wagter, A. H. Clarke, K. F. Taylor, P. A. W. Van Der Heide, and N. S. McIntyre, "Topographic correction of 3D SIMS images," *Surface and Interface Analysis*, vol. 25, no. 10, pp. 788–789, 1997.

[40] H. Gnaser, A. Brodyanski, and B. Reuscher, "Focused ion beam implantation of Ga in Si and Ge: fluence-dependent retention and surface morphology," *Surface and Interface Analysis*, vol. 40, no. 11, pp. 1415–1422, 2008.

[41] R. Levi-Setti, J. M. Chabala, and Y. L. Wang, "Aspects of high resolution imaging with a scanning ion microprobe," *Ultramicroscopy*, vol. 24, no. 2-3, pp. 97–113, 1988.

[42] E. De Chambost, P. Monsallut, B. Rasser, and M. Schuhmacher, "Depth scale calibration of SIMS depth profiles by means of an online crater depth measurement technique," *Applied Surface Science*, vol. 203-204, pp. 391–395, 2003.

[43] P. A. Ronsheim, R. Loesing, and A. Madan, "Sputtered depth scales of multi-layered samples with in situ laser interferometry: arsenic diffusion in Si/SiGe layers," *Applied Surface Science*, vol. 231-232, pp. 762–767, 2004.

[44] A. Wucher, J. Cheng, L. Zheng, D. Willingham, and N. Winograd, "Three-dimensional molecular imaging using mass spectrometry and atomic force microscopy," *Applied Surface Science*, vol. 255, no. 4, pp. 984–986, 2008.

[45] T. Akiyama, U. Staufer, and N. F. De Rooij, "Self-sensing and self-actuating probe based on quartz tuning fork combined with microfabricated cantilever for dynamic mode atomic force microscopy," *Applied Surface Science*, vol. 210, no. 1-2, pp. 18–21, 2003.

[46] I. Horcas, R. Fernández, J. M. Gómez-Rodríguez, J. Colchero, J. Gómez-Herrero, and A. M. Baro, "WSXM: a software for scanning probe microscopy and a tool for nanotechnology," *Review of Scientific Instruments*, vol. 78, no. 1, Article ID 013705, 2007.

Modeling Approach for Determining Equivalent Optical Constants of Plastic Shading Nets under Solar Radiation Conditions

A. M. Abdel-Ghany[1,2] and I. M. Al-Helal[1]

[1] *Department of Agricultural Engineering, College of Food and Agriculture Sciences, King Saud University, P.O. Box 2460, Riyadh 11451, Saudi Arabia*
[2] *Mechanical Power Engineering Department, Faculty of Energy Engineering, South Valley University, Aswan 81528, Egypt*

Correspondence should be addressed to A. M. Abdel-Ghany, aghany@ksu.edu.sa

Academic Editor: Fengqiang Sun

The radiative properties of several plastic shading nets were measured under natural solar radiation conditions. We found that the plastic nets behave as homogeneous translucent materials (e.g., plastic film, plastic sheets, and glass). Based on this behavior, we suggest that it is possible to treat plastic nets as translucent materials and to characterize them with equivalent optical constants (i.e., equivalent refractive indexes, n_{eq}, and equivalent extinction coefficients, σ_{eq}). Here a physical model to determine n_{eq} and σ_{eq} of plastic nets was described in analogy to homogeneous translucent materials. We examined three groups of nets based on their color (black, black-green, and beige). Each group consisted of nets with four or five different porosities. Nets of each group had almost the same texture structure. For each group, we derived an equation for n_{eq} as a function of the net porosity and determined an average value for σ_{eq}. Once values of n_{eq} and σ_{eq} were determined, the solar radiative properties of a net could then be calculated from n_{eq} and σ_{eq} for any incident angle of solar beam radiation without the need of measurements. The present model was validated by comparing the calculated with the measured radiative properties of three nets at different incident angle of solar beam radiation. The calculated radiative properties reasonably agreed with measured values.

1. Introduction

During the last decade, the use of plastic nets for shading agricultural industrial and residential structures in hot and sunny regions has steadily expanded because plastic nets offer many economical and environmental benefits [1–4]. These include (i) reduction or elimination of energy consumption used for heating or cooling structures, (ii) reduction of crop transpiration as well as water consumption for irrigation in agriculture structures in arid regions, and (iii) improvement in environmental and human-health conditions by reducing the need for pesticides. The diffusive nature of the net threads causes the incident beam radiation to be scattered forward (specularly and diffusively) during transmission [5]. This provides a uniform radiation distribution inside structures covered with nets to add more advantages to plastic nets applications. Various types of plastic nets with different colors and different shading powers are popular in the markets and are used for many purposes in many parts of the world. Commercial nets are usually defined by the following (i) the net porosity (ϕ), that is, the surface empty area (uncovered by the plastic threads) divided by the total surface area of the net. The value of ϕ is usually determined by image processing or solar radiation transmission [6]. (ii) The shading factor (SF), defined as the percentage of global solar radiation that the net is able to block. SF strongly depends on the incident angle of the beam radiation (θ), and to estimate the value of SF for a net correctly, the directional solar radiative properties (θ dependence) of the net should be determined. Accordingly, the directional solar radiative properties of nets are essentially needed to characterize different types of plastic nets according to their shading power. However, little information about how to determine these properties is available in the literature.

Some research groups have investigated the radiative properties of different types of plastic net or total net-covered structure models in laboratories under artificial lighting [2–4, 7]. In order to replace expensive measuring techniques in laboratories, we developed a simple method for measuring the radiative properties of nets under natural conditions of global, diffuse, and direct beam solar radiation, and under natural conditions of the photosynthetically active radiation (PAR) [8, 9]. The method was used to test nets with different colors, porosities, and texture structures. Even though, nets are three-dimensional, nonhomogeneous, and porous materials, the behaviors of the plastic nets under solar radiation conditions were found to be similar to those of translucent homogeneous materials (e.g., plastic films, glass and rigid plastic sheets such as polycarbonate and acrylic) [8, 9]. In other words, the changes in the global, diffuse, and direct beam transmittances and reflectances of nets with the incident angle (θ) were similar to those of different translucent materials that were measured under natural conditions of solar radiation [10–14]. This makes it possible to determine optical constants, in analogy to homogeneous transparent materials, to be used for calculating the radiative properties of nets. This is because calculation methods are preferable and easier than the measuring methods which are quite difficult and prohibitively expensive. Accordingly, the goal of the present study was to develop a theoretical model to investigate equivalent optical constants (i.e., equivalent refractive indexes and equivalent extinction coefficients) for plastic nets in analogy to homogeneous translucent materials. Three groups of nets (i.e., a black group, beige group, and black-green group) were selected to develop the model.

2. Materials and Methods

2.1. The Plastic Net Materials. The nets selected for testing were classified into three groups. Nets in each group have the same color, same texture structure, and different porosities (ϕ). The nets were locally produced by Saudi companies. Samples of the nets (5 cm × 5 cm) were scanned at a resolution of 4800 dpi with a flatbed scanner (HP-5590). Photos of the scanned nets were magnified several times and converted into high-contrast black (thread projection area) and white (empty area) colors using Adobe Photoshop software. The resulting photos showed the net threads as black on a white background. The porosity (the empty area divided by the total surface area of the net, ϕ) of each net was estimated. The three groups were the black group having porosities ϕ of 0.30, 0.36, 0.43, 0.54, and 0.64; the beige group having porosities ϕ of 0.32, 0.34, 0.36, and 0.45; the black-green group having porosities ϕ of 0.45, 0.46, 0.48, 0.50, and 0.51. In the latter group, the texture was interlaced threads made of black and green plastic robs knitted together. Photos of the tested nets (one from each group) are shown in Figure 1.

2.2. The Equivalent Optical Constants Model. In the case of homogeneous translucent materials (e.g., plastic films, plastic sheets, or glass), the measured radiative properties (reflectance, transmittance, and absorptance) are referred to as effective properties. This is because the incident beam over the material surface suffers an infinite number of multiple internal reflections inside the material itself (Figure 2). With each internal reflection, there are transmitted and reflected portions below and above the material surfaces. The summation of the reflected components above the material surface is the effective reflectance and of those below the material surface is the effective transmittance; both can be measured using radiation sensors. According to [5], transmission of direct beam radiation through a plastic net was characterized as (i) the unscattered beam that is transmitted through the net pores, (ii) the beam that is reflected forward on the thread surfaces specularly, and (iii) the beam that is diffused during transmission. On the other hand, the reflected beam will be a combination of specularly reflected beam, and backward scattered beam. For several types of nets (having different colors, porosities, and texture structures) tested in [5], the forward diffused beam was less than 20% of the incident beam radiation. For the modeling approach, the unscattered beam, the specularly reflected (forward or backward) beam and the diffused beam were assumed to have the same characteristic manner and were considered as direct beams transmitted or reflected from the net. In analogy to homogeneous materials, the effective transmittance and reflectance of a plastic net to the direct beam solar radiation (τ_r and ρ_r) can be expressed as [15] by

$$\tau_r = \frac{\tau^*(1-\rho^*)}{1-(\tau^*\rho^*)^2}, \qquad \rho_r = \rho^*(1+\tau^*\tau_r), \tag{1}$$

where τ^* is the transmittance of the beam radiation due to absorption only in the net material with a porosity ϕ and a solid matrix of $(1-\phi)$; ρ^* is the interface reflectance between the net surface and air.

By eliminating ρ^* from (1) a second-order algebraic equation with the unknown τ^* can be obtained. By solving the roots of this equation, the value of τ^* is given by [15] as

$$\tau^* = \left\{\frac{\tau_r^2-(1-\rho_r)^2}{2\tau_r}\right\} \pm \left\{\left(\frac{\tau_r^2-(1-\rho_r)^2}{2\tau_r}\right)^2+1\right\}^{1/2}. \tag{2}$$

Reasonable (i.e., positive) values for τ^* are obtained by considering the plus sign of (2). The input parameters to (1) and (2) are the transmittance and reflectance of the net to direct beam radiation (τ_r and ρ_r) that can be measured using a pyranometer underneath the net and an albedometer above the net. The value of ρ^* is obtained from the direct substitution of τ^* in (1). Once values of τ^* and ρ^* were determined one time (e.g., at normal incidence), the equivalent optical constants (n_{eq} and σ_{eq}) of the plastic net can be determined, in analogy to homogeneous translucent materials, as follows.

The interface reflectance ρ^* at normal incidence between the air ($n = 1$) and the net surface having the equivalent refractive index (n_{eq}) is given by

$$\rho^* = \frac{(n_{eq}-1)^2}{(n_{eq}+1)^2}. \tag{3}$$

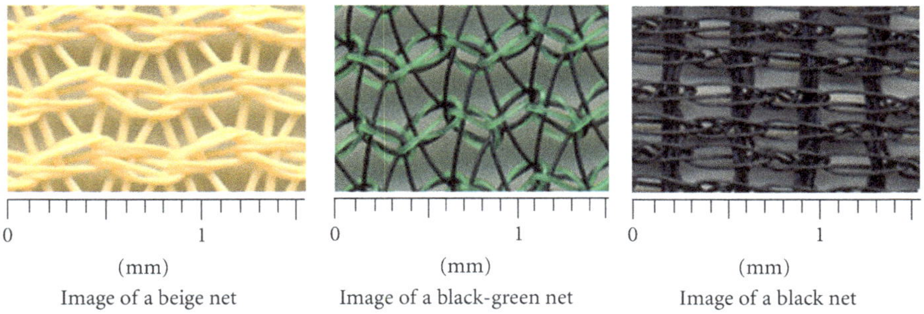

FIGURE 1: Scanner images of a net in each group tested in this study.

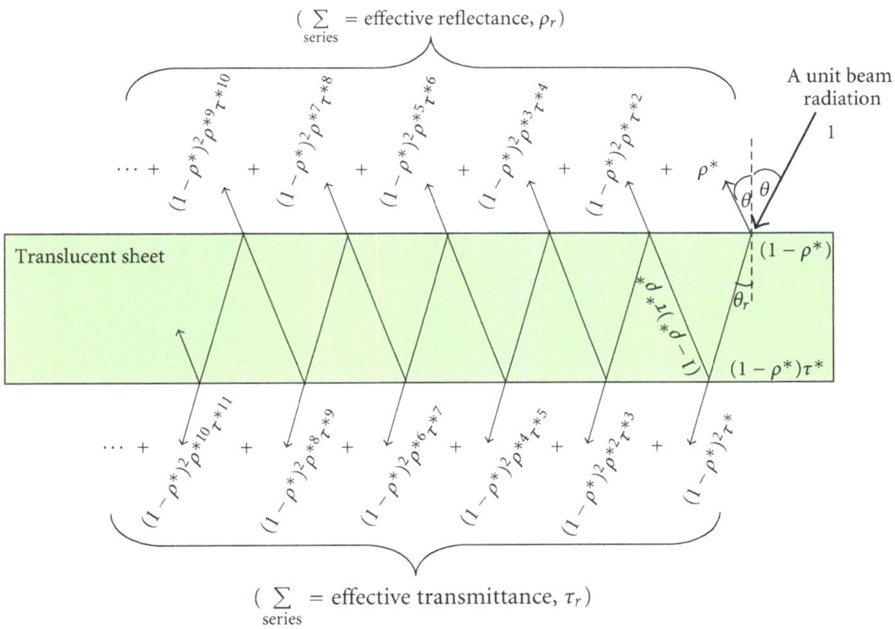

FIGURE 2: Tracing of a unit beam radiation through a translucent sheet to illustrate the concept of the effective reflectance and effective transmittance of the sheet.

Also at normal incidence, the transmittance due to absorption (τ^*) and the equivalent absorption coefficient (σ_{eq}) of the net material can be expressed in similar manner of the homogeneous translucent films by replacing the film thickness with the net density or net solidity $(1-\phi)$, ϕ is the net porosity. Thus, the relation between τ^* and σ_{eq} is given by

$$\tau^* = \text{Exp}\left\{-\sigma_{\text{eq}}(1-\phi)\right\}. \tag{4}$$

In (4), σ_{eq} is dimensionless whereas the unit of the absorption coefficient of a homogeneous translucent sheet is mm^{-1} if the sheet thickness is in mm.

Once the values of the equivalent optical constants (σ_{eq} and n_{eq}) are determined for a net, the directional radiative properties of this net can be calculated at any incidence angle (θ) of the beam radiation. In this case, the directional interface reflectance (ρ_θ^*) between the net surface and air is given by [12, 13] as

$$\rho_\theta^* = \frac{1}{2}\left\{\frac{\sin^2(\theta-\theta_r)}{\sin^2(\theta+\theta_r)} + \frac{\tan^2(\theta-\theta_r)}{\tan^2(\theta+\theta_r)}\right\},$$
$$\theta_r = \sin^{-1}\left(\frac{\sin\theta}{n_{\text{eq}}}\right), \tag{5}$$

where θ_r is the equivalent angle of refraction for a beam radiation penetrates through the net. The directional transmittance due to absorption (τ_θ^*) is given by

$$\tau_\theta^* = \text{Exp}\left\{\frac{-\sigma_{\text{eq}}(1-\phi)}{\cos\theta_r}\right\}. \tag{6}$$

The directional radiative properties, with respect to the direct beam radiation, of a plastic net can be obtained by

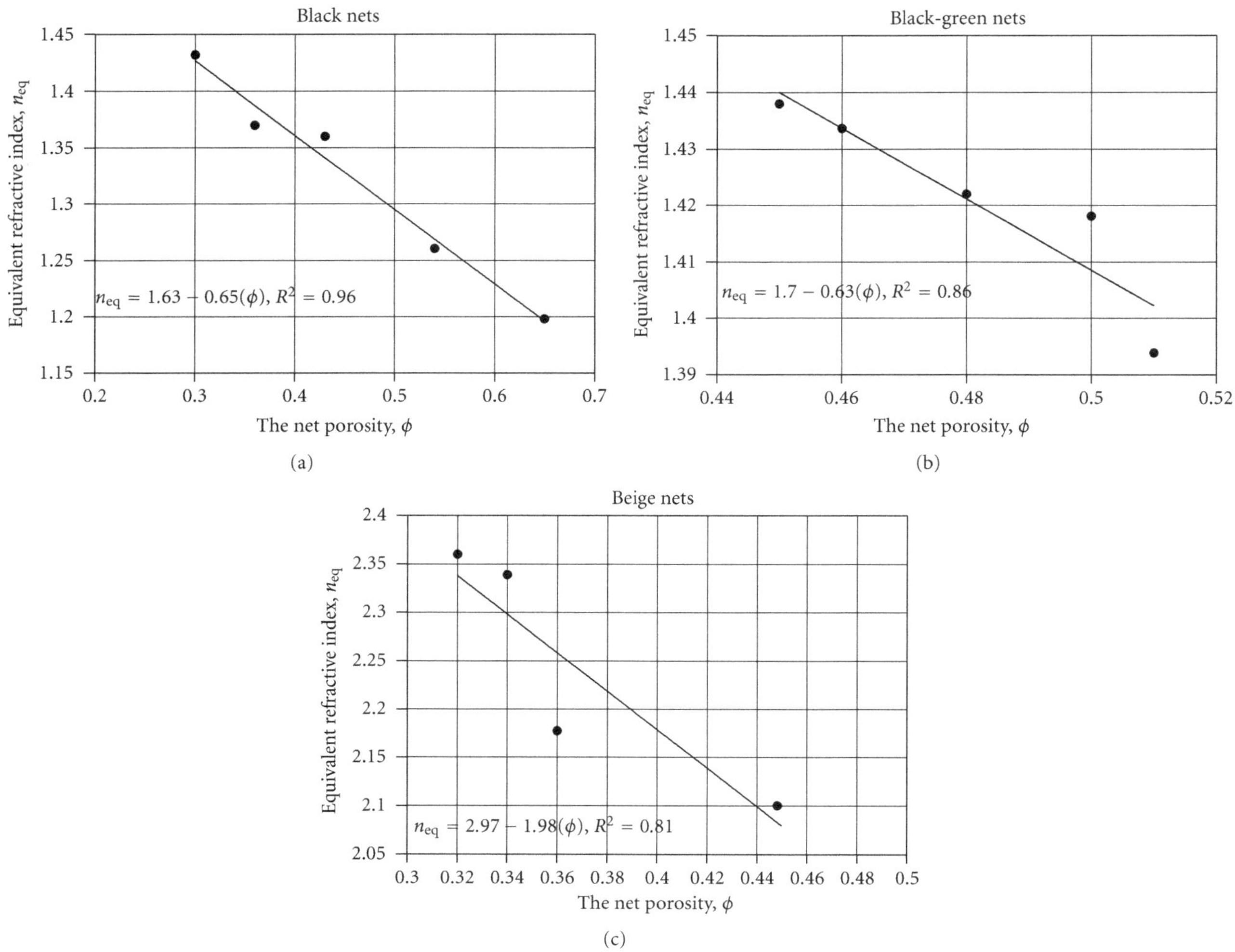

FIGURE 3: The equivalent refractive index (n_{eq}) as affected by the porosity of the tested nets: (a) is for the black group, (b) for the black-green group, and (c) for the beige group.

resubstituting τ_θ^* and ρ_θ^* into (1) to obtain τ_r and ρ_r for the desired direction. The diffuse solar radiation incident over a surface can be treated as a direct beam radiation with an incidence angle (θ) of 60° [12]. Thus the directional transmittance and reflectance of a net with respect to the global solar radiation (τ_g and ρ_g) are defined by

$$\begin{aligned} \tau_g &= \tau_{r,60^\circ} I_D + (1 - I_D)\tau_r, \\ \rho_g &= \rho_{r,60^\circ} I_D + (1 - I_D)\rho_r, \end{aligned} \tag{7}$$

where I_D is the diffuse index estimated as (D_i/S_i) at the desired direction θ.

2.3. Measuring the Required Parameters. For determining the transmittance and reflectance of the net to direct beam solar radiation (τ_r and ρ_r) to be used as input parameters to (1) and (2), the experiment was conducted on the roof of the building of the Agricultural Research and Experiment Station, Agriculture Engineering Department, King Saud University (Riyadh, Saudi Arabia, 46° 47′ E, longitude and 24° 39′ N, latitude). The measurements were conducted at around solar noon on clear, sunny days from September 10 to September 25, 2011. One pyranometer, CMP3 (Kipp & Zonen B.V. Inc., USA), and one albedometer, CMA-11 (Kipp & Zonen B.V. Inc., USA) were used. Each net sample was tacked onto a black painted wooden frame (50 cm × 60 cm × 30 cm). The frame was oriented toward the sun, so that incidence angle of the solar beam radiation (θ) was nearly zero. The transmitted global solar radiation (S_t) was measured below the net surface using the pyranometer. The incident and reflected global solar radiation (S_i and S_r) were measured above the net sample using the albedometer. This procedure was repeated while shading the frame with an opaque plate to measure the diffuse solar radiation components below (D_t) and above (D_i, D_r) the net sample. The measured parameters were recorded every 10 seconds, averaged over one-minute intervals and saved in a data logger (LI-1400, 9 channels, LI-COR, Inc.). For each parameter, five values were measured at normal incidence, and the average value was obtained. The sensors had been calibrated before

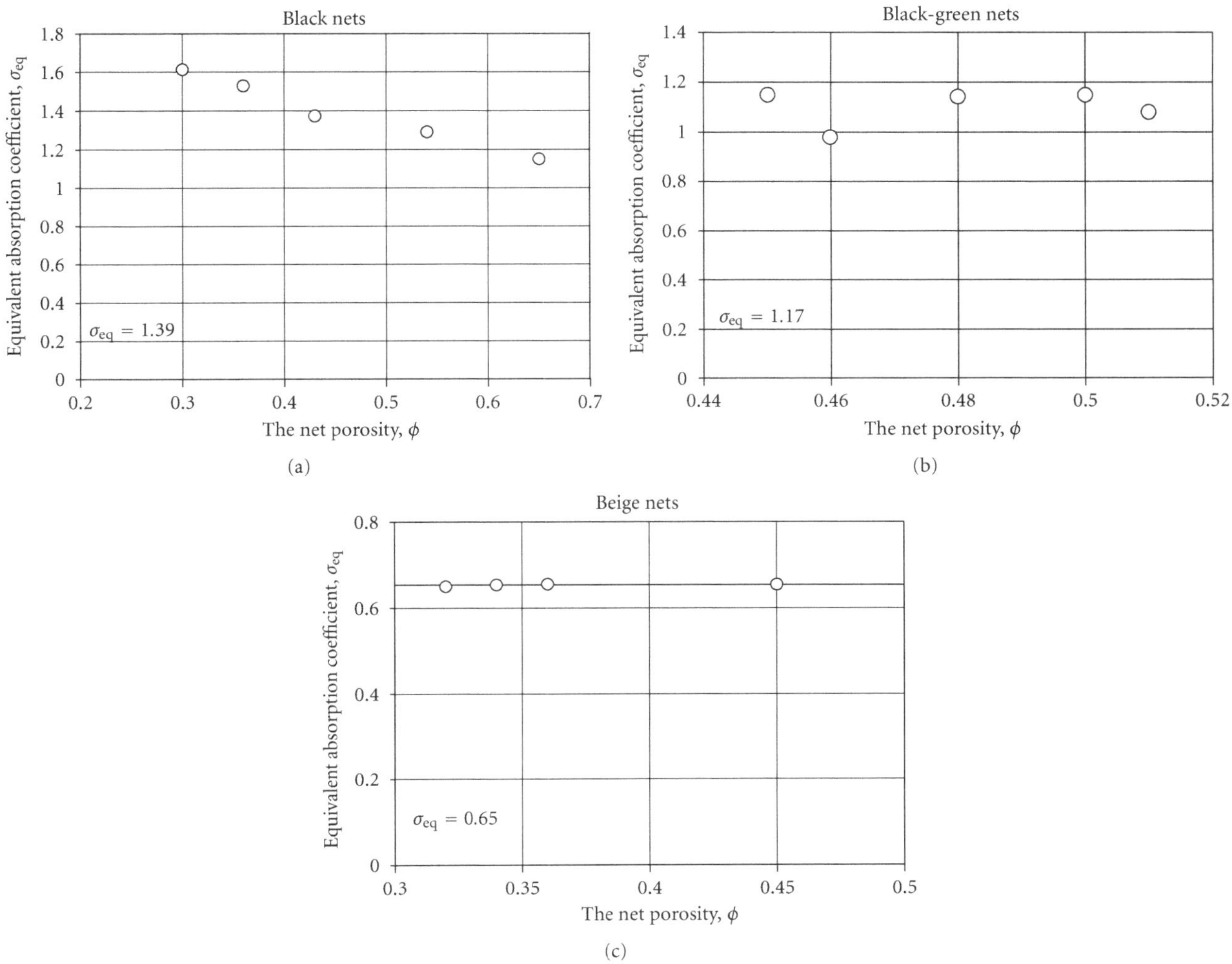

Figure 4: The equivalent absorption coefficient (σ_{eq}) as affected by the porosity of the tested nets: (a) is for the black group, (b) for the black-green group, and (c) for the beige group.

use by the supplier. The components of the direct beam radiation (B_i, B_t, and B_r) were determined as the difference between the global and diffuse radiation components. Thus at normal incidence, the net transmittance to direct beam radiation (τ_r) was estimated as (B_t/B_i) and the net reflectance to direct beam radiation (ρ_r) was estimated as (B_r/B_i).

3. Results and Discussion

For each net sample, the measured values of τ_r and ρ_r at normal incidence were substituted consequently into (2) and (1) to obtain the values of τ^* and ρ^*. Substituting the values of τ^* and ρ^* into (3) and (4) gives the equivalent optical constants (n_{eq} and σ_{eq}) of the selected net. The optical constants of a net did not depend on the direction of the incident beam radiation; therefore they are used to estimate the directional radiative properties of the net. The values of n_{eq} as affected by the net porosities (ϕ) are illustrated in Figures 3(a), 3(b), and 3(c), for the black nets group, the black-green nets group and the beige nets group, respectively. The data in Figures 3(a), 3(b), and 3(c) show that n_{eq} decreases when ϕ increases and correlations between n_{eq} and ϕ could be obtained with reasonable R^2 values in the following form:

$$n_{eq} = 1.63 - 0.65(\phi), \quad (R^2 = 0.96) \text{ for black nets,} \quad (8)$$

$$n_{eq} = 1.7 - 0.63(\phi), \quad (R^2 = 0.86) \text{ for black-green nets,} \quad (9)$$

$$n_{eq} = 2.97 - 1.98(\phi), \quad (R^2 = 0.81) \text{ for beige nets.} \quad (10)$$

Ideally, if the value of ϕ in (8), (9), and (10) increased to one (no net), then the values of n_{eq} should equal to one (i.e., the refractive index of air). However, there are residuals of +0.02 in (8), −0.07 in (9), and +0.01 in (10) that should be added to n_{eq} estimated form (8)–(10). These equations also show that value of n_{eq} increases with increasing brightness of the net color, which means more reflection is expected. The nets with the brightest color in this study (the beige nets) showed

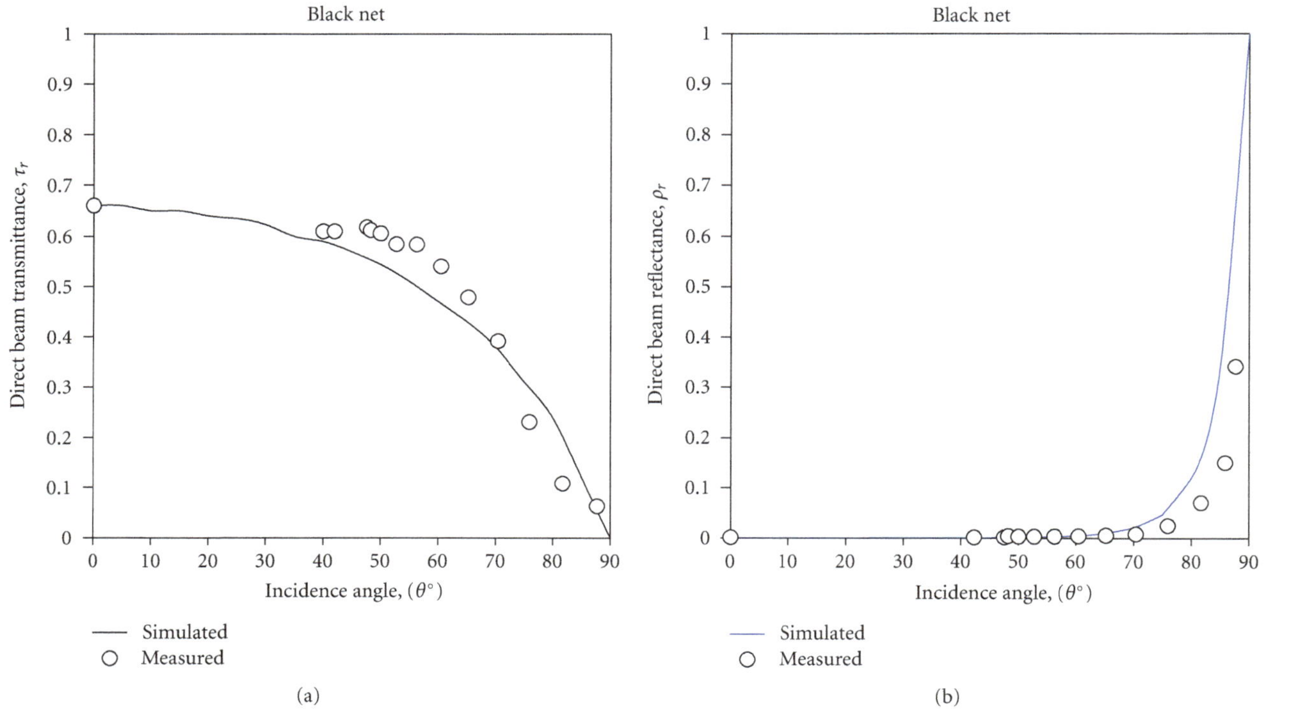

Figure 5: Comparison between the measured [8] and simulated values of (a) the direct beam transmittance (τ_r) and (b) the direct beam reflectance (ρ_r) for the black net.

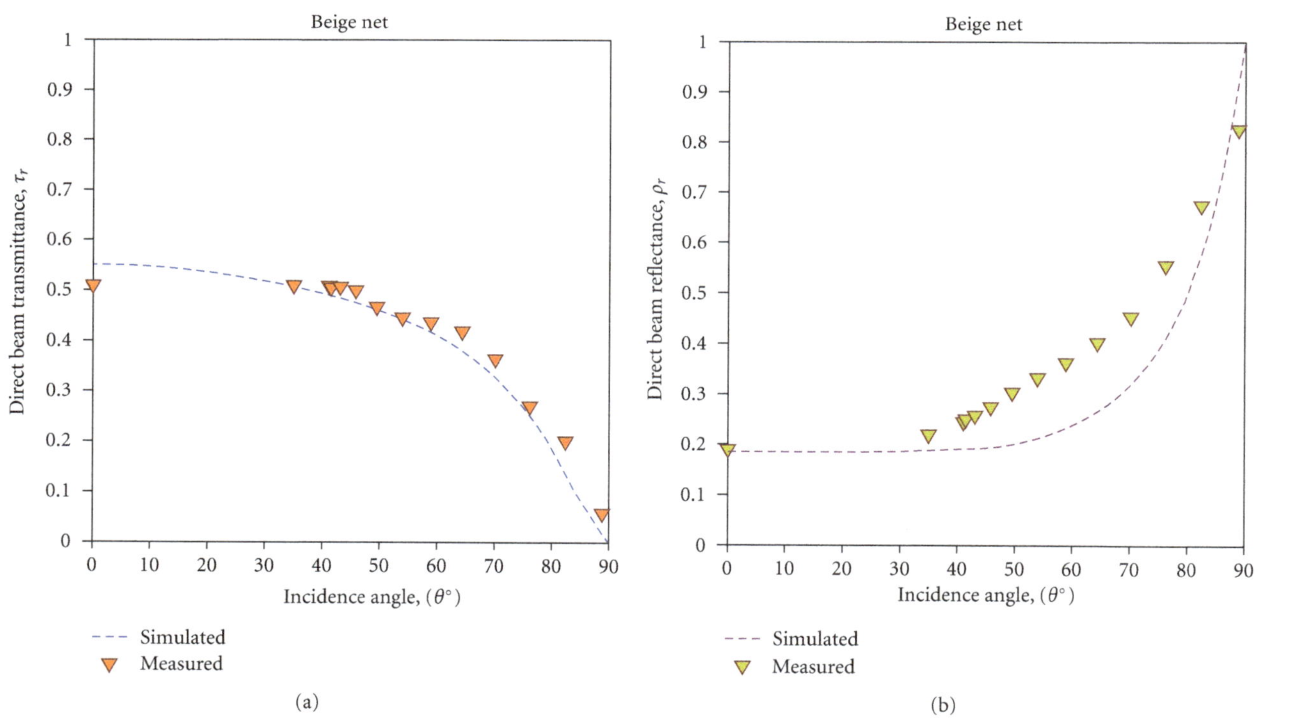

Figure 6: Comparison between the measured [8] and simulated values of (a) the direct beam transmittance (τ_r) and (b) the direct beam reflectance (ρ_r) for the beige net.

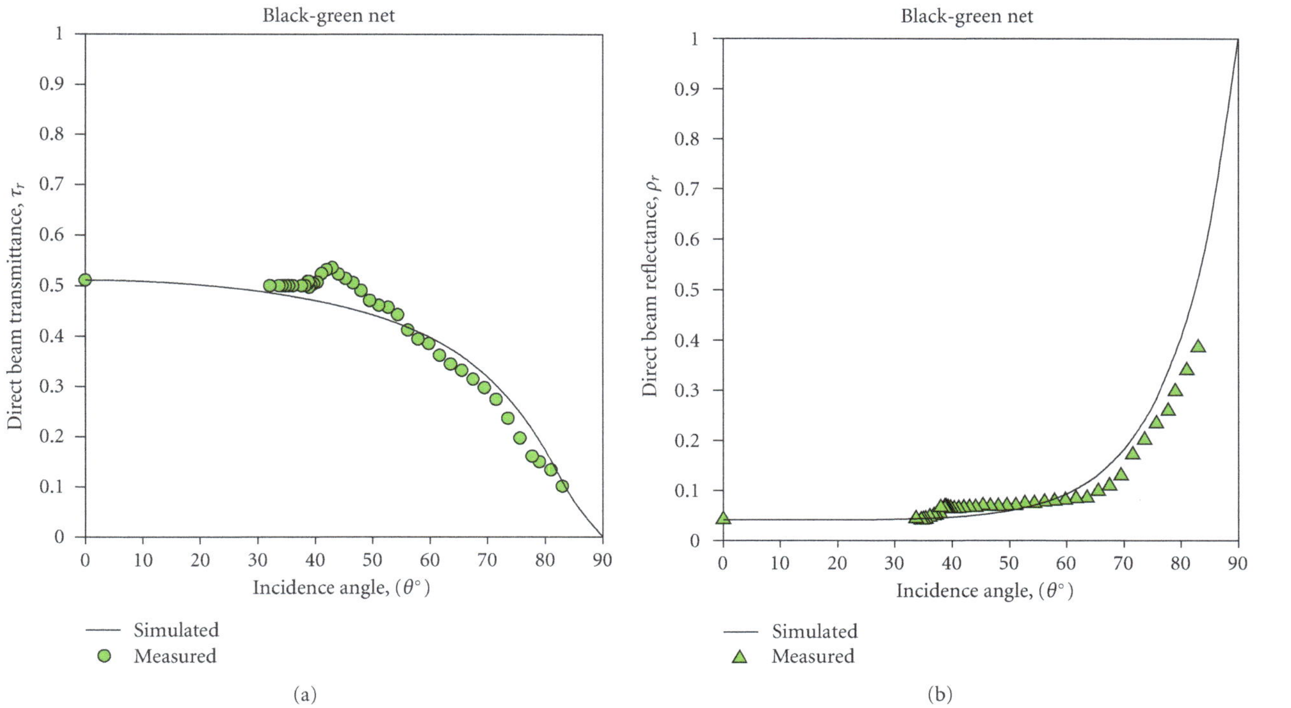

FIGURE 7: Comparison between the measured [8] and simulated values of (a) the direct beam transmittance (τ_r) and (b) direct beam reflectance (ρ_r) for the black-green net.

the highest value of n_{eq} whereas nets with the darkest color (the black nets) showed the lowest n_{eq}.

The effects of the net porosity ϕ on the value of σ_{eq} are illustrated in Figures 4(a), 4(b), and 4(c) for the black group, for the black-green group, and for the beige group, respectively. According to (4) and (6) and in analogy with homogeneous translucent materials, the value of σ_{eq} should be constant for each group of nets. And the value of τ_θ^* in (6) depends on the porosity, ϕ, of the net and on the direction, θ, of the incident beam radiation. The data in Figure 4(c) for the beige nets group satisfy the hypothesis that σ_{eq} is constant for each group of nets. However, the values of σ_{eq} for the black and black-green groups (Figures 4(a) and 4(b)) did not satisfy the hypothesis exactly. This was mainly due to the low number of data points as a result of the low availability of net groups in the local markets. Therefore, an average value for each net group was used.

4. Validation of the Model

The main purpose of the validation is to check the capability of n_{eq} and σ_{eq}, estimated using the present model, to calculate the directional radiative properties of nets under natural solar radiation conditions. In a previous study [8] we measured the direct beam transmittance (τ_r) and reflectance (ρ_r) for a black net ($\phi = 0.65$), for a beige net ($\phi = 0.12$), and for a black-green net ($\phi = 0.47$) at different incident angles (θ) of the direct beam solar radiation. Values of τ_r and ρ_r at normal incidence were measured for the three nets (see Section 2.2). The simulated values of τ_r and ρ_r using the present model at different θ comparing with the measured values of τ_r and ρ_r are illustrated, respectively, in Figures 5(a) and 5(b) for the black net, in Figures 6(a) and 6(b) for the beige net, and in Figures 7(a) and 7(b) for the black-green net. Reasonable agreement was shown between the simulated and the measured τ_r and ρ_r for the black, beige, and black-green nets.

Figures 5(a), 5(b), 6(a), 6(b), 7(a), and 7(b) do not show measured values of τ_r and ρ_r at low values of incidence angle ($0° < \theta < 33°$) because of the latitude and the time of the year at which the experiments were carried out (the lowest value of θ was estimated to be about 33° at solar noon). In addition, when θ was near or equal to 90°, the beam radiation incident on the thread surfaces reflects downwards. This increased τ_r to a value higher than zero and reduced ρ_r to a value much lower than one.

5. Conclusions

A modeling approach for determining the equivalent optical constants (n_{eq} and σ_{eq}) of plastic nets was described and validated. Three groups of nets (black, beige, and black-green) were tested. For each group, an equation for n_{eq} as a function of the net porosity was derived, and an average value for σ_{eq} was determined. Using values of n_{eq} and σ_{eq}, the directional solar radiative properties of the net can be calculated without the need of measurements which are expensive and time consumed. The resulting properties are equivalent for the whole net as a semitransparent material, not for the net material itself. The present model was validated by comparing the measured radiative properties of three of the nets, at different incident angle of solar beam radiation, with those calculated using the present model. The calculated radiative properties agreed with the measured values. Further research is needed to investigate n_{eq} and σ_{eq} for other nets communities with different colors (e.g., a blue group, an orange group, a white group, a green group, etc.).

Nomenclature

B_i: Direct beam solar radiation flux incident on the net surface ($W\,m^{-2}$)
B_r: Direct beam solar radiation flux reflected from the net ($W\,m^{-2}$)
B_t: Direct beam solar radiation flux transmitted through the net ($W\,m^{-2}$)
D_i: Diffuse radiation flux incident on the net surface ($W\,m^{-2}$)
D_r: Diffuse radiation flux reflected from the net surface ($W\,m^{-2}$)
D_t: Diffuse radiation flux transmitted through the net ($W\,m^{-2}$)
I_D: The diffuse index (i.e., ratio of diffuse to global solar radiation flux)
n_{eq}: Equivalent refractive index of the net (dimensionless)
S_i: Global solar radiation flux incident on the net surface ($W\,m^{-2}$)
S_r: Global solar radiation flux reflected from the net surface ($W\,m^{-2}$)
S_t: Global solar radiation flux transmitted through the net ($W\,m^{-2}$)
Φ: The net porosity (i.e., the empty area per unit area)
θ: Incident angle of solar beam radiation (degree)
θ_r: Equivalent refracted angle of solar beam radiation (degree)
ρ_g: Reflectance of the net to global solar radiation (dimensionless)
ρ_r: Reflectance of the net to direct beam solar radiation (dimensionless)
ρ^*: Interface reflectance between the net surface and air (dimensionless)
ρ_θ^*: Directional interface reflectance between the net surface and air (dimensionless)
σ_{eq}: Equivalent absorption coefficient of the net (dimensionless)
τ_g: Transmittance of the net to global solar radiation (dimensionless)
τ_r: Transmittance of the net to direct beam solar radiation (dimensionless)
τ^*: Transmittance of the net due to absorption only in the net threads (dimensionless).

Acknowledgments

This work was financially supported by the National Plan for Sciences and Technology (NPST) Program, King Saud University, Project no. 09-ENE912-02. The authors express thank to M. R. Shady for technical assistance during the experiments.

References

[1] S. Castellano, G. M. Scarascia, G. Russo et al., "Plastic nets in agriculture: a general review of types and applications," *Applied Engineering in Agriculture*, vol. 24, pp. 799–808, 2008.
[2] S. Castellano, G. Russo, and G. Scarascia Mugnozza, "The influence of construction parameters on radiometric performances of agricultural nets," *Acta Horticulturae*, vol. 718, pp. 283–290, 2006.
[3] D. Briassoulis, A. Mistriotis, and D. Eleftherakis, "Mechanical behaviour and properties of agricultural nets—part II: analysis of the performance of the main categories of agricultural nets," *Polymer Testing*, vol. 26, no. 8, pp. 970–984, 2007.
[4] D. Briassoulis, A. Mistriotis, and D. Eleftherakis, "Mechanical behaviour and properties of agricultural nets—part I: testing methods for agricultural nets," *Polymer Testing*, vol. 26, no. 6, pp. 822–832, 2007.
[5] A. M. Abdel-Ghany and I. M. Al-Helal, "Characterization of solar radiation transmission through plastic shading nets," *Solar Energy Materials and Solar Cells*, vol. 94, no. 8, pp. 1371–1378, 2010.
[6] A. M. Abdel-Ghany and I. M. Al-Helal, "Analysis of solar radiation transfer: a method to estimate the porosity of a plastic shading net," *Energy Conversion and Management*, vol. 52, no. 3, pp. 1755–1762, 2011.
[7] S. Hemming, G. L. A. M. Swinkels, S. Castellano, G. Russo, and G. M. Scarascia, "Numerical model to estimate the radiometric performance of net covered structures (AGRONETS)," in *Proceedings of the Agricultural and Biosystems Engineering for a Sustainable World (AgEng '08)*, Crete, Greece, June 2008.
[8] I. M. Al-Helal and A. M. Abdel-Ghany, "Measuring and evaluating solar radiative properties of plastic shading nets," *Solar Energy Materials and Solar Cells*, vol. 95, no. 2, pp. 677–683, 2011.
[9] I. M. Al-Helal and A. M. Abdel-Ghany, "Responses of plastic shading nets to global and diffuse PAR transfer: optical properties and evaluation," *Wageningen Journal of Life Sciences*, vol. 57, no. 2, pp. 125–132, 2010.
[10] A. M. Abdel-Ghany, T. Kozai, and C. Chun, "Evaluation of selected greenhouse covers for use in regions with a hot climate," *Japanese Journal of Tropical Agriculture*, vol. 45, pp. 242–250, 2001.
[11] L. He, T. H. Short, and X. Yang, "Solar radiation transmittance of a double-walled acrylic pellet-insulated greenhouse," *Transactions of the American Society of Agricultural Engineers*, vol. 34, no. 6, pp. 2559–2563, 1991.
[12] J. A. Duffie and W. A. Beckman, *Solar Engineering of Thermal Processes*, John Wiley & Sons, New York, NY, USA, 2nd edition, 1991.
[13] M. M. Elsayed, I. S. Taha, and J. A. Sabbagh, *Design of Solar Thermal Systems*, Scientific Publishing Center, King Abdulaziz University, Jeddah, Saudi Arabia, 1st edition, 1994.
[14] S. S. Soulayman, Z. Nassif, and A. Harfoush, "Transmittance, reflectance and absorptance of multi-plate planar window," *Renewable Energy*, vol. 1, no. 2, pp. 211–217, 1991.
[15] A. M. Abdel-Ghany, T. Kozai, C. Kubota, and I. S. Taha, "Investigation of the spectral optical properties of the liquid radiation filters for using in the greenhouse application," *Japan Journal of Agriculture Meteorology*, vol. 57, pp. 11–19, 2001.

22

Materials Selection, Synthesis, and Dielectrical Properties of PVC Nanocomposites

Youssef Mobarak,[1,2] M. Bassyouni,[3,4] and M. Almutawa[5]

[1] *Electrical Engineering Department, King Abdulaziz University, Rabigh 21911, Saudi Arabia*
[2] *Electrical Engineering Department, Faculty of Energy Engineering, Aswan University, Sahari City 81528, Egypt*
[3] *Chemical and Materials Engineering Department, King Abdulaziz University, Rabigh 21911, Saudi Arabia*
[4] *Chemical Engineering Department, Higher Technological Institute, Tenth of Ramadan City 11111, Egypt*
[5] *Faculty of Science, King Abdulaziz University, Jeddah 21589, Saudi Arabia*

Correspondence should be addressed to M. Bassyouni; migb2000@hotmail.com

Academic Editor: Mohd Sapuan Salit

Materials selection process for electrical insulation application was carried out using Cambridge Engineering Selector (CES) program. Melt mixing technique was applied to prepare polyvinyl-chloride- (PVC-) nanofumed silica and nanomontmorillonite clay composites. Surface analysis and particles dispersibility were examined using scanning electron microscope. Dielectrical properties were assessed using Hipot tester. An experimental work for dielectric loss of the nanocomposite materials has been investigated in a frequency range of 10 Hz–50 kHz. The initial results using CES program showed that microparticles of silica and clay can improve electrical insulation properties and modulus of elasticity of PVC. Nano-montmorillonite clay composites were synthesized and characterized. Experimental analyses displayed that trapping properties of matrix are highly modified by the presence of nanofillers. The nanofumed silica and nanoclay particles were dispersed homogenously in PVC up to 10% wt/wt. Dielectric loss tangent constant of PVC-nanoclay composites was decreased successfully from 0.57 to 0.5 at 100 Hz using fillers loading from 1% to 10% wt/wt, respectively. Nano-fumed silica showed a significant influence on the electrical resistivity of PVC by enhancing it up to 1×10^{11} Ohm·m.

1. Introduction

Nanocomposites represent a very attractive route to upgrade and diversify properties of the polymers. Nanofiller-filled polymers might be differentiated from microfiller-filled polymers in three major aspects that the nanocomposites normally contain smaller amounts, are in range of nanometers in size, and have tremendously large specific surface area. All these characteristics are reflected in their material properties [1–3]. Fillers are added to polymeric materials in order to enhance physical and mechanical properties [4–7]. Over the past few years, there have been few numbers of researches on the effect of fillers on di-electric properties of polymers [8–10]. The shift from ceramic electrical insulating materials (e.g., porcelain and glass) and from oil-paper insulations to polymeric materials has been the major change in the field of high voltage insulation technology [11–13]. Today polymers are widely used in most of the high voltage equipment, for example, power transformers, insulators, capacitors, reactors, surge arresters, current and voltage sensors, bushings, power cables, and terminations. The huge scenarios of new polymer composites in high voltage technology inspire the researchers of the field to innovate new materials and to study their properties [14–16]. There is a need for developing a range of compact devices and accessories, for both outdoor and indoor conditions, in which novel and more reliable insulation systems will play the key role [17–19]. Nanomaterials, in form of polymeric nanocomposites, are foreseen as excellent candidates which are able to fulfill the new requirements [20–22]. Elemental properties are usually integrated over macroscopic volumes to reach explanations for macroscopic properties [23–26]. The new developments in nanoscience

Table 1: Physical and mechanical properties of polyvinyl chloride.

Physical and mechanical properties of polyvinyl chloride (Sabic, Saudi Arabia)	
Young's modulus	3.2 MPa
Shear modulus	1.2 MPa
Bulk modulus	1.8 GPa
Poisson's ratio	0.49
Yield strength (elastic limit)	10 MPa
Tensile strength	10 MPa
Compressive strength	20 MPa
Flexural strength (modulus of rupture)	11 MPa
Hardness—Shore D	12
Heat deflection temperature at 455 KPa	−30°C

and technology stop short of the final integration and consider what special properties are present at the nanometric level and how they might be exploited [26, 27].

PVC is widely used in industrial applications. Chemically, PVC has a structure which is similar to that of PE but instead of several hydrogen atoms, it has chlorine atoms, which are attached to the molecular chains at the side in a random manner [28–30]. Also, it has excellent forming properties, is suitable for extreme thermoforming requirements, chemically resistant, UV-stabilized, weather resistant, and has increased impact resistance [31–33]. PVC is stronger and more rigid than other general purpose thermoplastic materials. It has a high tensile strength and modulus of elasticity. Additives are used to further specific end use, such as thermal stabilizers, lubricity, impact modifiers, and pigmentation. PVC is basically tough and strong, resists water and abrasion, and is an excellent electrical insulator [34–36].

2. Experimental

2.1. Materials Selection for Electrical Insulation and Mechanical Properties Using CES Program. Effect of silica and clay particles on the performance of PVC can be governed by the mechanical and electrical properties. CES program (Granta Design Company) was initially used to predict the desired properties of PVC composites using different fillers (microscale)/matrix mixing ratio. Synthesis and manufacturing of PVC composites were carried out based on the obtained results using CES program.

2.2. Raw Materials and Equipment. Polyvinylchloride (PVC) was received from petrochemical company (Sabic, Saudi Arabia). Physical and mechanical properties of PVC are listed in Table 1. Treated nano-montmorillonite clay was purchased from Sigma Aldrich. It is montmorillonite clay (Nanomer 1.30E), clay surface modified with 25–30 wt% octadecylamine. Spherical particle shape is the most important characteristic of nanoclay for polymer applications. The platy nature of clay fillers has a greater effect on properties such as viscosity, stiffness, and strength. Using clay as nanofiller gives high levels of flame retardancy to produced composite. Nano-fumed silica powder was obtained from Sigma Aldrich.

Hipot Tester (HIOKI 3522-50 LCR Hi-tester) device is used to measure electrical parameters of nanocomposites at various frequencies: $|Z|$, $|Y|$, θ, Rp (DCR), Rs (ESR, DCR),G, X, B, Cp, Cs, Lp, Ls, D (tan δ), and Q. Specification of LCR is power supply: 100, 120, 220, or 240 V(±10%) AC (selectable), 50/60 Hz, frequency: DC, 1 mHz to 100 kHz, and display. Voltage rating is 220 V AC 50 Hz, test voltage is 0~5 KV (adjustable), and output capacity is up to 1 KVA, cutoff current. PVC was mixed and pressed with nanofillers using two roll mills made of cast iron rolls, barrel dimensions 6″ × 16″, working length 14″, and motor capacity 7.5 HP. The samples were dried using vacuum-drying oven, at 100°C for 12 hours. Samples morphology was analyzed using scanning electron microscope (JEOL model JXA 840A (ADS+OM-Japan)). Monolayer of gold atoms was induced using gold sputter coater (S150 from Edwards for 6 min vacuum at 40 mA).

2.3. Synthesis of Polyvinylchloride Composites. Two sets of PVC composites were prepared. In the first set, PVC was composited with nano-fumed silica (1–10% wt/wt). Fumed silica was mixed and heated up to 200°C for 8 min using corotating twin-screw extruder (Berstorff ZE25A, Hannover, Germany) at 300 rpm. The compounded materials were ground and rolled at 185°C to obtain thin film (thickness of 1 ± 0.01 mm). PVC-fumed silica composites were obtained under 25 MPa and 185°C for 5 min using hot press. In the second set, polyvinylchloride nanomontmorillonite clay was obtained using typical techniques and operating conditions.

2.4. Characterization of PVC Composites. Electrical and surface analysis of PVC-nanostructured material specimen were carried out. This was achieved by measuring dielectric properties losses, electrical resistivity, and SEM analysis. These tests are able to identify the best combination of polymers—nanofillers in addition to the optimum fillers loading, in terms of improved dielectric strength and smaller space charge accumulation.

2.4.1. Electrical Properties. Dielectric spectroscopy is a powerful experimental method to investigate the dynamical behavior of a sample through the analysis of its frequency-dependent dielectric response. This technique is based on the measurement of the dielectric loss constants as a function of frequency of a sample sandwiched between two electrodes. The tan δ and susceptance (B) were measured as a function of frequency in the range from 10 Hz to 50 kHz at 25°C for all test specimens. The measurements were made using high resolution dielectric spectroscopy.

2.4.2. SEM Analysis. The morphology and dimensions of the PVC composites were tested using scanning electron microscopy. Specimens were cut in liquid nitrogen and then coated with nanogold layers using a sputter coater to make them conductive.

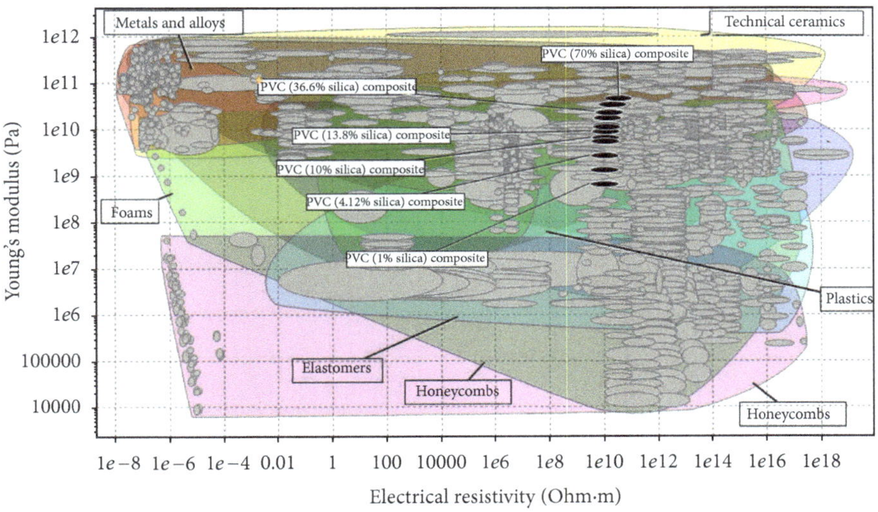

PVC-silica composites	Electrical resistivity (Ohm·m)
PVC (1% silica)	3.19×10^9–3.19×10^{10}
PVC (2.03% silica)	3.23×10^9–3.23×10^{10}
PVC (4.12% silica)	3.3×10^9–3.3×10^{10}
PVC (8.37% silica)	3.45×10^9–3.45×10^{10}
PVC (10% silica)	3.51×10^9–3.51×10^{10}
PVC (13.8% silica)	3.67×10^9–3.67×10^{10}
PVC (17% silica)	3.81×10^9–3.81×10^{10}
PVC (19.1% silica)	3.91×10^9–3.91×10^{10}
PVC (26.5% silica)	4.3×10^9–4.3×10^{10}
PVC (34.5% silica)	4.82×10^9–4.82×10^{10}
PVC (36.6% silica)	4.98×10^9–4.98×10^{10}
PVC (50.6% silica)	6.4×10^9–6.4×10^{10}
PVC (70% silica)	1.05×10^{10}–1.05×10^{11}

FIGURE 1: Electrical resistivity and mechanical properties of PVC-silica using CES program.

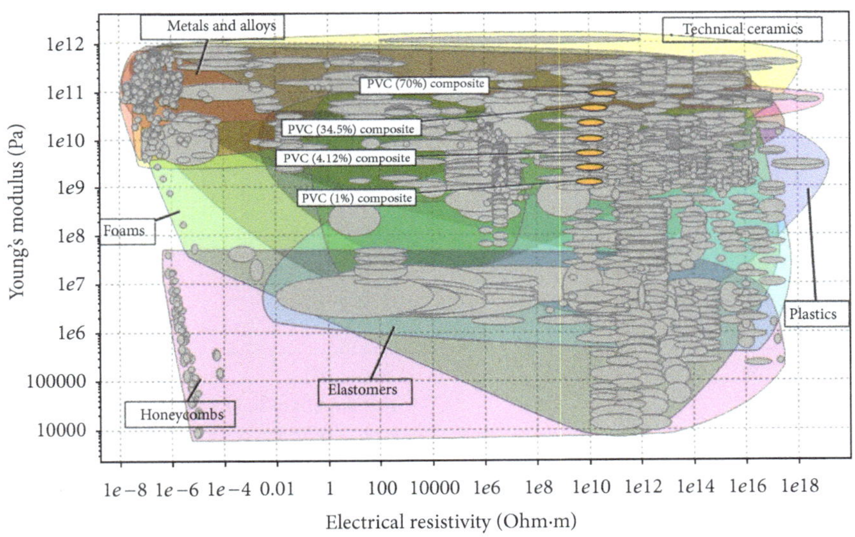

PVC-clay composites	Electrical resistivity (Ohm·m)
PVC (1% clay)	1.01×10^9–1.21×10^9
PVC (2.03% clay)	2.5×10^9–2.72×10^9
PVC (4.12% clay)	5.04×10^9–5.24×10^9
PVC (8.37% clay)	1.09×10^{10}–1.15×10^{10}
PVC (17% clay)	2.25×10^{10}–2.27×10^{10}
PVC (34.5% clay)	4.3×10^{10}–4.6×10^{10}
PVC (70% clay)	8.8×10^{10}–9.3×10^{10}

FIGURE 2: Electrical resistivity and mechanical properties of PVC-clay using CES program.

3. Results and Discussion

3.1. Predictable Mechanical and Electrical Behavior of PVC-Silica Composites. Figure 1 illustrates the electrical and mechanical properties of PVC-silica composites using CES software. Addition of silica to PVC leads to improving the electrical resistivity of PVC. It is detected that the electrical resistivity can be increased up to 5.0×10^{10} Ohm·m (mean value) using 70% silica wt/wt. This can be attributed to the high electrical resistivity of silica (1.0×10^{12}–1.0×10^{13} Ohm·m) comparing with lower electrical resistivity value of PVC matrix (3.16×10^9–3.16×10^9 Ohm·m). Modulus of elasticity of PVC was increased from 3 GPa to 72 GPa in the presence of 70% wt/wt silica.

3.2. Predictable Electrical and Mechanical Behavior of PVC-Clay Composites. Figure 2 illustrates the electrical and mechanical properties of PVC-clay composites using CES software. The initial results using the predictable model (CES software) showed that addition of clay particles to PVC can cause an increase in the electrical resistivity and modulus of elasticity. The clay fillers loading is between 1 and 70% wt/wt. Electrical resistivity was increased from 1.1×10^9 to 9×10^{10} Ohm·m (mean value). Modulus of elasticity was improved significantly from 3.190 GPa to 93.30 GPa with respect to clay fillers loading 1 to 70% wt/wt.

3.3. Effect of Nanofumed Silica on Electrical Insulation of PVC. Figure 3(a) shows loss tangent as a function of frequency

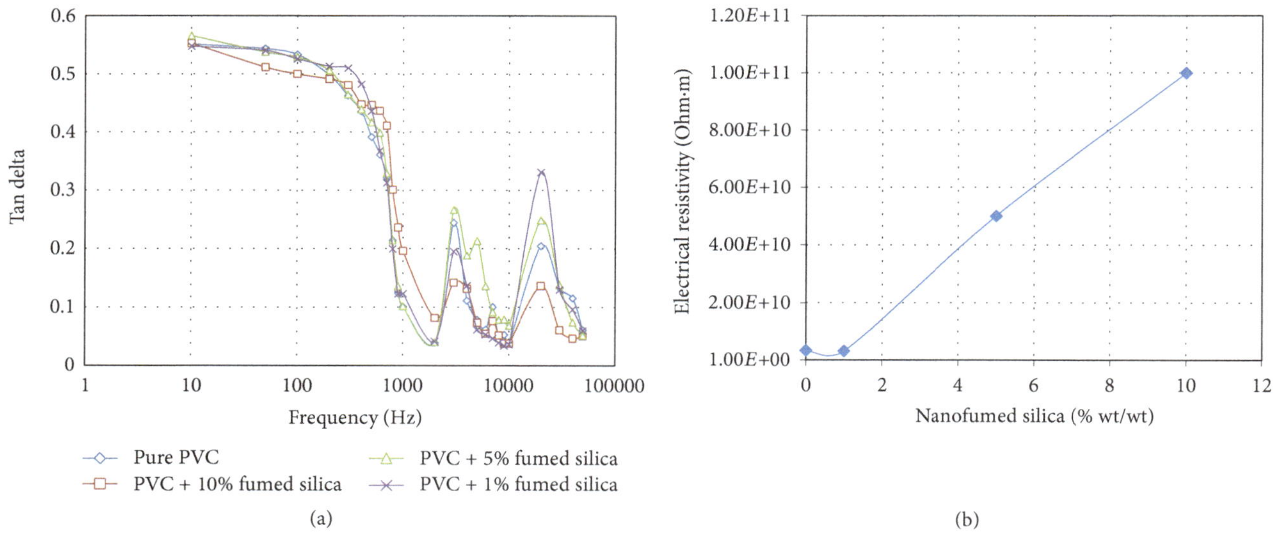

FIGURE 3: Measured loss tangent (a) and resistivity of PVC nano fumed silica.

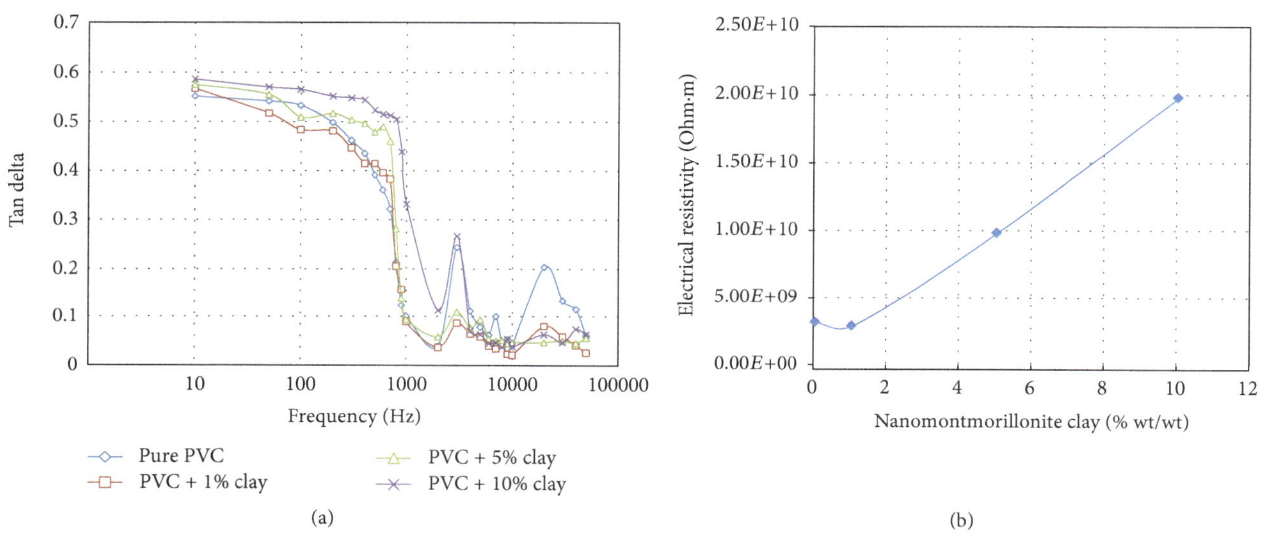

FIGURE 4: Measured loss tangent (a) and resistivity of PVC-nano montmorillonite clay.

PVC-nano fumed silica at room temperature (25°C). It is depicted that the loss tangent of PVC nanocomposites increased with increasing the fumed silica loading. The addition of nano-fumed silica from 5 to 10% wt/wt effectively improved the loss tangent values. However, it can be observed that the loss tangent does not enhance below 5 wt%. This can be attributed to the presence of voids at PVC-nano fumed silica interphase due to low melt flow around the nanofumed silica [36]. Electrical resistivity of PVC-nano fumed silica showed higher values than the CES program results (see Figure 3(b)). This can be attributed to the influence of using nanoscale fillers (larger specific area).

3.4. Effect of Nanomontmorillonite Clay on PVC Electrical Properties. Figure 4(a) shows loss tangent as a function of frequency for PVC-nanomontmorillonite clay composites at room temperature (25°C). It is clear that the loss tangent of PVC nanocomposites decreases with the increasing frequency, while it increases with increasing clay percentage of nanofillers up to 10% wt. This can be attributed to low response of the PVC dipole to follow the system variations at high frequency [37]. Figure 4(b) shows resistivity as a function of nano-montmorillonite clay loading at room temperature 25°C. The electrical resistivity of PVC-nanomontmorillonite clay composites increases with increasing nanoclay. It is noticed that the resistivity was decreased below 1% nanomontmorillonite clay which has the same behavior as PVC-nano fumed silica composite.

3.5. SEM Analysis. Microstructure studies were carried out in order to detect voids or agglomerates which can be formed through polymer composite processing. SEM images

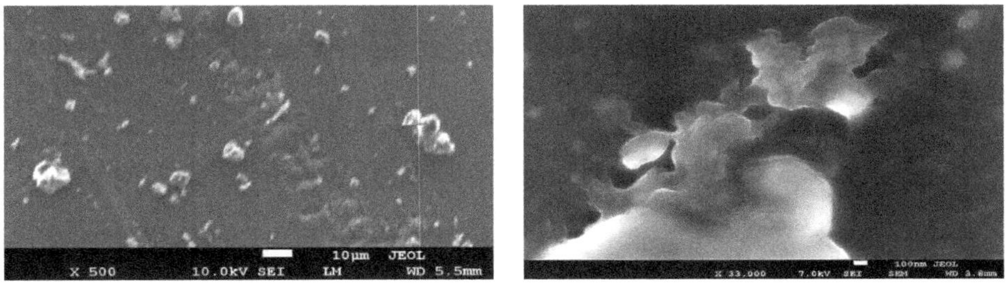

(a) (b)

FIGURE 5: SEM analysis of PVC nanocomposites (10% wt/wt): (a) nanoclay composites, (b) nanofumed silica.

for PVC composites with nano-montmorillonite and nanofumed silica fillers have been obtained as shown in Figures 5(a) and 5(b). It is depicted that nano fillers are homogenously dispersed up to 10% wt/wt. Nanoclay surface treatment using octadecylamine promotes better dispersion of the nanoclay fillers in PVC matrix.

4. Conclusions

As the electrical insulation of PVC composites contribute to its tan delta value, the variation of tan delta value in net PVC nanocomposites in lower frequency range may result in the electrical insulation of the nanocomposites having been affected by the presence of nanosize fillers. As this study was carried out under constant temperature, the influence of the relaxation time of the charge carriers on the electrical insulation of PVC nanocomposites can be ignored. Thus, the number of charge carriers and applied frequency become dominating factors of the electrical insulation of PVC nanocomposites. The presence of nanosize fillers inside PVC will restrict the chain mobility and result in increasing electric insulation as such restriction limited the generation of mobile charge and the movement of charge carriers in polymer dielectrics, especially at a lower frequency range where the insulation will play an important role. Thus, the variation of tan delta value at low frequency range may be due to the influence of inorganic fillers' electrical insulation.

Acknowledgment

The present work was supported by the Deanship of Scientific Research, King Abdulaziz University, Saudi Arabia, Project ID: 1432/829/388.

References

[1] T. Lan and T. J. Pinnavaia, "Clay-reinforced epoxy nanocomposites," *Chemistry of Materials*, vol. 6, no. 12, pp. 2216–2219, 1994.

[2] W. Feng, A. Ait-kadi, and B. Riedl, "Polymerization compounding: epoxy-montmorillonite nanocomposites," *Polymer Engineering and Science*, vol. 42, no. 9, pp. 1827–1835, 2002.

[3] M. Sh. Zoromba, M. Bassyouni, and A. Dahshan, "Preparation and mechanico-optical properties of ultraviolet-curable transparent polyurethane elastomer nanocomposites," *Journal of Nanotechnology in Engineering and Medicine*, vol. 3, no. 1, Article ID 011008, 5 pages, 2012.

[4] M. Bassyouni, S. A. Sherif, M. A. Sadek, and F. H. Ashour, "Synthesis and characterization of polyurethane—treated waste milled light bulbs composites," *Composites B: Engineering*, vol. 43, no. 3, pp. 1439–1444, 2012.

[5] M. Bassyouni, I. Taha, S. M.-S. Abdel-Hamid, and L. Steuernagel, "Physico-mechanical properties of chemically treated polypropylene rice straw bio-composites," *Journal of Reinforced Plastics and Composites*, vol. 31, no. 5, pp. 303–312, 2012.

[6] S. A. Gutub, M. Bassyouni, and S. M. S. Abdel-Hamid, "Dissolved solids adsorption of freshwater using synthesized biofoam composite," *Life Science Journal*, vol. 10, no. 2, pp. 464–471, 2013.

[7] S. A. Sheriff, M. A. Sadek, F. H. Ashour, and M. Bassyouni, "Effects of surface treatment of ground rice husk on the polyurethane based on castor oil," *Polymers and Polymer Composites*, vol. 17, no. 8, pp. 481–485, 2009.

[8] A. Z. Simeõs and C. S. Riccardi, "Dielectric spectroscopy analyses of SrBi4Ti4O 15 films obtained from soft chemical solution," *Advances in Materials Science and Engineering*, vol. 2009, Article ID 928545, 6 pages, 2009.

[9] C. Y. Lai, S. M. Sapuan, M. Ahmad, N. Yahya, and K. Z. H. M. Dahlan, "Mechanical and electrical properties of coconut coir fiber-reinforced polypropylene composites," *Polymer - Plastics Technology and Engineering*, vol. 44, no. 4, pp. 619–632, 2005.

[10] T. Imai, F. Sawa, T. Yoshimitsu, T. Ozaki, and T. Shimizu, "Preparation and insulation properties of epoxy-layered silicate nanocomposite," in *Proceedings of IEEE Conference on Electrical Insulation and Dielectric Phenomena (CEIDP '04)*, pp. 402–405, October 2004.

[11] D. A. Bolon, "Epoxy chemistry for electrical insulation," *IEEE Electrical Insulation Magazine*, vol. 11, no. 4, pp. 10–18, 1995.

[12] R. S. Gorur, E. A. Cherney, and J. T. Burnham, *Outdoor Insulators*, Ravi S Gorur, Phoenix, Ariz, USA, 1999.

[13] R. Hackam, "Outdoor HV composite polymeric insulators," *IEEE Transactions on Dielectrics and Electrical Insulation*, vol. 6, no. 5, pp. 557–585, 1999.

[14] L. H. Meyer, E. A. Cherney, and S. H. Jayaram, "The role of inorganic fillers in silicone rubber for outdoor insulation—alumina tri-hydrate or silica," *IEEE Electrical Insulation Magazine*, vol. 20, no. 4, pp. 13–21, 2004.

[15] T. J. Lewis, "Nanometric dielectrics," *Philosophical Transactions of the Royal Society, London*, vol. A247, pp. 529–551, 1994.

[16] C. Green and A. Vaughan, "Nanodielectrics—how much do we really understand?" *IEEE Electrical Insulation Magazine*, vol. 24, no. 4, pp. 6–16, 2008.

[17] B. Venkatasulu and M. J. Thomas, "Erosion resistance of silicone rubber nanocomposites at lower filler loadings," in *Proceedings of International Symposium on High Voltage Engineering (ISH '09)*, pp. 1-6, 2009.

[18] B. Venkatesulu and M. J. Thomas, "Studies on the tracking and erosion resistance of RTV silicone rubber nanocomposite," in *Proceedings of Annual Report Conference on Electrical Insulation and Dielectric Phenomena (CEIDP '08)*, pp. 204–207, October 2008.

[19] I. Ramirez, E. A. Cherney, S. Jayaram, and M. Gauthier, "Silicone rubber nanocomposites for outdoor insulation applications," in *Proceedings of Annual Report Conference on Electrical Insulation and Dielectric Phenomena (CEIDP '07)*, pp. 384–387, October 2007.

[20] G. Iyer, R. S. Gorur, R. Richert, A. Krivda, and L. E. Schmidt, "Dielectric properties of epoxy based nanocomposites for high voltage insulation," *IEEE Transactions on Dielectrics and Electrical Insulation*, vol. 18, no. 3, pp. 659–666, 2011.

[21] M. S. Naidu and V. Karamaju, *High Voltage Engineering*, Tata McGraw-Hill, Noida, India, 3rd edition, 2004.

[22] R. I. Buhl, J. Brandrup, M. Bitterner, W. Michaeli, and G. Menges, *Recyling and Recovery of Plastics*, Carl Hanser, Munich, Germany, 1996.

[23] M. E. Bassiouni, F. Al-Shamy, N. K. Madi, and M. E. Kassem, "Temperature and electric field effects on the dielectric dispersion of modified polyvinyl chloride," *Materials Letters*, vol. 57, no. 9-10, pp. 1595–1603, 2003.

[24] X.-L. Xie, R. K.-Y. Li, Q.-X. Liu, and Y.-W. Mai, "Structure-property relationships of in-situ PMMA modified nano-sized antimony trioxide filled poly(vinyl chloride) nanocomposites," *Polymer*, vol. 45, no. 8, pp. 2793–2802, 2004.

[25] M. M. Ueki and M. Zanin, "Influence of additives on the dielectric strength of high-density polyethylene," *IEEE Transactions on Dielectrics and Electrical Insulation*, vol. 6, pp. 876–881, 1999.

[26] G. Wypych, *Handbook of Fillers*, Transcontinental Printing, Quebec, Canada, 2nd edition, 1999.

[27] N. Fuse, Y. Ohki, and T. Tanaka, "Comparison of nano-structuration effects in polypropylene among four typical dielectric properties," *IEEE Transactions on Dielectrics and Electrical Insulation*, vol. 17, no. 3, pp. 671–677, 2010.

[28] M. Takala, H. Ranta, P. Nevalainen et al., "Dielectric properties and partial discharge endurance of polypropylene-silica nanocomposite," *IEEE Transactions on Dielectrics and Electrical Insulation*, vol. 17, no. 4, pp. 1259–1267, 2010.

[29] M. Ambid, D. Mary, G. Teyssedre et al., "Effect of filler concentration on dielectric behaviour and on charge trapping in PP/clay nanocomposites," in *Proceedings of Annual Report Conference on Electrical Insulation and Dielectric Phenomena (CEIDP '04)*, pp. 389–392, October 2004.

[30] L. Bois, F. Chassagneux, S. Parola et al., "Growth of ordered silver nanoparticles in silica film mesostructured with a triblock copolymer PEO-PPO-PEO," *Journal of Solid State Chemistry*, vol. 182, no. 7, pp. 1700–1707, 2009.

[31] A. Usuki, Y. Kojima, M. Kawasumi et al., "Synthesis of nylon 6-clay hybrid," *Journal of Materials Research*, vol. 8, no. 5, pp. 1179–1184, 1993.

[32] M. Ochi, R. Takahashi, and A. Terauchi, "Phase structure and mechanical and adhesion properties of epoxy/silica hybrids," *Polymer*, vol. 42, no. 12, pp. 5151–5158, 2001.

[33] P. O. Henk, T. W. Kortsen, and T. Kvarts, "Increasing the electrical discharge endurance of acid anhydride cured DGEBA epoxy resin by dispersion of nanoparticle silica," *High Performance Polymers*, vol. 11, no. 3, pp. 281–296, 1999.

[34] A. Thabet and Y. A. Mobarak, "A model for dielectric characterization of nanocomposite polymeric industrial materials," *Journal of Engineering Sciences*, vol. 40, no. 5, 2012.

[35] A. Thabet and Y. A. Mobarak, "Experimental study for dielectric strength of new nanocomposite polyethylene industrial materials," *International Journal of Electrical Engineering and Technology*, vol. 3, no. 1, pp. 553–564, 2012.

[36] A. Thabet and Y. A. Mobarak, "Dielectric characteristics of new nano-composite industrial materials," in *Proceedings of International Conference on High Voltage Engineering and Application (ICHVE '10)*, pp. 568–571, New Orleans, La, USA, October 2010.

[37] V. Raja, A. K. Sharma, and V. V. R. N. Rao, "Impedance spectroscopic and dielectric analysis of PMMA-CO-P4VPNO polymer films," *Materials Letters*, vol. 58, no. 26, pp. 3242–3247, 2004.

23

Recent Advances in the Deposition of Diamond Coatings on Co-Cemented Tungsten Carbides

R. Polini,[1] M. Barletta,[2] G. Rubino,[2] and S. Vesco[2]

[1] *Dipartimento di Scienze e Tecnologie Chimiche, Università degli Studi di Roma Tor Vergata, Via della Ricerca Scientifica, 00133 Roma, Italy*
[2] *Dipartimento di Ingegneria Meccanica, Università degli Studi di Roma Tor Vergata, Via del Politecnico, 00133 Roma, Italy*

Correspondence should be addressed to M. Barletta, barletta@ing.uniroma2.it

Academic Editor: Pavel Lejcek

Co-cemented tungsten carbides, namely, hard metals are largely used to manufacture high wear resistant components in several manufacturing segments. Coating hard metals with superhard materials like diamond is of utmost interest as it can further extend their useful lifespan. The deposition of diamond coatings onto WC-Co can be extremely complicated as a result of poor adhesion. This can be essentially ascribed to (i) the mismatch in thermal expansion coefficients between diamond and WC-Co, at the typical high temperatures inside the chemical vapour deposition (CVD) chamber, generates large residual stresses at the interface; (ii) the role of surface Co inside the WC-Co matrix during diamond CVD, which promotes carbon dissolution and diffusion. The present investigation reviews the techniques by which Co-cemented tungsten carbides can be treated to make them prone to receive diamond coatings by CVD. Further, it proposes interesting ecofriendly and sustainable alternatives to further improve the diamond deposition process as well as the overall performance of the coated hard metals.

1. Introduction

Co-(5.8% wt.) cemented tungsten carbides belong to the class of hard metals, widely used for the manufacturing of cutting tools, sliding bearings, drawing dies, and a large variety of wear-resistant components. High temperature chemical vapor deposition (CVD) of diamond coatings onto Co-cemented tungsten carbide (WC-Co 5.8% wt.) is a highly effective procedure to further increase the useful lifespan of hardmetal components by application on them of a superhard and chemically inert thin layer. Yet, the diamond deposition process is affected by several drawbacks. These are essentially ascribable to the onset of large residual stresses in the coating due to the mismatch in the thermal expansion coefficients between diamond and WC-Co and to the presence of the Co-based binder. Co can interact with the gas phase at high temperature during chemical vapor deposition (CVD), be quickly saturated with C [1] and promote the growth of an intermediate graphitic layer on which poorly adherent diamond can grow [2]. The deleterious effect of Co can be taken aback by removing it from the outermost layers of WC-Co using multistep etching procedures involving aggressive chemicals [3]. Ti-, Si-, and Cr-based interlayers deposited by physical vapor deposition (PVD) have also been reported to improve the growth of diamond coatings onto WC-Co [4]. The interlayers are deposited on the WC-Co substrate and act as C and Co diffusion barrier during the CVD process. They also interpose themselves between the WC-Co and diamond reducing the mismatch in their thermal properties [5]. Among the interlayers, the Cr-based ones play a special role. Metallic chromium (Cr) interlayers alone cannot sustain the diamond coatings and further mechanical treatments (i.e., shot or fluidized bed peening) to corrugate them and increase the diamond adhesion are of crucial importance [6]. Indeed, mechanically treated Cr/CrN and CrN interlayers seem to be very competitive although a very high time and cost consuming procedure is needed to deposit and, eventually, mechanically treat them. Thermal treatments of as-ground WC-Co are aimed at modifying the substrate morphology, reducing Co concentration at the interface and reducing some kind of surface reconstruction [7, 8]. Thermochemical treatments can be a viable alternative as they could promote the conversion of Co in, for example, CoB, which is chemically inert during CVD [9]. Among the large variety of thermal treatments, laser treatments play a very important role, too. Pulsed laser sources can reduce Co concentration in the outermost layers of the WC-Co, promote their surface corrugation, and improve the adhesion on them of the diamond coating by mechanical interlocking [8]. Nonetheless, pulsed source is operated at high laser irradiance (>80 MW/cm^2). Rather than selectively remove Co, they tend to remelt the substrate material and give rise to a sort of massive surface reconstruction [10]. The final results are a macrocorrugated morphology with rather flat peaks, composed of WC and hollow valleys in which some excess of Co is still present. Moreover, some Co and W oxides can be generated at such temperature, if any caution to protect the WC-Co from oxygen is not taken [10]. Accordingly, pulsed laser source has always been associated to further treatments like etching with dilute acid solutions to remove Co excess, eventually, most of the oxides and, thus, allow the deposition of diamond films [10]. Recently, continuous diode lasers in argon and nitrogen atmosphere are showed to be effective to selectively remove the Co-binder form WC-Co substrate, promote the establishment of a layer rich in WC and, thus, with thermal expansion coefficient close to diamond and avoid further substrate pre-treatments [11, 12]. Such laser treatments are also found to be very promising to prepare the WC-Co substrates to the application of diamond coatings by CVD [12].

This is the context in which the present investigation moves to report some of the recent advances of the technology to make the Co-cemented tungsten carbides ready to diamond deposition by CVD. Further, wear resistance and adhesive toughness of diamond coated Co-cemented tungsten carbides for some selected scenarios of growing industrial interest are also discussed into details.

2. Methodology

The most widespread technique to pre-treat Co-cemented tungsten carbides prior to diamond CVD is by single or multi-step etching procedures with aggressive chemicals [13–15]. Chemical etching is generally performed in aqueous solution of either strong oxyacids or hydrochloric acid as well as by multi-step procedures, which etch Co out of WC-Co surface and corrugate the substrate. A typical example of simple single step etching procedure involves the use of concentrated nitric acid for rather long time (10 to 15 min) to remove Co from the surface of the WC-Co substrates. Chemical etching is usually followed by abrasion of the etched substrate by 1–3 μm abrasive powders to increase the surface contact area between the substrate and the overlying coating, thus promoting their adhesion by mechanical interlocking [13]. An example of multi-step procedure is the so-called Murakami's Treatment [14]. The procedure involves a first step in which the Co-cemented tungsten carbides are etched with a solution composed of potassium iron cyanide and potassium hydroxide in water (10 g KOH + 10 g $K_3[Fe(CN)_6]$ + 100 mL water) for about 60 s. The second step is a chemical etching of the substrate by Caro's acid, that is, a solution of sulfuric acid in hydrogen peroxide (3 mL 96 wt.% H_2SO_4 + 88 mL 40% w/v H_2O_2) for very short time (10 s or less). This way, Co removal and corrugation of the WC-Co substrate can be concurrently achieved.

The application of interlayers between Co-cemented tungsten carbides and diamond coatings is also widely spread in the scientific literature, although it is rather unusual in the industry practice for the very large time and cost consuming procedures they need to be applied. A wide range of interlayer materials (amorphous carbon, metallic Cr, Ti, Ta, and so forth and ceramic CrN, TiC, TiN, SiC, and Si_3N_4) can be successfully applied by physical or chemical vapor deposition on WC-Co substrates. PVD arc with chamber pressure of 1-2 Pa and deposition temperature of 350–450°C has been reported to deposit Ti and Cr-based interlayers [16]. In this case, typical gas phase includes pure N_2 for TiN and CrN interlayers, mixture rich in Ar (>70%) with some CH_4 (~15–30%) for TiC and CrC interlayers and mixture rich in N_2 (~85%) with some CH_4 (~15%) for Ti(C, N) interlayers. Typical thicknesses of the PVD films range from few to 10 microns. Ti-based interlayers on WC-Co substrate deposited by unbalanced magnetron sputtering have also been reported in the literature [17]. SiC interlayers applied on WC-Co substrates by chemical vapor deposition, with tetramethylsilane (TMS) as single-source precursor for both silicon and carbon, are often reported in the literature [18]. Typical gas phase involves 0.1% TMS in hydrogen, chamber pressure of 500 Pa and substrate temperature of 780°C. The deposition was therefore performed at high temperature, although, recently, novel low temperature deposition process was introduced [19].

Corresponding coating thicknesses can range from 0.5 to 1 μm. TiN, Si_3N_4, a-SiC, a-SiC_xN_y, a-C, TiB_2, Mo, Ti and W deposited by plasma assisted chemical vapor deposition (PACVD) on Co-cemented tungsten carbide have also been documented in the scientific literature [20–22].

Thermal or thermochemical treatments are viable and industrially attractive alternatives to pretreat Co-cemented tungsten carbides before diamond CVD. For example, decarburizing of tungsten carbides by plasma CVD in H_2-O_2 atmosphere followed by recarburizing in CH_4-H_2 atmosphere is one of the simplest techniques to promote the growth and improve the adhesion of diamond coatings on WC-Co substrates. Such treatments are usually performed at temperature of 900 to 1000°C with chamber pressure of 4-5 kPa [23]. Further approaches involve heat treating the WC-Co substrate for a time, at a temperature and in a protective atmosphere that produces WC grain growth at the surface, increases roughness without forming subsurface porosity, and virtually eliminates surface cobalt by evaporation. The specific heat-treating conditions must be tailored to the cobalt content of the WC-Co alloy [24]. Thermochemical treatments involve plasma nitriding and carbonitriding as well as plasma boronitriding. Plasma nitriding and carbonitriding can be carried out in r.f. induction plasma. The usual conditions of plasma nitriding and carbonitriding are as follows: 45 Pa gas pressure,

$N_2:C_2H_2$ = 100:0∼90:10 gas composition, 450–900°C substrate temperature, 5 kW plasma power [25]. Plasma boronitriding can be carried out in a microwave plasma. Boron oxide was generally used as the boron source. The plasma boronitriding conditions are as follows: substrate temperature, 800°C; H_2-NH_3 atmosphere; gas pressure, 5.4 kPa; temperature of boron oxide is approximately lower 200–250°C than that of the substrate; boronitriding time, 0.5–2 h; microwave power, 700–1500 W [26]. Laser treatments belong to the class of thermal treatments. Pulsed laser sources are mostly applied to modify the surface of the Co-cemented tungsten carbides, using a repetitive pulsed excimer laser irradiation process (λ = 250–300 nm, τ = 20–25 ns) [27–29]. Incident laser energies are usually varied from 2 to 8 J/cm^2 and the number of pulses from 100 to 800 [27, 28]. Etching with chemical solutions is often needed to remove oxides and/or contaminants which are generated during laser treatments [29]. 940 nm continuous high power diode laser can be also used to pretreat the Co-cemented tungsten carbides prior to diamond deposition [11, 12]. In order to prevent Co and W oxidation, the substrates can be placed in a protective chamber and flushed with pure Ar or N_2 (99.999%). A glass window at the top of the chamber allows delivery of laser irradiation and inspection. Laser power can be in the range of 500–1500 W, and scan speed is typically set at 1 mm/s [11]. The beam is moved back and forth along linear and parallel patterns to uniformly cover the entire substrate surface with an overlap of ∼0.5–1.0 mm between two consecutive linear patterns. The total sample treatment can last up to 1 min [12].

3. Experimental

Selected case studies concerning the deposition of diamond coatings on Co-cemented tungsten carbides are hereby reported. The test cases deal with the different pretreatments, which can be used to favor the nucleation and growth of the diamond coatings and improve their adhesion to the substrate and the overall wear performance. The scenarios here investigated refer to the following class of pretreatments of the WC-Co substrates: (i) combined chemical etching and mechanical peening (i.e., fluidized bed treatment, FBT); (ii) application of a large variety of intermediate layers by PVD (intermediate layer treatment, ILT); (iii) thermal processing by 940 nm continuous high power diode laser (diode laser treatment, DLT). The pretreatments will be detailed in the incoming section.

The diamond coatings were always applied on 10 × 10 × 3 (ILT and DLT) or 4 (FBT) mm^3 WC-5.8 wt.%Co as-ground flat slabs, whose average grain size was 0.7–0.8 (ILT and DLT) or 5-6 μm (FBT), Rockwell A hardness 90.0 HRA, and transverse rupture strength (TRS, according to ISO 3327) 2900 MPa. For comparative purposes, as-ground WC-Co substrates submitted to the well-known two-step chemical etching (Murakami's treatment) were also reported. During Murakami's treatment, the WC-Co substrate was first etched for 15 min with Murakami's reagent (10 g $K_3[Fe(CN)_6]$ + 10 g KOH + 100 mL of H_2O) to corrugate the WC grains. Then, after rinsing in deionised water, the substrate was etched with Caro's acid (3 mL 96 wt.% H_2SO_4 + 88 mL 30% (w/v) H_2O_2) for 10 s to wash surface Co out, and rinsed again in H_2O. After all the pretreatments except for FBT, the substrates were seeded in an ultrasonic bath with a slurry of diamond particles (250 nm in diameter) in ethanol for 15 min.

The deposition of diamond films was performed in a stainless steel HFCVD reactor. The gas phase, a mixture of hydrogen and methane with a 1% CH_4/H_2 volume ratio, was activated by a set of linear tungsten filaments (0.79 mm in diameter), whose temperature was kept at about 2500 K and monitored by a two-colour pyrometer (by IRCON). Total flow rate was 4000 standard cm^3 min^1 (sccm), and pressure was 10 kPa. Several samples were positioned on the substrate holder (a Mo grid) just below the filaments, being 12 mm the distance between the filament and the center of their top surface. The substrate temperature was ∼650°C and it was measured by placing a thermocouple in the bulk of a WC-Co substrate located under the filament. Deposition rates, under those CVD conditions, were 0.5–0.7 μm/h, whilst deposition duration was varied from 3 to 15 h.

Sample roughness before and after diamond deposition was measured by a contact gauge surface profiler (Taylor Hobson Surface Topography System Taly-Surf CLI 2000). 100 surface profiles with a lateral spacing of 10 μm and with a resolution of 0.5 μm were recorded. A very low scan speed of 50 μm/s was adjusted during the measurements in order to minimize the sliding speed of the stylus tip and ensure measurement accuracy. TalyMap software Release 3.1 was used for data analysis and image processing.

Field emission gun scanning electron microscopy (FEG-SEM, LEO Supra 35) and energy dispersive X-ray spectroscopy (EDS, Oxford Instruments Ltd. Inca 300) were used to characterize the surface morphology after pre-treatments, CVD and wear tests, and to perform chemical analyses of the WC-Co substrates before and after each pre-treatment. Dry rotating and "linear reciprocating" tribological tests (Optimol SRV Tribometer) were performed in order to check wear resistance and film adhesion to differently pre-treated WC-Co substrates. The samples were tested at 10 N in rotational mode (radius 3 mm, frequency 10 Hz, duration 60 min, WC-5.8%Co ball 6 mm diameter) and 60 N in back-and-forth sliding (stroke length 2 mm, frequency 50 Hz, steel ball 10 mm diameter, duration 20 min). Wear resistance of the coatings was assessed by contact gauge profiler. Volume, dimension and average depth of the wear tracks were obtained by using the features of the TalyMap software.

The adhesion of the diamond films was assessed by Brale indentation tests. The tests were performed using a Wolpert Rockwell hardness tester with a Brale diamond indenter having a cone of 120° and a tip radius of 0.2 μm. According to Jindal et al. [30], when the applied load overcomes the so-called critical load, P_{cr}, the plastic deformation of the substrate induces the delamination of the film and cracks propagate between the film and the substrate. In this study, lateral crack diameter, X, was measured by SEM according to the procedure described in [31]. The results are plotted as a function of the loads, P, applied at the indenter tip. The

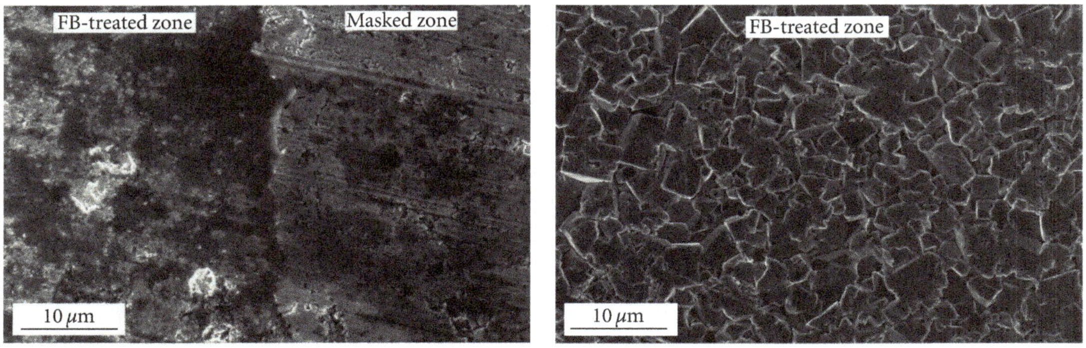

(a) (b)

FIGURE 1: Morphology of WC-Co: (a) 16 h FB and masked; (b) FB after etching.

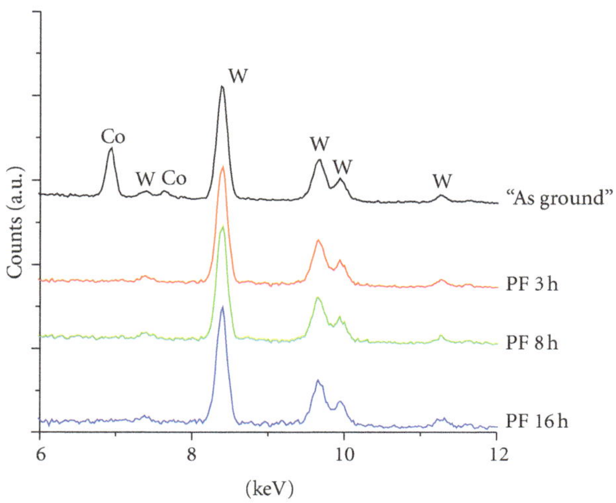

FIGURE 2: EDS after etching with Caro's acid and 3 h long FB treatment.

reciprocal of the slope obtained from the crack diameter-indentation load plot, $(dX/dP)^{-1}$, was used as an estimation of the adhesion.

4. Results and Discussion

4.1. Mechanical Peening and Co Etching. A FB, 40 × 40 mm^2 cross-section and 100 mm static bed height, of diamond powder (120 mesh size, 0.67 factor shape), in fast regime (3.5–5 m^3/h flow rate), was used to pretreat the WC-Co slabs. The slabs dipped in the fluidized powder were held on the shaft of a d.c. electrical motor and, rotated at moderate speed, they were exposed to the repeated impacts of incoming powders. Three slabs (namely., PF samples) were submitted to binder etching with Caro's acid (3 mL 96 wt.% H_2SO_4 + 88 mL 30% w/v H_2O_2) for 10 s and to fluidized bed processing for 3, 8, or 16 h. The fourth slab (viz., FP sample) was exposed to FB treatment prior to binder etching. Then, an etching with Caro's acid was performed. The fifth slab (MPS) was submitted to the Murakami's treatment [14].

Figure 1 shows the WC-Co substrate after 16 h FBT and after combined FBT and etching with Caro's acid. Significant changes in surface morphology can be noted with respect to as-ground substrate. Etching with Caro's acid removed surface Co, generating a slightly more roughened morphology (S_a = 0.14 μm). When Co etching was performed prior to FBT, the surface binder removal led to the embrittlement of the WC-Co external layer. FB machining of such embrittled surfaces contributed to remove WC grains and to establish more corrugated morphology. The duration of the FBT significantly affected the final surface roughness of PF samples. In fact, S_a of 0.17 μm, 0.25 μm, and 0.27 μm were measured for samples submitted to 3, 8, and 16 h FBT, respectively. This was because longer FBT had much more time to complete the removal of embrittled external layers of carbide grains, thereby establishing a rougher surface morphology. Figure 2 shows the EDS spectra of as-ground WC-Co and of substrates submitted to etching with Caro's acid and FBT for 3, 8 and 16 h. Binder concentration was below the detection limit. Figure 3 reports the diamond films. The morphology of coated samples becomes progressively rougher according to the FBT time. Spatial roughness S_a varied from 0.19 μm for PF sample 3 h FBT to 0.25 and 0.26 μm, respectively, for PF samples 8 h and 16 h FBT. An even rougher diamond film was achieved on MPS sample (S_a = 0.35 μm). FBT allowed attaining diamond coatings with finer grain size. Being the grains size related to nucleation density according to the model of "evolutionary selection" [32], FBT caused an enhancement of diamond nucleation density. Such effect can be ascribed to defects at the WC-Co surface induced by the grinding and FBT. Indentation tests confirmed that all diamond films were well adhered, with no noteworthy differences in film adhesion arising. Figure 4 shows the results in terms of crack radius versus indentation load. The $(dX/dP)^{-1}$ parameter averaged on PF-pretreated WC-Co substrates was around 0.216 ± 0.03 kg/μm and very close to the value of 0.213 kg/μm achieved on MPS.

4.2. Interlayers. Interlayers consisting of TiN, TiC, Ti(C, N), CrN, or CrC were deposited by PVD-arc technology on the WC-Co flat slabs. In Table 1 the thickness of

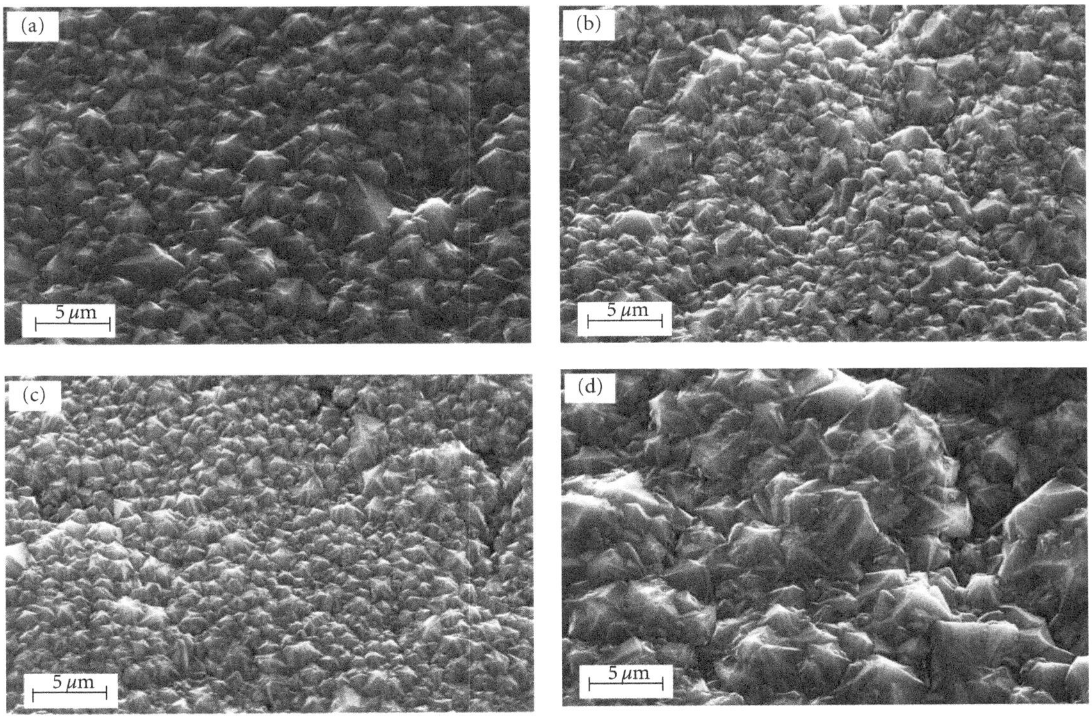

FIGURE 3: Diamond films after 8 h HF-CVD (a), PF 16 (b), PF 8 (c), and PF 3 (d) MPS.

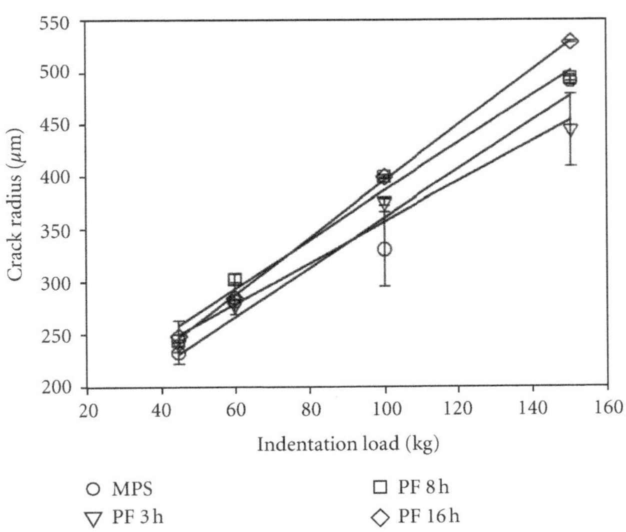

FIGURE 4: Crack radius versus indentation load function.

the PVD coatings, gas phase composition, and deposition temperatures are summarized.

Figure 5 shows the surface morphology of arc-evaporated interlayers. All PVD films showed the presence of globules (or "macros"). During cathode evaporation, the vacuum arc discharge ejects microscopic droplets from the cathode towards the substrates which cause a roughness increase on the coated parts [33]. Both size distribution function and surface density of droplets incorporated into the growing film depend, under given process conditions, on deposition time, and, therefore, on film thickness. SEM micrographs show that larger macroparticles are formed at the surface of thicker TiN films (sample TiN_a, Figure 5(a)) with respect to both thinner TiN films (TiN_b, Figure 5(b)), which were obtained by multiple rotation of the substrates during PVD, and 3-μm thick Ti(C, N) coatings (Figure 5(d)). Figure 5(c) shows the morphology of 2-μm thick TiC coatings. TiC coating exhibited a larger macroparticles density than TiN_b films with comparable thickness. CrN interlayers (Figure 5(e)) exhibited smaller droplets than TiN coatings with comparable thickness (Figure 5(a)), and thin (0.6 μm) CrC films were very smooth, being the surface morphology identical to that of as-ground WC-Co substrates.

Figure 6 shows the SEM micrographs of diamond coatings deposited onto different interlayers. A continuous diamond film was obtained after 3 h deposition. In the case of TiN_b interlayered sample, the diamond film underwent spontaneous delamination on cooling from deposition temperature to room temperature. After 3 h growth process, the surface morphology of the substrates was not significantly modified by the presence of the thin (~2 μm) diamond film. In fact, diamond coated macros were clearly visible at the surface of WC-Co substrates with Ti-based interlayers (Figures 6(a)–6(c)). The smooth surface morphology of the thin CrC interlayer was maintained after 3 h CVD (Figure 6(d)). Figures 6(e) and 6(f) show the morphology of the coating after 3 h and 9 h CVD, respectively. After 9 h deposition and consequently to the increase of film thickness, macros were embedded in the diamond coating, thus, allowing to obtain a smoother surface of the sample.

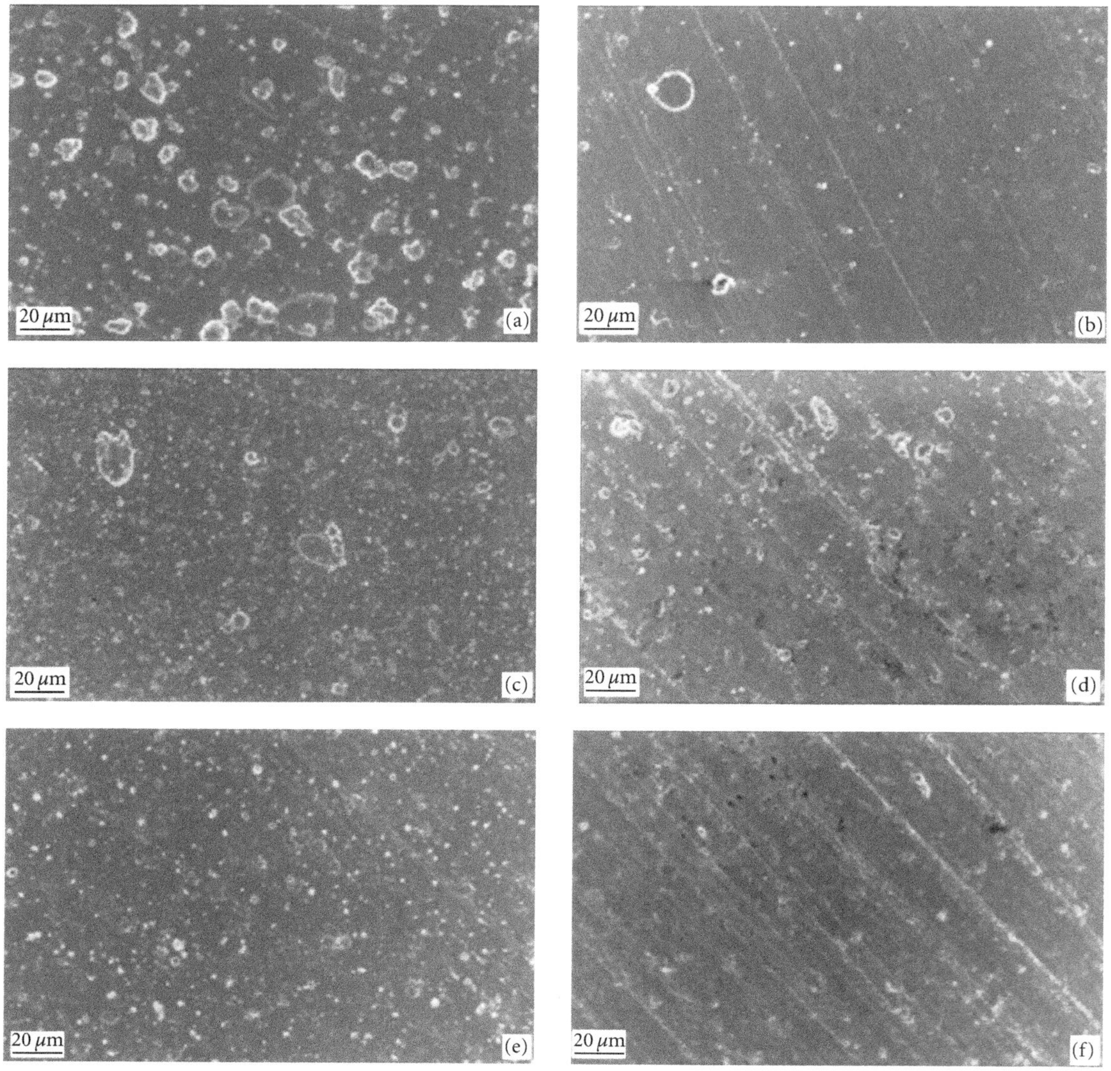

FIGURE 5: Surface morphology of arc-evaporated interlayers: (a) TiN_a, (b) TiN_b, (c) TiC, (d) Ti(C, N), (e) CrN, and (f) CrC.

TABLE 1: PVD parameters and thickness of arc evaporated Cr- and Ti-based interlayers.

Interlayer	Bias (V)	Gas phase			Chamber pressure (Pa)	Deposition temperature (°C)	Thickness (μm)
		Ar	N_2	CH_4			
CrN	−80	—	100%	—	2.0	350	8
TiN_a	−135	—	100%	—	1.5	450	7
TiN_b	−135	—	100%	—	1.5	450	3.3
CrC	−70	84%	—	16%	1.0	450	0.6
TiC	−70	70%	—	30%	1.0	450	2
Ti(C, N)	−70	—	84%	16%	1.0	450	3

Figure 7 shows the groove morphology obtained for the uncoated, PVD-coated, and HF-CVD-coated specimens after 30 min dry sliding wear tests. Figure 8 shows the volume losses as determined by three-dimension profilometry. The volume losses of samples submitted to 3 h CVD were systematically less than the ones measured on PVD-coated substrates. Only Ti(C, N) samples showed comparable wear.

The presence of macros, although was deleterious for getting high levels of wear resistance, played some positive effect on the adhesion of diamond films. In fact, the adhesion of diamond on the smooth TiN_b substrate was practically nil, while the diamond-coated TiN_a sample showed a better wear resistance (wear volume = $4.2 \cdot 10^5\ \mu m^3$) compared to bare WC-Co (wear volume = $1.6 \cdot 10^6\ \mu m^3$). The overall

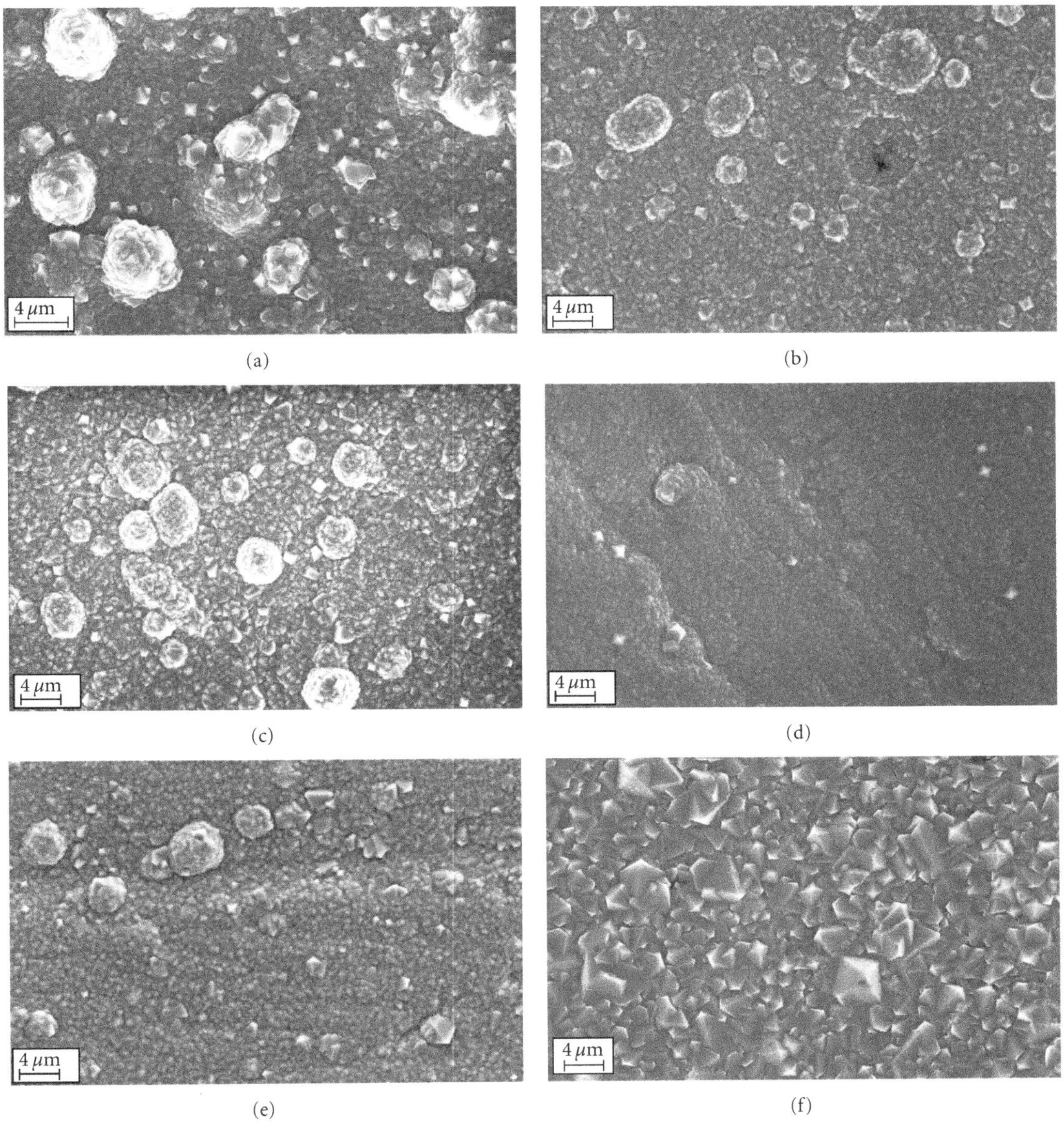

FIGURE 6: SEM micrographs of diamond films grown by HFCVD on different interlayered WC-Co samples: (a) 3 h CVD on TiN_a, (b) 3 h CVD on TiC, (c) 3 h CVD on Ti(C, N), (d) 3 h CVD on CrC, (e) 3 h CVD on CrN, and (f) 9 h CVD on CrN.

results indicate that the chemical bonds between diamond and TiN were not strong enough, and that the satisfactory adhesion level in the case of TiN_a sample was mostly attributable to the three-dimensional (3D) interface, which is expected to increase, according to Singh et al. [8], the adhesive strength by modifying the stress distribution in the diamond coating. In the case of the CrC interlayer, the good adhesion has to be ascribed to stronger chemical bonds between the diamond film and chromium carbides, rather than to interface morphology issues. The role of surface morphology of the diamond-coated sample on wear behaviour was demonstrated in the case of CrN-interlayered substrates. In fact, the wear volume of the sample coated with 9 h CVD diamond was practically nil. This fact was attributable to the more regular surface morphology of the thicker diamond coating, where most macros were completely embedded (Figure 5(f)). Therefore, the surface roughness was determined by the faceted morphology of diamond crystallites, and no removal of macros and consequent local damage could occur during the sliding test.

4.3. Laser Treatments. WC-Co slabs were treated by a 940 nm continuous wave high-power diode laser (cw-HPDL) in a protective environment (Ar) to prevent Co and W oxidation. Laser power was varied in the range of 600 to 1500 W, and scan speed was set at 1 mm/s. Laser beam was focused (elliptical spot size 3.8 × 1.2 mm) on the sample and, then, moved back and forth along three linear and parallel patterns

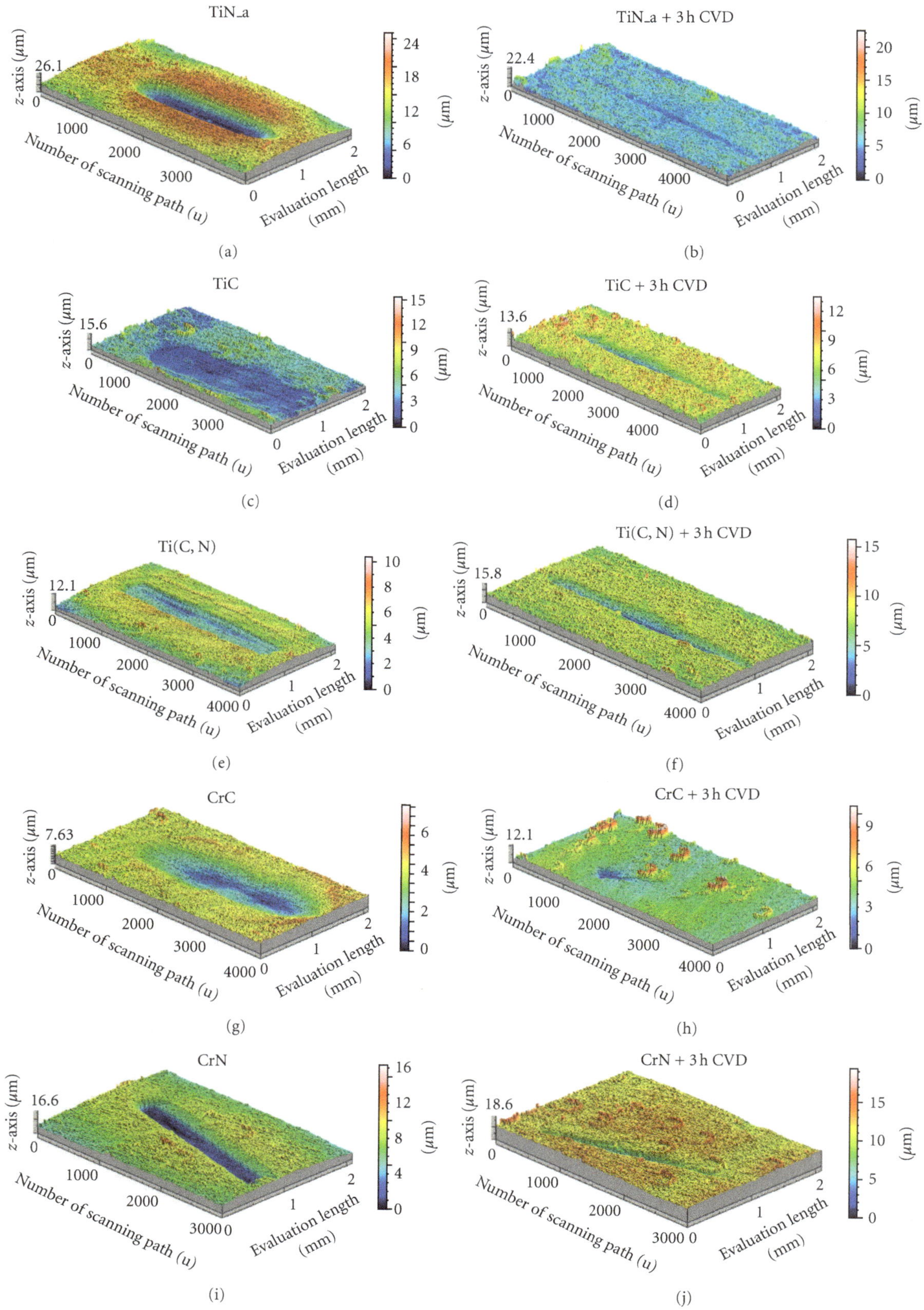

FIGURE 7: Continued.

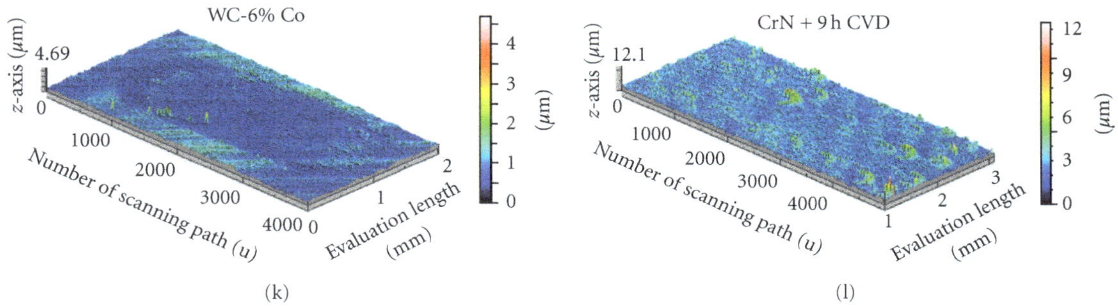

Figure 7: Wear tracks of uncoated, PVD- and CVD-coated samples.

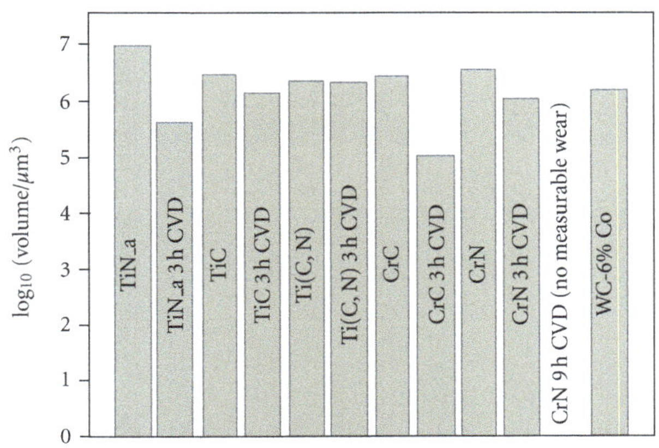

Figure 8: Wear volume of different samples submitted to dry sliding.

up to uniformly cover the entire substrate surface with an overlap of 0.7 mm among the different laser passes. As-ground WC-Co submitted to two-step chemical etching (MP treatment) were also investigated.

Figure 9 reports the evolution of the WC-Co morphology by varying the laser power from 600 to 1500 W. Figure 9(a) shows the starting morphology of the as-received WC-Co. WC grains are not clearly distinguishable, and Co matrix is spread over the surface. The morphology is also characterized by several macrostripes spreading over the surface as a result of the previous grinding process. Laser processing modifies the morphology of the WC-Co (Figures 9(b)-9(i)). Basically, laser selectively removes Co and allows WC grains sticking out from the "bulk" of the substrate. Such effect is even more apparent at any time laser power is increased (Figures 9(e)–9(i)). Co binder is reduced at the outermost layer of the WC-Co. The starting surface texture of the as-ground surface is not visible anymore, and the individual WC grains protrude from the core of the material. Increasing the power causes the increase in the average size of the WC grains and this contributes to the establishment of a rougher profile on the WC-Co. EDS analysis of the WC-Co after laser and MP treatments is reported in Figure 10. Increasing laser power from 800 to 1500 W, Co is progressively reduced up to nearly completely disappear. Similarly, Co after MP treatments is nearly completely disappeared. MP and some laser-treated samples can be therefore submitted to diamond CVD without any further treatment.

Figure 11 reports the morphology of the diamond coating deposited on MP-and laser- (1250 W) treated WC-Co substrate. It can be argued that after laser treatment the resulting diamond coating presents the same morphological characteristic of the coating deposited after the state-of-art MP treatment. Under both circumstances, small diamond crystallites due to secondary nucleation grow onto large-sized crystallites.

Both diamond films are continuous and uniformly cover the underlying WC-Co substrates. Wear resistance and adhesion of diamond films deposited on MP-and laser-treated WC-Co substrates were comparatively evaluated, too.

Figure 12 and Table 2 summarize the wear performance of the diamond films deposited by 10 h CVD onto the laser and MP-treated WC-Co substrates. The circular wear pattern was chosen to allow checking wear endurance of the diamond films by continuously varying the direction of the counterpart, being the laser treatment unidirectional. Diamond films on laser-treated substrates behave better at any time laser power is increased. The diamond films deposited on WC-Co treated at laser power of 1200 W or lower are characterized by the presence of uneven wear patterns, with the occurrence of some delamination phenomena of progressively bigger extent when lower laser power is set (Figures 12(a)–12(e)).

Table 2 shows that the size of the wear patterns due to delamination is progressively decreasing when laser power is increased from 800 to 1200 W (worn volume from ~0.0582 to 0.0019 mm^3, worn surface from ~21.6 to 0.0179 mm^2). However, the diamond film is mostly locally affected, with partial delamination occurring and not spreading too much in the surroundings. The diamond films deposited on the WC-Co laser treated at 1300 W and over are characterized by the lack of any visible wear damage (Figure 12(f)). Such better behaviour can be ascribed to the better removal of Co and the increase in the contact surface between the diamond and the underlying substrate itself. As seen before, laser treatments at high power tend to corrugate the WC-Co substrates and promote a better adhesion with the overlying diamond coating by mechanical interlocking. Wear performance of WC-Co after proper laser and MP treatments are comparable, too. In fact, even the diamond film deposited

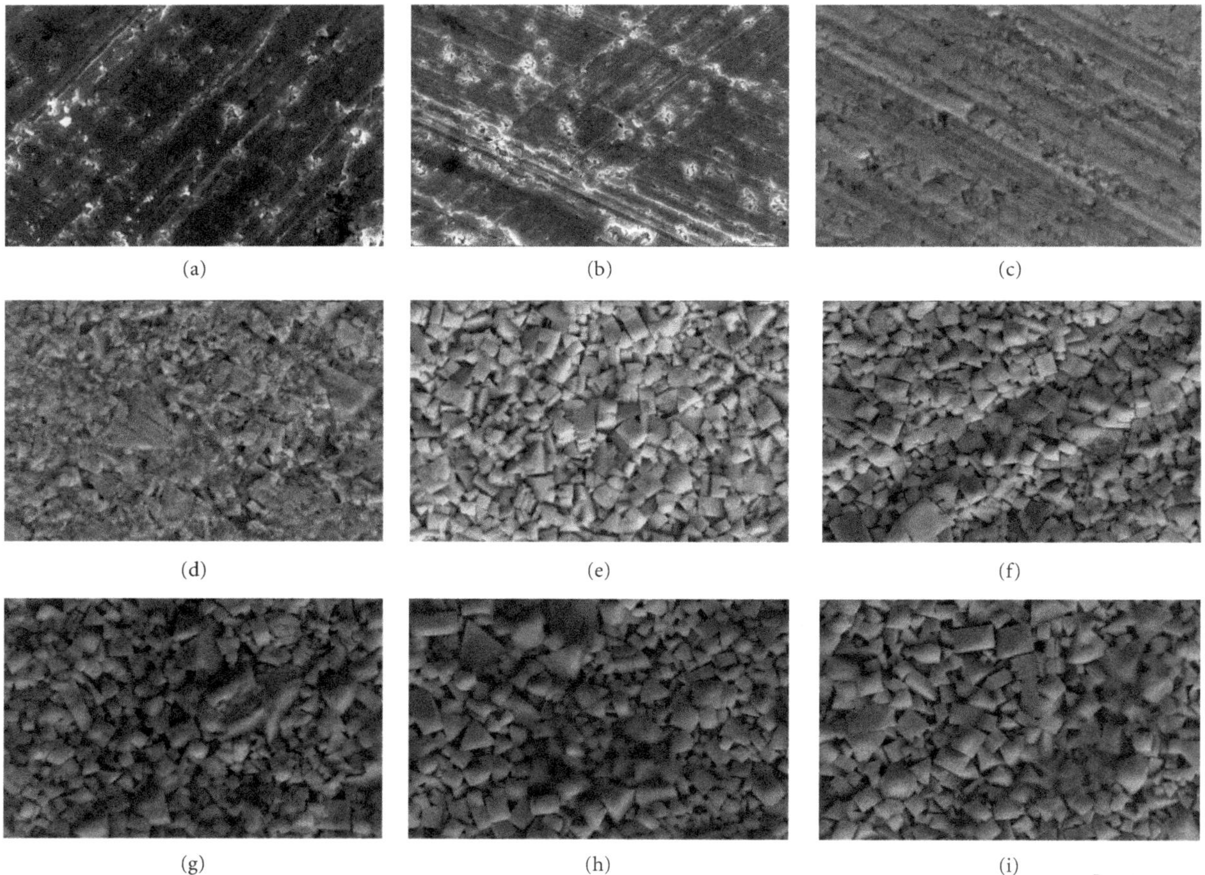

FIGURE 9: WC-Co substrate after and before laser processing: (a) as-received, (b) 600 W, (c) 800 W, (d) 900 W, (e) 1000 W, (f) 1100 W, (g) 1200 W, (h) 1300 W,(i) 1400 W.

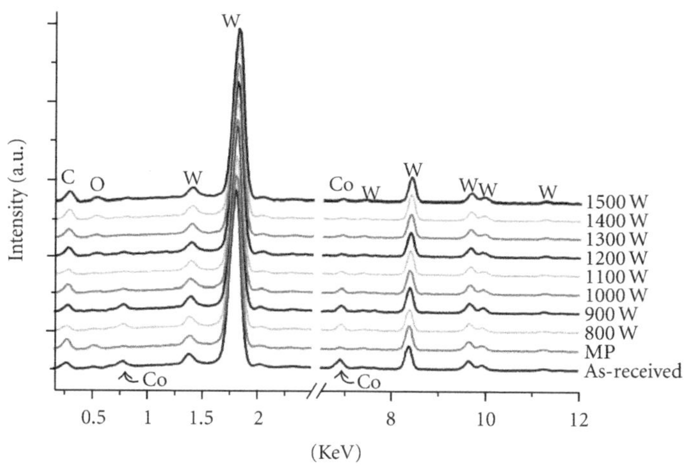

FIGURE 10: EDS of as-received, MP-treated and laser-treated WC-Co substrates.

on WC-Co after MP treatment withstands the tribological test, without any occurrence of delamination phenomena (Table 2).

In order to rank the performance of diamond films deposited onto laser (1250 W) and MP-treated WC-Co substrate, indentation tests with progressively increasing loads were performed, too. The experimental findings show that both diamond films were characterized by a good adhesion to the substrate, with no noteworthy differences regardless the applied load.

FIGURE 11: SEM morphology of diamond-coated WC-Co substrate after MP and 1250 W laser processing.

TABLE 2: Wear resistance of the diamond coatings after laser and MP treatments.

Sample	Worn surface, mm^2	Worn volume, mm^3	Max depth, μm	Mean depth, μm
800 W	21.6	0.0582	12.5	2.69
900 W	14.7	0.0342	11.8	2.32
1000 W	1.32	0.0185	5.77	1.4
1100 W	1.29	0.0181	5.83	1.4
1200 W	0.179	0.0019	4.64	1.08
1300 W	0	0	0	0
1400 W	0	0	0	0
1500 W	0	0	0	0
MP	0	0	0	0

TABLE 3: Cracks radius on diamond coatings after laser (1250 W) and MP treatments.

Load	Crack radius, μm	
	After laser treatment at 1250 W	After MP treatment
28 kg	195 ± 23	204 ± 29
45 kg	253 ± 31	239 ± 28
62 kg	290 ± 30	308 ± 36
105 kg	388 ± 59	370 ± 38
154 kg	483 ± 41	495 ± 46
	$(dX/dP)^{-1}$, $\mu m/kg$	
	0.437 ± 0.03	0.452 ± 0.04

Table 3 reports the experimental data in terms of crack radius versus indentation load. Average $(dX/dP)^{-1}$ parameter measured on diamond deposited on laser-treated WC-Co substrate was ~0.437 ± 0.03 kg/μm and very close to the value of 0.452 ± 0.03 kg/μm achieved on diamond deposited on MP-treated WC-Co substrate. Such results are in accordance with results measured for films of similar thickness and grown onto MP-treated WC-Co substrate [34, 35]. Laser treatments of WC-Co can be therefore as effective as the widely recognized Murakami's treatment insofar as the adhesion of diamond coatings on WC-Co substrates is concerned. These findings also agree with several data reported in the relevant literature, that is, that proper

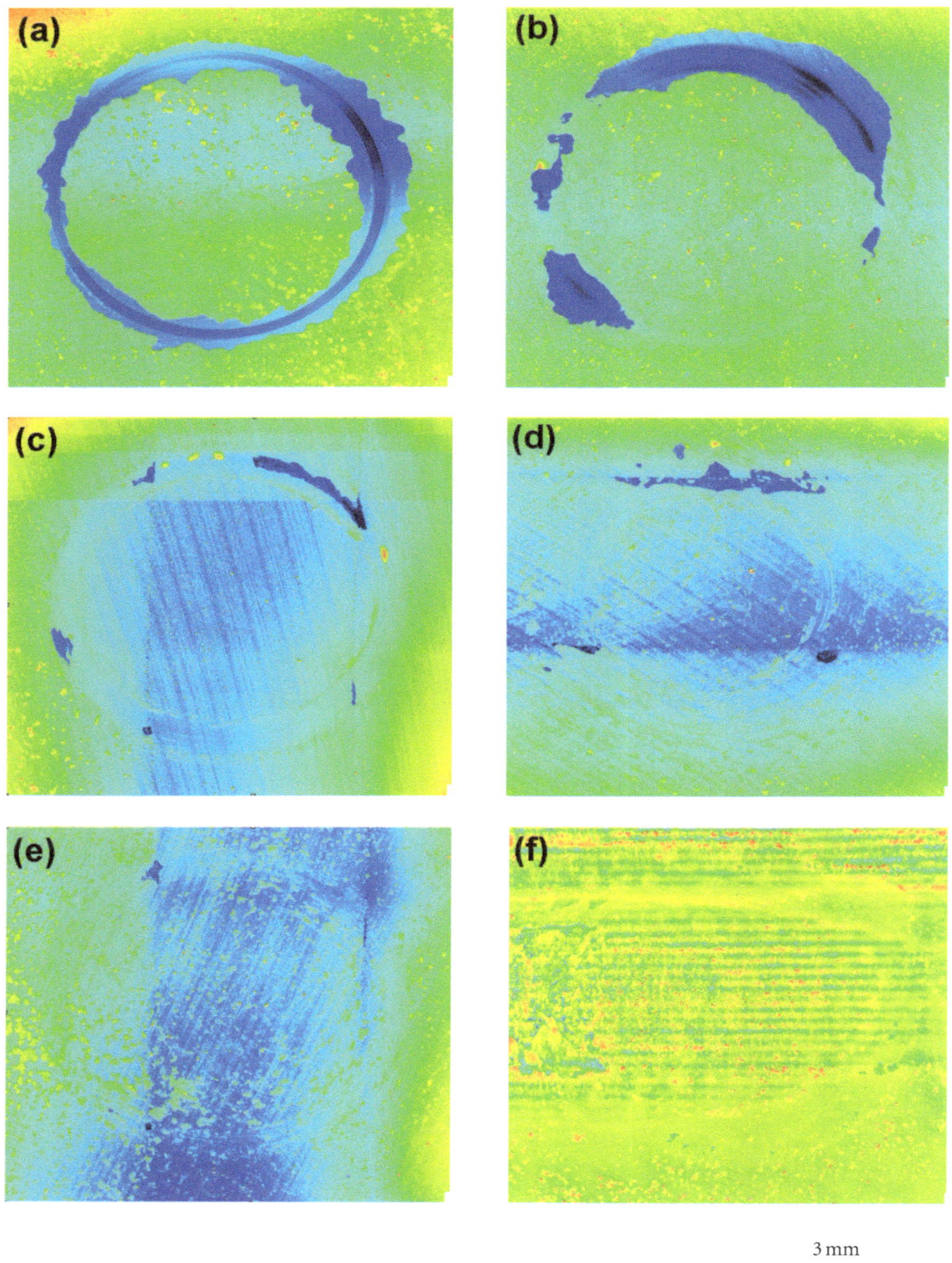

Figure 12: Wear patterns of diamond coated WC-Co substrate: (a) 800 W, (b) 900 W, (c) 1000 W, (d) 1100 W, (e) 1200 W, (f) 1300 W.

substrate surface roughening is a crucial issue affecting diamond film adhesion [8, 34, 35].

5. Conclusions

Some conclusions can be drawn as following.

(1) FB treatment after chemical etching with Caro's reagent caused (i) a perceptible modification of the surface morphology of the substrate with the 16 h long FB-pretreatment producing the more roughened WC-Co substrate, (ii) to deposit by HFCVD a continuous and well-adhered diamond film all over the substrates with a very high nucleation density without any further nucleation enhancement pretreatment, and, (iii) consequently, to obtain smooth and fine-grained polycrystalline diamond coatings. If the FB treatment of hardmetal substrates was performed prior to binder etching, that is, directly on as-ground WC-Co substrates, a smoother surface morphology was attained. Rockwell indentation tests showed that the adhesion of diamond films deposited on FB-pretreated substrates was as good as the one obtained for Murakami pre-treated substrates.

(2) The interlayer surface morphology played an important role in determining the performance of thin diamond films. Good wear resistance was exhibited by TiN and TiC interlayers characterized by large macros, whereas the adhesion of diamond on smooth TiN interlayers was very low. The wear resistance of diamond films on Ti(C,N) interlayered sample was poor, independently of the diamond film thickness. Better results were obtained in the case of Cr-based interlayers. The sample obtained by combining a thin CrC interlayer with a thin diamond film exhibited a wear volume about one order of magnitude less than bare WC-Co substrate. The use of CrN also proved to be effective in enhancing the properties of diamond films. In fact, diamond coatings thick enough to completely surround CrN globules, showed an excellent behaviour with no measurable wear volume after 30 min dry sliding against steel counterface. The good adhesion between diamond and Cr-based interlayers might be ascribed to the interfacial formation of crystalline chromium carbides during HFCVD.

(3) High power diode laser allows the selective removal of Co from the outermost layer of the WC-Co, leaving substantially unchanged the configuration and microstructure of the WC grains. The achieved morphology and the elementary chemical composition of the outermost layer of the laser-treated materials is very close to those achievable by the state-of-art of chemical treatment of the WC-Co substrate, that is, the MP treatment. The main mechanism by which the "low melting and boiling point" Co is selectively removed is by local melting and vaporization in the surrounding of the hotter "high melting and boiling point" WC. In fact, WC grains can reach higher temperature under laser source, being better energy absorber in NIR region and possessing lower thermal conductivity and high thermal capability than Co. A fluence of 134 J/mm^2 can be considered the threshold to significantly activate the removal of Co, whilst a fluence of 218 J/mm^2 is enough to activate first modification of the WC grains by melting and resolidification in the form of larger agglomerate. Fluence in excess of 250 J/mm^2 initiates a sort of WC-Co surface reconstruction, with significant melting and resolidification phenomena. In that case, Co cannot be selectively removed without affecting the structure of the WC.

References

[1] X. Chen and J. Narayan, "Effect of the chemical nature of transition-metal substrates on chemical-vapor deposition of diamond," *Journal of Applied Physics*, vol. 74, no. 6, pp. 4168–4173, 1993.

[2] K. Shibuki, M. Yagi, K. Saijo, and S. Takatsu, "Adhesion strength of diamond films on cemented carbide substrates," *Surface and Coatings Technology*, vol. 36, no. 1-2, pp. 295–302, 1988.

[3] J. Oakes, X. X. Pan, R. Haubner, and B. Lux, "Chemical vapour deposition diamond coatings on cemented carbide tools," *Surface and Coatings Technology*, vol. 47, no. 1–3, pp. 600–607, 1991.

[4] I. Y. Konyashin and M. B. Guseva, "Thin films comparable with WC-Co cemented carbides as underlayers for hard and superhard coatings: the state of the art," *Diamond and Related Materials*, vol. 5, no. 3–5, pp. 575–579, 1996.

[5] I. Endler, A. Leonhardt, H. J. Scheibe, and R. Born, "Interlayers for diamond deposition on tool materials," *Diamond and Related Materials*, vol. 5, no. 3–5, pp. 299–303, 1996.

[6] X. Xiao, B. W. Sheldon, E. Konca, L. C. Lev, and M. J. Lukitsch, "The failure mechanism of chromium as the interlayer to enhance the adhesion of nanocrystalline diamond coatings on cemented carbide," *Diamond and Related Materials*, vol. 18, no. 9, pp. 1114–1117, 2009.

[7] K. Saijo, M. Yagi, K. Shibuki, and S. Takatsu, "Improvements in adhesive strength and cutting performance of diamond-coated tools," *Surface and Coatings Technology*, vol. 47, no. 1–3, pp. 646–653, 1991.

[8] R. K. Singh, D. R. Gilbert, J. Fitz-Gerald, S. Harkness, and D. G. Lee, "Engineered interfaces for adherent diamond coatings on large thermal-expansion coefficient mismatched substrates," *Science*, vol. 272, no. 5260, pp. 396–398, 1996.

[9] S. Kubelka, R. Haubner, B. Lux, R. Steiner, G. Stingeder, and M. Grasserbauer, "Influences of WC-Co hard metal substrate pre-treatments with boron and silicon on low pressure diamond deposition," *Diamond and Related Materials*, vol. 3, no. 11-12, pp. 1360–1369, 1994.

[10] T. Li, Q. Lou, J. Dong, Y. Wei, and J. Liu, "Modified surface morphology in surface ablation of cobalt-cemented tungsten carbide with pulsed UV laser radiation," *Applied Surface Science*, vol. 172, no. 3-4, pp. 331–344, 2001.

[11] M. Barletta, G. Rubino, and A. Gisario, "Co removal and phase transformations during high power diode laser irradiation of cemented carbide," *Applied Surface Science*, vol. 257, no. 9, pp. 4239–4245, 2011.

[12] M. Barletta, G. Rubino, and A. Gisario, "Adhesion and wear resistance of CVD diamond coatings on laser treated WC-Co substrates," *Wear*, vol. 271, no. 9-10, pp. 2016–2024, 2011.

[13] N. M. Everitt, R. F. Silva, J. Vieira, C. A. Rego, C. R. Henderson, and P. W. May, "Friction measurements on hot filament CVD diamond films deposited on etched tungsten carbide surfaces," *Diamond and Related Materials*, vol. 4, no. 5-6, pp. 730–734, 1995.

[14] M. G. Peters and R. H. Cummings, European Patent 0519587 A1, 1992.

[15] J. Miao, J. Song, Y. Xue, Y. Tong, W. Tang, and F. Lu, "Effect of a two-step pretreatment method on adhesion of CVD diamond coatings on cemented carbide substrates," *Surface and Coatings Technology*, vol. 187, no. 1, pp. 33–36, 2004.

[16] R. Polini, F. P. Mantini, M. Barletta, R. Valle, and F. Casadei, "Hot filament chemical vapour deposition and wear resistance of diamond films on WC-Co substrates coated using PVD-arc deposition technique," *Diamond and Related Materials*, vol. 15, no. 9, pp. 1284–1291, 2006.

[17] S. K. Sarangi, A. Chattopadhyay, and A. K. Chattopadhyay, "Effect of pretreatment, seeding and interlayer on nucleation and growth of HFCVD diamond films on cemented carbide tools," *International Journal of Refractory Metals and Hard Materials*, vol. 26, no. 3, pp. 220–231, 2008.

[18] G. Cabral, J. Gäbler, J. Lindner, J. Grácio, and R. Polini, "A study of diamond film deposition on WC-Co inserts for graphite machining: effectiveness of SiC interlayers prepared

by HFCVD," *Diamond and Related Materials*, vol. 17, no. 6, pp. 1008–1014, 2008.

[19] F. Fendrych, A. Taylor, L. Peksa et al., "Growth and characterization of nanodiamond layers prepared using the plasma-enhanced linear antennas microwave CVD system," *Journal of Physics D*, vol. 43, no. 37, Article ID 374018, 2010.

[20] S. D. Wolter, J. T. Glass, and B. R. Stoner, "Bias induced diamond nucleation studies on refractory metal substrates," *Journal of Applied Physics*, vol. 77, no. 10, pp. 5119–5124, 1995.

[21] E. R. Kupp, W. R. Drawl, and K. E. Spear, "Interlayers for diamond-coated cutting tools," *Surface and Coatings Technology*, vol. 68-69, pp. 378–383, 1994.

[22] C. R. Lin, C. T. Kuo, and R. M. Chang, "Improvement in adhesion of diamond films on cemented WC substrate with Ti-Si interlayers," *Diamond and Related Materials*, vol. 7, no. 11-12, pp. 1628–1632, 1998.

[23] K. Saijo, M. Yagi, K. Shibuki, and S. Takatsu, "Improvements in adhesive strength and cutting performance of diamond-coated tools," *Surface and Coatings Technology*, vol. 47, no. 1–3, pp. 646–653, 1991.

[24] E. J. Oles, A. Inspektor, and C. E. Bauer, "The new diamond-coated carbide cutting tools," *Diamond and Related Materials*, vol. 5, no. 6–8, pp. 617–624, 1996.

[25] T. Sato, Y. Hosokawa, S. Ito, and K. Akashi, "Plasma carbonitriding of cemented carbide substrate as an effective pretreatment process for diamond CVD," *Surface and Coatings Technology*, vol. 112, no. 1–3, pp. 189–193, 1999.

[26] W. D. Man, J. H. Wang, Z. B. Ma, and C. X. Wang, "Plasma boronitriding of WC(Co) substrate as an effective pretreatment process for diamond CVD," *Surface and Coatings Technology*, vol. 171, no. 1–3, pp. 241–246, 2003.

[27] D. G. Lee, J. M. Fitz-Gerald, and R. K. Singh, "Novel method for adherent diamond coatings on cemented carbide substrates," *Surface and Coatings Technology*, vol. 100-101, no. 1–3, pp. 187–191, 1998.

[28] T. Li, Q. Lou, J. Dong, Y. Wei, and J. Liu, "Selective removal of cobalt binder in surface ablation of tungsten carbide hardmetal with pulsed UV laser," *Surface and Coatings Technology*, vol. 145, no. 1–3, pp. 16–23, 2001.

[29] L. Tiejun, L. Qihong, D. Jingxing et al., "Improved adhesion of diamond coating on cobalt-cemented tungsten carbide hardmetal by using pulsed-UV-laser substrate surface pretreatment," *Applied Surface Science*, vol. 193, no. 1–4, pp. 102–119, 2002.

[30] P. C. Jindal, D. T. Quinto, and G. J. Wolfe, "Adhesion measurements of chemically vapor deposited and physically vapor deposited hard coatings on WCCo substrates," *Thin Solid Films*, vol. 154, no. 1-2, pp. 361–375, 1987.

[31] R. Polini, P. D'Antonio, S. Lo Casto, V. F. Ruisi, and E. Traversa, "Cutting performance and indentation behaviour of diamond films on Co-cemented tungsten carbide," *Surface and Coatings Technology*, vol. 123, no. 1, pp. 78–83, 2000.

[32] C. Wild, N. Herres, and P. Koidl, "Texture formation in polycrystalline diamond films," *Journal of Applied Physics*, vol. 68, no. 3, pp. 973–978, 1990.

[33] T. Schuelke, T. Witke, H. J. Scheibe et al., "Comparison of DC and AC arc thin film deposition techniques," *Surface and Coatings Technology*, vol. 120-121, pp. 226–232, 1999.

[34] R. Polini, M. Santarelli, and E. Traversa, "Nucleation and adhesion of diamond films on Co cemented tungsten carbide," *Journal of the Electrochemical Society*, vol. 146, no. 12, pp. 4490–4498, 1999.

[35] R. Polini, M. Barletta, and M. Delogu, "Fluidized bed micromachining and HFCVD of diamond films onto Co-cemented tungsten carbide (WC-Co) hardmetal slabs," *Thin Solid Films*, vol. 515, no. 1, pp. 87–94, 2006.

Permissions

The contributors of this book come from diverse backgrounds, making this book a truly international effort. This book will bring forth new frontiers with its revolutionizing research information and detailed analysis of the nascent developments around the world.

We would like to thank all the contributing authors for lending their expertise to make the book truly unique. They have played a crucial role in the development of this book. Without their invaluable contributions this book wouldn't have been possible. They have made vital efforts to compile up to date information on the varied aspects of this subject to make this book a valuable addition to the collection of many professionals and students.

This book was conceptualized with the vision of imparting up-to-date information and advanced data in this field. To ensure the same, a matchless editorial board was set up. Every individual on the board went through rigorous rounds of assessment to prove their worth. After which they invested a large part of their time researching and compiling the most relevant data for our readers. Conferences and sessions were held from time to time between the editorial board and the contributing authors to present the data in the most comprehensible form. The editorial team has worked tirelessly to provide valuable and valid information to help people across the globe.

Every chapter published in this book has been scrutinized by our experts. Their significance has been extensively debated. The topics covered herein carry significant findings which will fuel the growth of the discipline. They may even be implemented as practical applications or may be referred to as a beginning point for another development. Chapters in this book were first published by Hindawi Publishing Corporation; hereby published with permission under the Creative Commons Attribution License or equivalent.

The editorial board has been involved in producing this book since its inception. They have spent rigorous hours researching and exploring the diverse topics which have resulted in the successful publishing of this book. They have passed on their knowledge of decades through this book. To expedite this challenging task, the publisher supported the team at every step. A small team of assistant editors was also appointed to further simplify the editing procedure and attain best results for the readers.

Our editorial team has been hand-picked from every corner of the world. Their multi-ethnicity adds dynamic inputs to the discussions which result in innovative outcomes. These outcomes are then further discussed with the researchers and contributors who give their valuable feedback and opinion regarding the same. The feedback is then collaborated with the researches and they are edited in a comprehensive manner to aid the understanding of the subject.

Apart from the editorial board, the designing team has also invested a significant amount of their time in understanding the subject and creating the most relevant covers. They scrutinized every image to scout for the most suitable representation of the subject and create an appropriate cover for the book.

The publishing team has been involved in this book since its early stages. They were actively engaged in every process, be it collecting the data, connecting with the contributors or procuring relevant information. The team has been an ardent support to the editorial, designing and production team. Their endless efforts to recruit the best for this project, has resulted in the accomplishment of this book. They are a veteran in the field of academics and their pool of knowledge is as vast as their experience in printing. Their expertise and guidance has proved useful at every step. Their uncompromising quality standards have made this book an exceptional effort. Their encouragement from time to time has been an inspiration for everyone.

The publisher and the editorial board hope that this book will prove to be a valuable piece of knowledge for researchers, students, practitioners and scholars across the globe.

List of Contributors

Toshihiro Inami, Masaru Yamaguchi and Kazutaka Kasai
Department of Orthodontics, Nihon University School of Dentistry at Matsudo, 2-870-1 Sakaecho Nishi, Matsudo, Chiba 271-8587, Japan

Yasuhiro Tanimoto
Department of Dental Biomaterials, Nihon University School of Dentistry at Matsudo, 2-870-1 Sakaecho Nishi, Matsudo, Chiba 271-8587, Japan

Yo Shibata
Department of Oral Biomaterials and Technology, Showa University School of Dentistry, 1-5-8 Hatanodai, Shinagawa-ku, Tokyo 142-8555, Japan

Satoshi Hirayama
Department of Dental Caries and Aesthetic Dentistry, Nihon University School of Dentistry at Matsudo, 2-870-1 Sakaecho Nishi, Matsudo, Chiba 271-8587, Japan

Masayuki Ueda
Department of Orthodontics, Nihon University School of Dentistry at Matsudo, 2-870-1 Sakaecho Nishi, Matsudo, Chiba 271-8587, Japan
Department of Oral Biomaterials and Technology, Showa University School of Dentistry, 1-5-8 Hatanodai, Shinagawa-ku, Tokyo 142-8555, Japan

T. Siva Rao, Teshome Abdo Segne, T. Susmitha and A. Balaram Kiran
Department of Inorganic and Analytical Chemistry, School of Chemistry, Andhra University, Visakhapatnam 530003, India

C. Subrahmanyam
Department of Chemistry, Indian Institute of Technology Hyderabad, Hyderabad 500007, India

Hamad R. Jappor
Department of Physics, College of Education, University of Babylon, P.O. Box 4, Hilla, Iraq

Zeyad Adnan Saleh
Department of Physics, College of Science, Al-Mustansiriyah University, P.O. Box 46010, Baghdad, Iraq

Mudar A. Abdulsattar
Directorate of Materials Research, Ministry of Science and Technology, P.O. Box 8012, Baghdad, Iraq

Arturo López-Barraza and Sonia E. Ruiz
Coordinacion de Mecanica Aplicada, Instituto de Ingenierıa, Universidad Nacional Autonoma de Mexico, Ciudad Universitaria, Coyoacan, 04510 Mexico City, Mexico

Edén Bojórquez and Alfredo Reyes-Salazar
Facultad de Ingenierıa, Universidad Autonoma de Sinaloa, Calzada de las Americas y Boulevard Universitarios S/N, Ciudad Universitaria, 80040 Culiacan, Sinaloa, Mexico

Rasoul Khandan, Siamak Noroozi, Philip Sewell, John Vinney and Mehran Koohgilani
School of Design, Engineering and Computing, Bournemouth University, Dorset Bournemouth BH12 5BB, UK

Enrique Lima and Marıa Elena Villafuerte-Castrejon
Instituto de Investigaciones en Materiales, Universidad Nacional Autonoma de Mexico, Circuito Exterior, A. P. 70-360, 04510 Mexico, DF, Mexico

Jose Saniger
CCADET, Universidad Nacional Autonoma de Mexico, Circuito Exterior, A. P. 70-188, 04510 Mexico, DF, Mexico

Victor Lara
Departamento de Quımica, Universidad Autonoma Metropolitana, Iztapalapa, Avenida San Rafael Atlixco No. 186, 09340 Mexico, DF, Mexico

Jorge E. Sanchez-Sanchez and Luis Javier Alvarez
Laboratorio de Simulacion, Instituto de Matematicas, Unidad Cuernavaca, Universidad Nacional Autonoma de Mexico, AP-273-3 Admon. 3, Cuernavaca, Morelos 62251, Mexico

Alex Elías-Zúñiga, Beatriz Montoya, Wendy Ortega-Lara, Ciro A. Rodríguez, Hector R. Siller and Oscar Martínez-Romero
Centro de Innovacion en Diseno y Tecnologıa, Tecnologico de Monterrey—Campus Monterrey, Avenida E. Garza Sada 2501 Sur, 64849 Monterrey, NL, Mexico

Eduardo Flores-Villalba
Centro de Innovacion en Diseno y Tecnologıa, Tecnologico de Monterrey—Campus Monterrey, Avenida E. Garza Sada 2501 Sur, 64849 Monterrey, NL, Mexico
Escuela de Medicina y Ciencias de la Salud, Tecnol´ogico de Monterrey—Campus Monterrey, Avenida E. Garza Sada 2501 Sur, 64849 Monterrey, NL,Mexico

José A. Díaz-Elizondo
Escuela de Medicina y Ciencias de la Salud, Tecnologico de Monterrey—Campus Monterrey, Avenida E. Garza Sada 2501 Sur, 64849 Monterrey, NL, Mexico

Jorge A. Gomez
Instituto Potosino de Investigacion Cientıfica y Tecnologica (IPICyT), Camino a la Presa San Jose 2055, Colonia Lomas 4 Seccion, 78216 San Luis Potosı, SLP, Mexico

Alfredo Marquez, Antonino Perez and A. Duarte-Moller
Centro de Investigacion en Materiales Avanzados (CIMAV), Miguel de Cervantes 120, Complejo Industrial Chihuahua, 31109 Chihuahua, CHIH, Mexico

Megumu Inaba, KazuhisaMurata, Isao Takahara and Ken-ichiro Inoue
BTL Catalyst Group, Energy Technology Research Institute, National Institute of Advanced Industrial Science and Technology (AIST), AIST Tsukuba Central 5, 1-1-1 Higashi, Ibaraki, Tsukuba 305-8565, Japan

M. M. S. Dezfouli, Mohd Hafidz Ruslan and Sohif Mat
Solar Energy Research Institute (SERI), National University of Malaysia (UKM), 43600 Bangi, Selangor, Malaysia

Mohd Roshdi Hassan
Department of Mechanical and Manufacturing Engineering, University Putra Malaysia, 43400 Serdang, Selangor, Malaysia

B. Bakhtyar
SEFB, COB, University Utara Malaysia (UUM), 6010 Sintok, Kedah, Malaysia

Eltjani-Eltahir Hago
Department of Chemistry and Chemical Engineering, Southeast University, Nanjing 210096, China

Xinsong Li
School of Chemistry and Chemical Engineering, Southeast University, Nanjing 211189, China

Sungnam Hong and Sun-Kyu Park
Department of Civil and Environmental Engineering, Sungkyunkwan University, Suwon 440-746, Republic of Korea

In-Seok Yoon
Department of Construction Information Engineering, Induk University, Seoul 139-749, Republic of Korea

G. Kermouche
Ecole Nationale Superieure des Mines de Saint-Etienne, SMS Division, LGF UMR CNRS 5307, 158 Cours Fauriel 42023 Saint-Etienne Cedex, France

N. Aleksy and J. M. Bergheau
Universite de Lyon, ENISE, LTDS UMR CNRS 5513, 58 Rue Jean Parot, 42023 Saint-Etienne Cedex 2, France

M. N. Mohammed, M. Z. Omar, M. S. Salleh and K. S. Alhawari
Department of Mechanical and Materials Engineering, Faculty of Engineering and Built Environment, Universiti Kebangsaan Malaysia, 43600 Selangor, Malaysia

Asim Hassan Ali, Nuha S. Mashaan and Mohamed Rehan Karim
Center for Transportation Research, Faculty of Engineering, University of Malaya, 50603 Kuala Lumpur, Malaysia

Y. Prawoto
Faculty of Mechanical Engineering, Universiti Teknologi Malaysia, Skudai, 81310 Johor, Malaysia

K. Sumeru
Faculty of Mechanical Engineering, Universiti Teknologi Malaysia, Skudai, 81310 Johor, Malaysia
Department of Refrigeration and Air Conditioning, Politeknik Negeri Bandung, Bandung 40012, Indonesia

W. B. Wan Nik
Faculty of Maritime Studies and Marine Science, University Malaysia Terengganu, 21030 Kuala Terengganu, Malaysia

P. K. Choudhury
Institute of Microengineering and Nanoelectronics (IMEN), Universiti Kebangsaan Malaysia, Bangi, 43600 Kuala Lumpur, Malaysia

Roham Rafiee and Reza Pourazizi
Composites Research Laboratory, Faculty of New Sciences and Technologies, University of Tehran, Tehran 1439955941, Iran

Timon Rabczuk, Junhua Zhao and Yancheng Zhang
Institute of Structural Mechanics, Bauhaus University Weimar, Marienstraße 15, 99423Weimar, Germany

James A. Whitby, Peter Horvath, Mihai Gabureac, Ivo Utke and Johann Michler
EMPA, Swiss Federal Laboratories for Materials Science and Technology, Laboratory for Mechanics of Materials and Nanostructures, Feuerwerkerstrasse 39, 3602 Thun, Switzerland

Fredrik Ostlund
EMPA, Swiss Federal Laboratories for Materials Science and Technology, Laboratory for Mechanics of Materials and Nanostructures, Feuerwerkerstrasse 39, 3602 Thun, Switzerland
TOFWERK AG, Uttigenstrasse 22, 3600 Thun, Switzerland

Jessica L. Riesterer
EMPA, Swiss Federal Laboratories for Materials Science and Technology, Laboratory for Mechanics of Materials and Nanostructures, Feuerwerkerstrasse 39, 3602 Thun, Switzerland
Idaho National Laboratory, Idaho Falls, ID 83415, USA

Markus Hohl
TOFWERK AG, Uttigenstrasse 22, 3600 Thun, Switzerland

Libor Sedlacek and Jaroslav Jiruse
Tescan a. s., Libusina trıda 21, 623 00 Brno, Czech Republic

Vinzenz Friedli
EMPA, Swiss Federal Laboratories for Materials Science and Technology, Laboratory for Mechanics of Materials and Nanostructures, Feuerwerkerstrasse 39, 3602 Thun, Switzerland
Specs Zurich GmbH, Technoparkstrasse 1, 8005 Zurich, Switzerland

Mikhael Bechelany
EMPA, Swiss Federal Laboratories for Materials Science and Technology, Laboratory for Mechanics of Materials and Nanostructures, Feuerwerkerstrasse 39, 3602 Thun, Switzerland
Institut Europeen des Membranes (UR CNRS 5635), Universite Montpellier 2, Place Eugene Bataillon, 34095 Montpellier, France

A. M. Abdel-Ghany
Department of Agricultural Engineering, College of Food and Agriculture Sciences, King Saud University, P.O. Box 2460, Riyadh 11451, Saudi Arabia
Mechanical Power Engineering Department, Faculty of Energy Engineering, South Valley University, Aswan 81528, Egypt

I. M. Al-Helal
Department of Agricultural Engineering, College of Food and Agriculture Sciences, King Saud University, P.O. Box 2460, Riyadh 11451, Saudi Arabia

Youssef Mobarak
Electrical Engineering Department, King Abdulaziz University, Rabigh 21911, Saudi Arabia
Electrical Engineering Department, Faculty of Energy Engineering, Aswan University, Sahari City 81528, Egypt

M. Bassyouni
Chemical and Materials Engineering Department, King Abdulaziz University, Rabigh 21911, Saudi Arabia
Chemical Engineering Department, Higher Technological Institute, Tenth of Ramadan City 11111, Egypt

M. Almutawa
Faculty of Science, King Abdulaziz University, Jeddah 21589, Saudi Arabia

R. Polini
Dipartimento di Scienze e Tecnologie Chimiche, Universita degli Studi di Roma Tor Vergata, Via della Ricerca Scientifica, 00133 Roma, Italy

M. Barletta, G. Rubino and S. Vesco
Dipartimento di Ingegneria Meccanica, Universita degli Studi di Roma Tor Vergata, Via del Politecnico, 00133 Roma, Italy